Designing for Kids

Designers, especially design students, rarely have access to children or their worlds when creating products, images, experiences and environments for them. Therefore, fine distinctions between age transitions and the day-to-day experiences of children are often overlooked.

Designing for Kids brings together all a designer needs to know about developmental stages, play patterns, age transitions, playtesting, safety standards, materials and the daily lives of kids, providing a primer on the differences in designing for kids versus designing for adults. Research and interviews with designers, social scientists and industry experts are included, highlighting theories and terms used in the fields of design, developmental psychology, sociology, cultural anthropology and education.

This textbook includes more than 150 color images, helpful discussion questions and clearly formatted chapters, making it relevant to a wide range of readers. It is a useful tool for students in industrial design, interaction design, environmental design and graphic design with children as the main audience for their creations.

Krystina Castella is a professor of industrial design and business at ArtCenter College of Design in Pasadena, California, USA. Krystina has practiced as a designer and has taught at ArtCenter College of Design across disciplines for almost three decades. Her research and teaching center around designing for play, the intersection between design and ethical business, designing for social innovation, and sustainable materials and manufacturing innovation.

Designing for Kids
Creating for Playing, Learning, and Growing

Krystina Castella

Routledge
Taylor & Francis Group

NEW YORK AND LONDON

First published 2019
by Routledge
711 Third Avenue, New York, NY 10017

and by Routledge
2 Park Square, Milton Park, Abingdon, Oxon, OX14 4RN

Routledge is an imprint of the Taylor & Francis Group, an informa business

Library of Congress Cataloging-in-Publication Data
Names: Castella, Krystina, author.
Title: Designing for kids : creating for playing,
learning, and growing / Krystina Castella.
Description: New York : Routledge, 2018. |
Includes bibliographical references and index.
Identifiers: LCCN 2018027089 | ISBN 9781138290754 (hardback) |
ISBN 9781138290761 (pbk.) | ISBN 9781315266015 (ebook)
Subjects: LCSH: Design–Human factors. | Child consumers.
Classification: LCC NK1520 .C37 2018 | DDC 745.4–dc23
LC record available at https://lccn.loc.gov/2018027089

ISBN: 978-1-138-29075-4 (hbk)
ISBN: 978-1-138-29076-1 (pbk)
ISBN: 978-1-315-26601-5 (ebk)

Typeset in Avenir by
Out of House Publishing

For my partners in play Brian and Sequoia

Contents

Preface

In the field of design we follow our hearts. That is how I became a designer who enjoys creating with and for kids and teenagers. I enjoy their sense of humor and seeing the world through fresh eyes. I enjoy researching, playtesting and co-designing with kids and teens and helping them interpret their ideas into real-world design solutions. I enjoy using play as a tool for creativity.

This has led me to practicing as a product and environmental designer for companies on projects with a youth and family focus, including Walt Disney Imagineering (theme park design), Fox Network (designing sets for children's programming), RTKL (designing public and community spaces) and The Brooklyn Children's Museum (exhibits). I have designed digital games and toys, board games, apparel, furniture, clothing, stationery, housewares and soft goods for children. I have written and photographed children's books and have written many cookbooks with family and fun food as the focus. In my consulting business and the Design Entrepreneur Network I founded, I have worked with many individuals and businesses on developing their designs, products, experiences, services and business strategies for children and teens.

I also work with kids directly and encourage anyone who likes creating for kids to do this as much as you can. Align the work with your own interests. My passion for the outdoors led me to be a volunteer park ranger for the National Parks Service and lead daily nature hikes and educational programs for kids of all ages, engaging them in nature. I volunteered at homeless shelters to keep kids who lived there on top of the schoolwork. I teach the design process to kids and teens in public schools and outreach programs in underserved communities. I volunteer at the local public grammar school and work with the students on project-based and design-based learning initiatives.

Years after being trained as an industrial designer at Rhode Island School of Design, to understand children's development and behavior even better, I took social science courses, including development psychology and cultural studies, at UCLA. Throughout my coursework I wondered: how had I advanced this far into the field of designing for kids without knowing all of this valuable and inspirational information? From that point on, I have included research in these fields throughout my design development process and in my teaching.

The translation of the research and direct experiences with children and youth into tangible insights and then viable business solutions is what I do in my coursework almost every day. For almost three decades, I have been a professor at ArtCenter College of Design and also at UCLA (Business) and Otis College of Art and Design (Toy Design). I have used my knowledge and expertise in designing for children to educate students in Graduate Industrial Design, Product Design, Interaction Design, Environmental Design, Graphic Design, Entertainment Design, Toy Design, Illustration, Film, Motion Design and

Animation, Photography and Advertising, and every design discipline you can imagine. I have taught research methods, the design process, ethics, sustainability, design for social impact, materials and manufacturing, and business. In my courses, students develop children's educational programs, products, medical devices, furniture, home goods, interactive media, entertainment properties, apps, books, character brands, environments, toys and games, wearables, and anything that supports child development. In each one of these projects, child development and current topics relevant to childhood form a premise for design research, inspiration and reaching a viable solution.

While writing much of this book, I was not sitting at a desk but on the road exploring America with my husband and 5-year-old son. We drove from Los Angeles to NY and back over many months through 35 states, Canada and 32 National Parks (tent camping) and explored every major city. The purpose was to experience play across America, interview people for this book, advocate for children, and watch kids and teens play in diverse environments. I also attended and spoke at several conferences across design disciplines. I traveled to Finland to present and work on ethical guidelines on designing for children, and to Denmark to be the keynote speaker and participate in a design camp on play at Design Kolding, where companies proposed business challenges about play to students. I spoke at the US Coalition of Play conference about the most important topics for designers to know about kids.

Throughout my journey in this field, I have met and worked with many interesting people who are working in diverse fields advocating for children. All have something to share about their experience. This book captures and shares some of that information and those insights for all designers to use for inspiration.

Enjoy! – I can't wait to see what you create!
Krystina Castella
Glendale CA 2018

Acknowledgments

Nathan Allen, Scott Allen, Dr. Mariana Amatullo, Mark Atkinson, Liliana Becerra, Melinda Beck, Jay Beckwith, Fridolin Beisert, Safir Bellali, Katherine Bennett, Kevin Bethune, Dr. Ann Bingham Newman, Nis Bojin, Ania Borysiewicz, Penny Bostain, Brian Boyl, Russell Chong, Robert Cron, Mihaly Csikszentmihalyi, Leslie Davol, Douglas Day, Rena Deitz, Eames Demetrios, Marie-Catherine Dube, Ana Dziengel, Karen Feder, Marla Frazee, Terri Fry Kasuba, Patricio Fuentes, Alice Fung, Adriana Galvez, Shuli Gilutz, Bill Goodwin, Paula Goodwin, Natalie Hampton, Roleen Heimann, Maggie Hendrie, Cas Holman, Shri Jambhekar, Shellie Kalmore, Petter Karlsson, Michelle Katz, Jessie Kawata, Lauren Kaye, Richard Keyes, Jin Hyung Kim, Yesim Kunter, Ken Kutska, Angela Kyle, Michael Lanza, Paul Levine, Amy Levner, Chris Lindgren, Richard Louv, Michael Mallory, Tessa Mansfield, Dr. Francois Martin, Jason Mayden, Devin Montes, Terry Montimore, Tom Mott, Heidrun Mumper-Drumm, Mari Nakano, Christopher Noxon, Ali Otto, Hsinping Pan, Eric Poesch, Valerie Poliakoff Struski, Caitlin Pontrella, Mariana Prieto, Claudia Puig, Erin Rechner, Shirley Rodriguez, Martin Sanders, Michael Shanklin, James Siegal, Andy Sklar, Alena St. James, Neeti Strittmatter, Cameron Tiede, Matthew Urbanski, Susan Weinschenk, Deb White, Dan Winger, Dana Wiser, Dice Yamaguchi, Jini Zopf.

Introduction

This book is for designers creating for kids and teens. Think of it as a quick course in child development and childhood today. It offers catalysts of inspiration from research and interviews with designers, social scientists and industry experts. Examples of design work in various fields show approaches that we can learn from to further develop a point-of-view and methods to approach our work.

There are areas within our design process that we think about differently when creating for children and teens as individuals who are continually playing, learning and growing. Our approach to research, design, co-design, prototyping, marketing, distribution and the user experience can prioritize the benefits to the child throughout the project lifecycle. Questions we ask as designers are: Which research methods are the best to use with children at different ages? What are the ethical and sustainability considerations? How do you conduct playtesting? How do you integrate co-design with children into our process? Which materials, colors and forms influence our choices in kids' products? How do we enhance the usability of an environment? How do we design with considerations for safety and law? What are the legal and business considerations for children's markets? Asking these questions throughout our process can strengthen the professional viability of a project for children and teens as users.

Our role and responsibility as designers for kids and teens in the products we make, the spaces we design, and the services we offer can strongly influence their childhood, growth and development. Understanding the developmental and cultural needs of children and teens helps designers make more informed decisions, regardless of the discipline we practice. Use this book when working on a product design, an environment, a retail experience, a children's brand, an entertainment property, a UX design, a poster, a toy or game, or any project for which you need to know more about children and teens as users and the audience for a creation.

This book is divided into three sections with nine chapters.

Part I The Design Process

In this section, you will learn about designing for kids and youth from concept to completion. Included are a variety of research methods and topics to choose from to create a research strategy and methods to implement it. There are suggestions for ideation, design development, materials and manufacturing, and prototyping. In addition, there are practical business topics to consider, including intellectual property, marketing and distribution. There are many references to design ethics, sustainable materials and design for social impact, as these are the areas where designers are now leading the conversation, and which can have a substantial impact specifically for children today and in the future.

Part II Child Development

While the basic needs of children and teens remain pretty stable throughout time and around the globe, we are continually learning more about how to address the needs of children at different stages and further customize them for individuals. This section covers developmental stages and the three main pillars of development: physical, cognitive, and social and emotional development. For this book, I have selected common topics within each of these domains of development that come up over and over again when sitting down and discussing projects with students or clients or on projects that I am designing directly. By understanding the cognitive, physical, and social and emotional development of our audience, we can better tailor our designs towards them.

Part III Childhood Today

There are many considerations related to the daily experience of children and teens in today's world that influence development and how they experience childhood. As designers, we consider what kids need in their daily life to give them a childhood filled with health, enrichment and promise. Knowing how our project fits into a broader social context helps us to make decisions that are culturally relevant and shapes innovation for the future. Growing up today is different than any adult, no matter what age, has experienced. Play, education, spaces, and media and technology are at the center of how children experience the world today and this glimpse into the current world of childhood, and there is a chapter covering each of these essential topics and how it relates back to what we think about as designers. In our practice, knowing the issues around these changes helps us make informed decisions that follow the research and our belief system.

This book melds research from the academic, design and business worlds. While developing the content of this book, I conducted nearly 100 interviews with people I have worked with or have admired over the years. They are all advocates for children and youth and range from researchers, to industry experts, to designers. They include psychologist Mihaly Csikszentmihalyi (*Flow*) and behavioral scientist Susan Weinschenk (*100 Things Designers Need to Know about People*); authors Richard Louv (*Last Child in the Woods*) and Marla Frazee (multiple recipients of the Caldecott medal) and Eames Demetrios (*An Eames Primer*); designers for children and social impact, including Mari Nakano (UNICEF), educator Rena Deitz (International Rescue Committee) and Mariana Prieto speaking about a project from IDEO.org; play experts Shuli Gilutz (Israeli UX researcher in children's technology), Dan Winger and Martin Sanders (LEGO

Innovation), Jay Beckwith (Gymboree Play and Music), Karen Feder (Danish play researcher), and James Siegal and Amy Levner (KaBOOM!); illustrators and artists Melinda Beck (Nickelodeon), Doug Day (Warner Brothers) and Andy Sklar (Universal Studios); designers and business professionals at lifestyle companies Ali Wong (The Gap/Old Navy), Safir Bellali (Vans) and Deb White (Pottery Barn Kids). There are invaluable insights from professors Heidrun Mumper-Drumm (sustainability), Katherine Bennett (design research), Fridolin Beisert (creativity), Richard Keyes (color), Nathan Allen (prototyping), Shri Jambhekar (human factors) and Brian Boyl (UX design). And there are many more who have shared their thoughts and knowledge throughout the book. These are all remarkable and amazing people doing great things in the field of design for kids and youth.

Incorporate what you learn on these pages to inspire your process and create products, services and experiences that add value to children's lives. Inspire them to create their own world too.

Part I | The Design Process

Chapter 1

The Research and Design Process

Design is a journey. Designers tend to think of our process as linear, a step-by-step method starting with research, setting goals, making decisions, achieving milestones and culminating with the project release. Having a clear process that we can rely on is helpful, although designing is closer to jumping and skipping around following different paths based on insights, making unexpected connections and considering future aspects of the design process before it is time to dig into them more deeply. In fact, often the most innovative solutions come from stopping, reflecting, playing and breaking the linear path. An insight can bring us very close to the end more quickly than we had hoped or open up a world of solutions we never thought about at the outset of the project. Continuous back and forth between researching, considering, playing and designing encourages innovation throughout the whole adventure.

When designing for children and teens, we use the same processes that we use when we are creating for a general user. Art Director Scott Allen points out the similarities: "Designing for kids is like designing for any age: Tell your story. Know your audience. Discover an awesome design solution."[1] However, there are unique factors that make designing for children different, and we consider these factors in our research methods and project development.

What Makes Designing for Children Different from Aiming for a General User?

- Children and teens change quickly. Designers consider these changes and provide age-appropriate messages, materials, products, services and experiences. They are growing physically, cognitively, socially and emotionally.
- Children are dynamic, not static. Their brains and bodies are powered up for enrichment.
- Learning is attached to everything kids do. We continually question what the child will learn and build on for future learning. How will they be enriched?
- They are always experimenting and change interests frequently. We ask: how does our product, service, system or environment adapt to evolving interests?
- The user is often not the consumer. The designer considers multiple perspectives. Who will use it and who will buy it?

◀ Figure 1.1

Include play within the design journey.
Source: Terri Fry Kasuba (used with
permission).

- Kids don't have the same life or contextual experience as adults. We are always questioning how much they already know and what skills they will be bringing to the topic.
- In research and testing, it is hard to get from them exactly what they are thinking. Younger kids cannot yet verbalize their thoughts. Older kids just may not tell you. To uncover insights we use multiple approaches.
- Kids are erratic and chaotic. You never can predict how they are going to interact with or interpret your project.
- Kids are usually up for a challenge. They thrive on a sense of accomplishment from taking it on and succeeding or just trying.
- Kids love repetition. They can do things over and over again when most adults might get bored.
- Many things are new to them. There are a lot of first-time experiences throughout childhood.
- Safety standards are tight. An extra level of effort needs to be addressed to satisfy safety requirements.
- Kids are unforgiving users; they get frustrated easily, while adults are more forgiving. The design has to be frustration free.
- Kids love interaction and feedback. They often like being critiqued on what they are doing wrong and want to learn how to improve.
- Kids have a level of trust we outgrow as adults. We can use this to educate and protect them when we need to.
- Kids are a vulnerable population. We need to always question the ethics behind our interactions with them, for research, what we make, how we make it and how we market it to them.
- There is more freedom to create with fun/humor and play.

There are also many differences from child to child, in families, in culture and throughout the world. Designers try to understand those similarities and differences and use our creativity to nurture each child so they can thrive and achieve to their fullest potential.

Design Research

Defining the Problem

Sometimes we start projects with what seems to be a clear mission. For example, if we design applications for a software company, we begin with brainstorming new apps. If we plan exhibits for a children's museum, we begin with developing ideas for new installations. It is smart to take a step back and ask, "What is the real problem that we are trying to solve?"

Kids know how to do it. As we get older, we lose that sense of questioning. Since questioning is a starting point for innovation, it is worth taking time to frame the problem to make sure we are working on the right issue. According to Warren Berger, author of *A More Beautiful Question: The Power of Inquiry to Spark Breakthrough Ideas*, "A beautiful question is an ambitious yet actionable question that can begin to shift the way we perceive or think about something – and that might catalyze to bring about change."[2] The right question sets up the framework for everything else that follows. For example, instead of starting with "What kind of toy should we make?" go broader with a question such as "What are potential solutions for play?" Strong questions do not know or imply the solution.

Goals and Objectives

Once we have our question, we outline measurable goals and objectives as the guiding force that drives the design process. Goals are the overall context and "what" of the project. What does this project hope to accomplish? Objectives are the "how" of the project. How will we accomplish the goal with specific tangible results? Sometimes these are solid from the beginning; others might change as priorities shift and we learn where we can create the most value and impact. Examples of goals and objectives may look like this for different types of projects:

Product Design

> Goal: Provide illumination at night so a child can get out of bed safely to walk to the bathroom.
> Objective: Create a lighting solution that allows a child to see when he awakes and not be scared.

Graphic Design

> Goal: Communicate to children the social problems surrounding gun violence.

Objective: Create promotional pieces including books, videos and educational materials that inform children about gun ownership and its risks.

Environment

Goal: Offer inner-city kids opportunities for active play.
Objective: Create play environments and programs on unused city property.

Interactive

Goal: Teach kids phonetic awareness.
Objective: Create a system of products that prioritize sound to learn the alphabet to be used by 2–5-year-olds.

Project Brief

A brief gets everyone involved on the same page. Sometimes a project starts with a brief that is very clear in setting up the goals, objectives and outcomes expected. Other projects are freeform, and the objectives, goals and constraints are developed and shaped throughout the process. The origin of the brief also forms the approach to a project. A brief that starts in the design studio may be more design driven and include defining the user experience and the materials and processes. A brief written by education professionals might outline learning outcomes and assessment criteria. One written by the marketing department would be business focused, such as defining the customer, the price point and the positioning. A brief developed from the legal team could include legal criteria to meet, such as patentability or outlining the licensing strategy. Where there are multiple interests involved, it is best to work on the brief as a team and prioritize competing objectives.

A typical brief includes:

- Company profile or description of the client
- Title of project: Reflecting the purpose of the project in its title
- Project description: Goals, design objectives, marketing objectives, target audience, purpose, function and format
- Scope and technical requirements: Regulatory, safety and sustainability, with a detailed list of deliverables
- Project budget: Provides an idea of how much money is spent on different aspects of the project
- Project timeline and deadlines: List the project milestones. What needs to be complete and who is responsible? Review date/s and completion date
- Measures of success: How will you ensure the design is appropriate for your objectives and audience? How will you test the qualitative features?

Play! Exhibition Autry Museum of the American West

Interview with core team of developers Carolyn Brucken, senior curator and curator of Western Women's History, Alban Cooper, exhibition designer, and Sarah Wilson, education curator, on their exhibit development process.

Can you explain the exhibit development process in depth?

GOALS AND OBJECTIVES

Once an exhibition proposal has been approved an exhibition core team forms to develop initial content and concepts. The Creative Brief is prepared which will guide all aspects of the exhibition, from design to label writing, programs, marketing, etc.

Our Creative Brief:

WHY

> Play matters because it is universal and creates a shared experience across cultures, time and place.

GOALS

> Have fun. Engage local families, and build goodwill with the surrounding community.
> Discover social and generational commonalities.
> Have visitors understand the social messaging within toys and games.

HOW

> Design spaces, and displays in an entertaining, and informative manner.
> Favor experience and haptic learning over didactic explanation.
> Create fellowship and dialogue among visitors.

TOOLS

> Present objects that span a diverse time frame, that also express or represent different cultures.
> Place the objects in environments that hint how they could be interpreted.
> Give visitors a voice.
> Provide comfortable places to sit and play.
> Provide opportunities for amusement, and encourage visitors of all ages to play in the gallery, on the south lawn, and other spaces.

MOOD

> Curious
> Nostalgic
> Engaged
> Whimsical/Fun
> Surprised
> Fondness

Research: How did you use design research methods in creating the project?
We utilized an in-house, collaborative, interdepartmental strategy when designing participatory elements. We engaged in multiple prototyping activities with various museum constituencies. For exhibition elements including case design, graphics, wayfinding, and visitor experience, we pulled from a variety of influences including film, amusement parks, children's museums, toy stores, and child development research.

Strategy: How did research drive your strategy?
Our research illuminated a variety of insights about children's cognitive development, visitor behavior in museums, and visitors' interests that led us to user-centric approaches to design strategies. Research indicated that few visitors would read label text – particularly younger audiences. To achieve multigenerational agency and participation, we incorporated interactive components in each exhibition section. Each interactive element was designed and built to reflect and reinforce the themes and learning objectives of its particular section so that visitors could intuitively recognize and absorb those themes.

Design: How did you prototype ideas?
We prototyped a variety of exhibition components including object selection, label text, case design, media, and interactive elements. We utilized focus groups with children and adults to test interest and interaction with types of objects. We took advantage of public programs to test label types, language, and interactive media. We built multiple scaled and full-sized casework to check function, accessibility for collections staff, structural integrity, lighting, aesthetics, and viewing angles for visitors of all ages and abilities. We tested outdoor games for logistic feasibility and staff buy-in.

Implementation: How did you decide on the final solution?
The core exhibition team continuously met to review content and design ideas, building consensus during the concept and development phases. Final design solution required a series of approvals that led to implementation,

including Creative Brief approval, Preliminary Design approval by department representatives, Feasibility approval, and Final Design review.

How does intergenerational engagement play a role in the informal learning that takes place at the Play exhibit?

The topic lends itself to intergenerational engagement and conversation; anyone can be a teacher: children can instruct parents, and grandparents can share stories with grandchildren.

We understand that different audience constituencies approach exhibitions with a variety of interests, learning styles, knowledge, and expectations. By building displays with historical objects, art, pop culture, media, graphics, poetry, and sound, visitors – regardless of age and experience – can find entry points into understanding the exhibition content and finding relevance in the storytelling.

Once the exhibit opened, what were the most significant surprises?

How visitors engaged with the exhibition space provided surprises. When creating the Reading Cave, we identified it as the space in which visitors who prefer tranquil, quiet play. In reality, the Reading Cave can also be a space of loud, animated, high-energy play, which surprised the team.

At the *Play!* exhibition, we are experiencing dwell times over 2 and 3 hours. We were also surprised by the amount of food brought in and consumed by visitors, an outcome tied to the more prolonged dwell times inside the exhibition. The team was pleasantly surprised by how the show has weathered consistent use.[3]

◀ Figure 1.2

Mother and daughter enjoying a tea party at the *Play!* Exhibit. Source: *Play!* Exhibition (used with permission Autry Museum of the American West).

Develop and Recognize a Point-of-View

As adults, we all have one key advantage when designing for kids compared with, say, if we are aiming for people over 100. We were all kids at one time and can use that experience to help develop a baseline perspective. Self-reflection exercises help us to understand our point-of-view on the topic and give us a starting point for design. Basing our decisions on our own childhood experiences highlights what was important to us and how our unique expertise can inform a project.

Artists and illustrators often capture this point-of-view in a style or visual approach to their work. Author and animation historian Michael Mallory, who co-write a book with Japanese American animator, television producer and film director Iwao Takamoto, explained what he would tell young creatives: "Iwao would say to young artists, Come up with your own ideas. Do something from your life. Give it a personal spin on it. Make it come from within you."[4]

Most people have a nostalgic perspective on their childhood and want today to be just like yesterday. It is a different world today than when we were kids. Technology has transformed the way we all live. We are connected to others around the globe. Educational methods have diversified, and parenting methods continue to change. When we talk to families with multiple children, we see that many things have changed in raising their kids since their last child just 2 or 3 years ago. As designers, we need to step outside of ourselves and expand our perspective to understand and meet the needs of today's audiences. Fridolin Beisert, author of *Creative Strategies*, suggests: "Read about other designers' points-of-view. Talk to others, learn about theirs and try out your own. Point-of-view is never static. It should evolve. Experiences you have, traveling and many things you do will influence it."[5]

Before you start researching the project topic, it is helpful to visualize your knowledge and biases. Mind mapping is a useful process that allows us to structure our ideas to help with analysis and recall visually. This intuitive framework around a central concept is more in line with the divergent thinking processes of the brain than making lists. Design researcher and strategist Katherine Bennett suggests:

> I prefer to do mind mapping before the research, so all the ideas are put down. Acknowledge and stay honest about when your bias comes into play. We usually pull out from our own experience first and then figure what we need to learn. Throughout the process, the mind map can be reorganized.[6]

Through mind mapping, writing, making and sketching, record all of your ideas on the topic. What are some quick thoughts that come to mind? Discover what you know, think and don't know. This understanding will drive your research plan.

Working in Teams

When working on a team project, we play to the strengths and the dynamic of the team. Start out by getting to know about your team members' childhood and their attitudes towards it. What type of unique experiences can each person share? Highlight each team member's perspective. It is also helpful to work with a group of people on an individual project, taking turns to focus on each project concept. Art director Scott Allen suggests: "When leading a team – I first start with a group discussion, followed by individual exploration, followed by a regroup for further discussion. It is important to have time alone to explore ideas and to be a good listener."[7]

Research Plan

Solutions can only be as good as our depth of understanding. To gain knowledge, we dive into research. A design research plan helps us to focus on topics and methods that will uncover opportunities to influence the design process. Why don't we just start developing ideas? We can, and we do. As a design practitioner, I am a firm believer that research and design occur simultaneously. Designers think and invent as we research shaping and guiding the type of research we do. Designers research to come up with insights and need to capture and interpret those immediately. Always research with a design journal in hand. If you enjoy building, make physical models that are inspired by the research. These impulsive gems made early on in the process might not be the most resolved ideas, but recording and working through them can offer a greater understanding of the project and may turn out to be insightful now and at later phases. Play researcher Karen Feder emphasizes:

> Design research has a lot to offer in the intersection between knowledge and practice – not only in design but also in the world. A significant aspect of research in the design process is that you are never done. It goes on all the way throughout the process, as a way to constantly qualify your data, insights, ideas, decisions, solutions and final purpose and impact. Not to say that you need to collect information all the time, but it is a constant humbleness and awareness of your interpretations and conclusions, to make sure your design stays relevant.[8]

Design research:

- Teaches us what we need to know and about what we don't
- Tests our assumptions
- Gives us a greater understanding of the points of view of the users and other stakeholders
- Helps us to avoid missed opportunities and make sure we are solving the correct problem
- Establishes the novelty, desirability, viability, feasibility, marketability and profitability of our project
- Helps us to understand the story we want to tell to connect to people emotionally
- Calumniates in insights and opportunities for design

In the research phase, we:

- Develop goals and objectives for the research
- Create a research plan by choosing the best methods to use that align with the goals, objectives and constraints of the project.
- Establish questions and experiences for collaborations
- Prepare research tools to use, such as surveys, interview questions, prototypes

- Adhere to ethical guidelines and create and use consent and release forms for participants
- Dig into the project space and uncover wants and needs; define problems, opportunities and improvements on existing work in this area
- Look into the history, the future, the context, the objects, the systems, the experience, the business, safety, sustainability and the people involved
- Examine the results to understand patterns and develop insights and opportunities to build on
- Disseminate the research to the participants and other sources that will find it useful

Putting Together a Design Research Strategy

1. Choose a topic area and research objectives.
   Decide on clear knowledge goals. What do you want to learn in your research? Are you looking for behaviors, audience size or usability? What are the best methods to find this out? Brainstorm a list based on what you want to know. Break it down into action areas. Edit it to form focused objectives for knowledge goals to look for in your research. Three is a good number of goals to start with. These are topic specific. Develop exercises for each of the knowledge goals. Create a timeline with crucial decision-making points throughout the process of project development. Align with the chosen methods. Include idea generation throughout your plan.
2. Methods and recruitment.
   Decide on the research methods you will use. Different generative tools can help with different types of questions. How might you best get the information? Which techniques will you use? There are many methods, and it is nearly impossible to learn them all. To recruit kids to observe or interview, go into the field. Who are you going to talk to? Have their parents fill out informed consent forms and model releases.
3. Conduct research with the different methods you chose. Co-design a lot.
4. Analyze your findings. Understanding comes from connecting the past (the research) to the future (the upcoming design). Conclude the analysis with insights that lead to viable design opportunities. Generate as many ideas as possible, and from those insights list as many opportunities as you can generate. Then refine this list to the few best. Throughout the rest of the design process, always present the idea in a way that connects it to the insight that produced it. This helps to remain focused on the original vision for the project – to remember the knowledge gained from the research that forms the foundation of the idea and communicates the information in presentations.[9]

Ethics and Research with Kids

Children are considered a vulnerable population, and therefore have special protections from research risks. Just like medical and psychological researchers in pediatrics, designers often encounter conflicts between protecting children and developing knowledge about children's needs. Depending on what designers are creating, we sometimes face the same challenges and often adhere to similar guidelines to those that are followed by research scientists, mostly in the social sciences fields. When working within the context of an academic institution, a project might need to be approved by an Institutional Review Board. They will review your research methods and procedures and ensure that they are following all federal, institutional and ethical guidelines. An academic institution might work in collaboration with non-profits or corporations that have concrete policies on how to research with children. Check with your institution to understand the process. If there is no policy in place, it is up to individuals to decide on best practices in conducting research, always keeping in mind the benefits to the child. Mariana Prieto, a design for social impact consultant, explains:

> The ethics of doing design research, in general, are essential to not only keep in mind but to address them as a team directly. When designing for children, it's even more critical especially on projects looking to address sensitive issues. On several projects, we have worked with an ethics board to guide us through the most challenging projects, but that's not always an option. If a team doesn't have that option, I would strongly recommend the group gather and do a roundtable of an ethics discussion and write guidelines on how their team will approach the research. Once they have this, share it with someone that has experience in ethics for research and get their feedback. Starting with a rough document will be more productive than calling them with nothing specific to discuss.[10]

Informed Consent

Informed consent describes the obligation of researchers to allow subjects to be active participants in decision-making with regard to research in which they play a role.

1. Disclosure (of information to the subject)
2. Comprehension (by the subject of the information disclosed)
3. Voluntariness (of the subject in making his/her choice)
4. Competence (of the subject to make a decision)
5. Consent (by the subject)

In the United States, for kids under 18, parents need to be informed and sign an informed consent release. If they are in an institutional setting, such as hospitals and schools, we need permission from the venue and the parents. If you are recording the session and plan on reusing, publishing and distributing it, be sure they know about it, and that you gain permission to use the quotes and photos with the proper release forms.

Children's Rights in Design

Kids have the right to have a say in what they will use. With all the good that can come from including kids in the process, however, there is also potentially a negative side, which individuals and companies can use to exploit children. To create ethically rich designs throughout every stage of the process, from research and co-design through manufacturing, materials and marketing, consider the benefits to the children.

Framework to address when researching with and designing for kids:

1. Take the time to understand your values and what you believe is appropriate for children within your user base. Voice your opinion through the work you create.
2. Ask: What type of project is it? Starting off with goals and objectives to sell sugary snacks to kids may have different ethical considerations from a project that engages kids in learning.
3. Think about: How can the research process benefit the child? Consider the short-term and long-term benefits for the child and the company.

Design Prompt: Outline the Benefits

Make a chart that shows a continuum. Establish the short-term and long-term benefits to the child on one side and the benefits to the company on the other. Do both equally benefit, or does the company gain more or the child? If the company benefits more, develop methods in your approach that can bring more benefits to the child. Ideally, it would be nice if the child could benefit significantly from the process and the project created.

I was invited to a Talkoot in Helsinki, Finland to be part of a team to work on a guide that integrates children's rights and ethics into the design process with specific considerations for technology. Daniel Kardefelt-Winther, research coordinator for UNICEF Office of Research, who was one of the sponsors of the event, explained:

> We decided to try a new approach where we talk to the "doers" – designers of digital products – directly and get them involved to think about what child rights would mean in their work. They are the best experts of their own work and through dialogue and co-creation we can both learn from each other and do something together. It's important to consider how we can design from a child rights perspective because policy solutions sometimes take a long time to implement and evaluate, so by the time policy is implemented, technology has moved on and is no longer applicable. So, if you can encourage designers to design products that are child rights compliant from the beginning, you can avoid some of the issues that we would otherwise need policy solutions for later on – it's a preventative approach that can complement the policy approach well.
> Talkoot event, Helsinki, Finland January 19–21 2018

Four pillars were created:

1. Supporting well-being and healthy psychological and cognitive development
2. Encouraging self-expression, creativity, learning and play
3. Nurturing the child as a social being and a citizen
4. Ensuring safety and privacy

The key principles in the voice of the child/teen user:

1. Everyone can use (NON-DISCRIMINATION)
   I need a product that does not discriminate against characteristics such as gender, ability, language, ethnicity and socioeconomic status. Support this diversity in all aspects of your company's design (including advertising).

2. Give me room to explore and support my growth (RIGHT TO DEVELOPMENT)
   I need to experiment, take risks and learn from my mistakes. If/when there are mistakes, support me to fix them by myself, or together with an adult. Encourage my curiosity but consider my capabilities based on age and I need support to acquire new skills and encouragement to try self-driven challenges.

3. I have purpose so make my influence matter (RIGHT TO PARTICIPATE)
   Help me understand my place and value in the world. I need space to build and express a stronger sense of self. You can help me do this by involving me as a contributor (not just a consumer).

4. Offer me something safe (RIGHT TO BE PROTECTED)
   Make sure your products are safe for me to use and do not assume anyone else will ensure my safety. A marked path or "lifeguard" can tell me why something is unsafe, and informs me on how to stay safe. Provide me also with a model for healthy behavior. Make sure you equip my guardians with an understanding of this as well.

5. Do not misuse my data (RIGHT TO PRIVACY)
   Help me keep control over my data by giving me choices about what data to share and let me know how my data is used. Do not take any more than you need, and do not monetize my personal data or give it to other people.

6. Create space for play (including a choice to chill) (RIGHT TO DEVELOPMENT, RIGHT TO LEARN, RIGHT TO LEISURE AND PLAY)
   When using your product or service, consider different moods, views and contexts of play. I am active, curious and creative, but do not forget to also offer me some breathing space. Foster interactive and passive time and encourage me to take breaks. Make it easy to set my own limits and help to develop and transform them as my understanding of the world around grows.

7. Encourage me to be active and play with others (RIGHT TO DEVELOPMENT, RIGHT TO PARTICIPATE)
   My well-being, social life, play, creativity, self-expression and learning can be enhanced when I collaborate and share with others. Provide me with experiences to help me build relationships and social skills with my peers and community, but also give me the tools to distance myself from those I do not want to have contact with.

8. Help me recognize and understand commercial activities (RIGHT TO INFORMATION)
   Label advertising clearly so I do not confuse it with other information. Make sure that I fully understand all purchases before I am paying for those in or through your product/service.

9. Use communication I can understand (RIGHT TO INFORMATION)

Make sure that I understand all relevant information that has an impact on me. This includes the terms and conditions of your product or service. Consider all forms of communication (visuals, sound, etc.) and make it accessible to all. Keep in mind that age, ability, culture and language impact my understanding.

10. You don't know me, so make sure you include me (RIGHT TO PARTICIPATE, RIGHT TO BE HEARD)

You should spend time with me, my friends and my guardians before you design a product or service. We have good ideas that could help you. Also ensure that you talk with people who are experts on my needs.

For more on the Children's Design Guide and Methods and Practices, visit https://childrensdesignguide.org/

▶ Figure 1.3

Brainstorming for the Children's Design Guide.
Source: Children's Design Guide (used with permission).

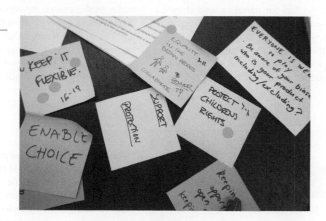

Social Innovation and Design Responsibility

Dr. Mariana Amatullo, associate professor, Strategic Design & Management at Parsons School of Design – The New School, shares her thoughts on designing for children within the social impact field.

What do you see as the most significant challenges in social innovation projects when designing for kids?
All design for social innovation projects inherently carries significant challenges because they call for creating something novel and useful that will also be measured by its positive impact on society. The focus on designing for kids only amplifies the scope of complexity a designer will encounter in the social innovation field. Suddenly your self-awareness about the influence you command with the choices you make as a designer becomes paramount.

What are the most important things designers could do for kids over the coming years?
Innovations that support the education sector, one's access to it, one's experience with it, that we probably see the lasting toll of inequity, both in a country like the US and around the world. What might be ways to open up new spaces for design to address the information poverty many children find themselves in inner-city environments and rural communities is a crucial challenge.

On the other end of the spectrum, in privileged circumstances where access to education and information is not an issue, we are seeing a different set of problems that stem from the technocratic environment that many kids today are growing up in. I am thinking of the classic picture of an entire family plugged to their mobile devices instead of conversing and relating with one another over a meal in a restaurant for example. In many settings, kids who are continuously wired live with a false sense of connection to others. The relational aspect of what it means to develop friendships can get distorted or lost. So how we might design a humanistic core into the technologies, we use and understand their limitation is vital. Bringing back whimsy and play into kids' lives beyond devices seems essential too.

What do you see as the role of a designer in this future designing for children?

I believe that designing for children is an opportunity to tap into the power design has for narrative and storytelling but also the emotional connection. Bridging children's innate curiosity with a sense of purpose and responsibility instilled through your design holds great promise. Focus on teaching children human relations and values while connected to a global society. Give

them a sense of the world. Design open-ended products and tools that can adjust and have a long lifespan. Invest in your role to collaborate with others and in teams. Don't underestimate your capacity to engage in responsible production and consumption. Ask yourself what I am putting out in the world?

Are there specific sustainability issues and ethical issues when designing for kids? How do you address them? What would you like to see change?

When designing for kids, ethical issues relating for example to how a designer might develop the appropriate research protocols to engage with a protected community, which minors are, is fundamental. Beyond questions of the research process, issues of right, wrong, justice, fairness, duty, obligation, responsibility, which are all encompassed by a rigorous understanding of applied ethics when designing for kids are all essential to take into account for a new generation of design leaders who will be savvy about the choices and decisions they make. In this sense, the design for social innovation field is one where we are confronted with ethical issues head-on. The great French writer Sartre once stated that "to be a leader is to be responsible." Designing for kids and in a context of social innovation represents an opportunity to exercise that leadership muscle.[11]

◀ Figure 1.4

"Designmatters Healing Tree project, a storybook system and environmental intervention in the COANIQUEM rehab clinic for pediatric burn victims in Chile, is a beautiful example of a design system for kids that I was part of, and that demonstrates how much research, empathy, understanding the cultural context, and co-creation pays off when designing for kids." Dr. Mariana Amatullo.
Source: Yixuan Liu (used with permission). Project team: Alvin Oei, Belle lee, Lori Nishikawa, Lauren Medina.

Research Methods

There are many types of design research to choose from. Some methods include learning from experience. Others are based on learning from data or the wisdom of others. Although incorporating as many types of research as possible might lead to a more significant knowledge base to spark solutions, it is not always the best use of time or money. For a research plan, we choose methods that allow us to focus on our priorities. A research plan usually includes both primary and secondary research.

The researcher collects primary research for the specific purpose of the project. It is fresh and new. We include all of the stakeholders, kids, parents, caregivers and industry experts at every step of the way from the beginning to the end to get into today's world of childhood. A strong emphasis on primary user research with kids shifts the way we think about our project and helps us to see it through their eyes. Using different methods throughout the process with kids and other stakeholders gives us multiple points of view, deepens empathy and strengthens our depth of knowledge. Play researcher Karen Feder explains:

> I am very inspired by Edith Ackermann and her thoughts. If you want to know someone, you have to "be in it together," to understand what's driving them and what's at stake. With children, you can't just stay on the outside and look in and believe that your perception is the truth. You have to go all in to be sure that you are capable of taking on a child's perspective. To be able to do that, you have to spend time with children, try to understand their everyday lives and respect their opinions and passions. Even if it doesn't fit into your idea of how they should live their life. If you are humble and honest in your engagement with children, you can build up a trustful and equal relationship.[12]

Secondary research was initially created by someone else, and it happens to be useful for our purpose. We immerse ourselves in the most up-to-date research on child development and specifics of children's culture. Information published about kids helps us better understand kids' development, their capabilities, and the cultural relevancy and timeliness of our projects. You will find that there are many things about childhood that stay the same across cultures and throughout time. As Mariana Prieto, a design for social impact consultant, said to me,

> Children are children, no matter where you go. They will want to laugh, play and feel safe and loved. Even teenagers that you might think would be different based on different cultures, religions, upbringings, or access to technology. By spending time with them, you quickly find out that they are not. Everywhere in the world teenagers are interested in the same things. Like what their friends are doing, their social media accounts or hanging out in parking lots after school. They are also scared and uncomfortable about talking about sex to adults. Just like we all were. And they all want something to look forward to.[13]

You may find existing research currently available that can be applied to your project.

Designers use ethnographic research, observing people and behaviors in their natural setting often. We study the context, actions, experiences and objects used to collect data and analyze our notes for themes and variables.

Qualitative research helps us gain an understanding of underlying reasons and motivations, and uncover trends in thoughts and opinions. The methods, usually with a small sampling pool, include focus groups, individual interviews and participation/observations.

It is exploratory, provides insights into the problem or helps to develop hypotheses for potential quantitative research.

Quantitative research is used to quantify the problem, attitudes, opinions and behaviors by generating data that can be transformed into usable statistics. Surveys, interviews, longitudinal studies and systematic observations are all methods used.

Build a Research Team

Include all stakeholders (or as many as you can gather) to be part of your research team. Prepare a list of your ideal collaborators and consider what you need to learn from users. Besides parents, teachers are the adults who spend the most significant amount of time with children and adolescents. Would they be helpful for research on your project? Include experts in the industry your project is in; consultants, creatives, scientists, researchers, business professionals, child development experts and psychologists can all contribute on some level.

- Decide which members of the team will become participants as part of the co-design team and have an active, hands-on role in molding and shaping the final design of the project. The others can participate in interviews, focus groups and other forms of collaboration.
- Ask yourself what role you play as a designer and what is your point-of-view. How about the visions of the members of the team? How can each participant contribute based on their expertise and the level of involvement they will have?
- Will you work directly and collaboratively through the entire process, or have you selected points of participation that are relevant to different stages of the process?
- Prepare the tools necessary for interviews, focus groups, teaching research and co-creation sessions. Develop appropriate brainstorming methods for the co-design team to use throughout the project.

Observational Research: Look and Listen

This process of observing, documenting and interpreting a child's goals and strategies gives designers insight into their thinking. Eric Poesch, senior vice president, Design and Development, at Uncle Milton Toys, recommends:

> If you have no kids of your own, via friends and family, observe kids playing. Observing a child or children involved in unstructured play

gives you insight that you can't possibly gain in any other way. You are reminded of your childhood and how you played, your favorite toys and how you explored your world. Consider these things and compare and contrast them to your current observations. Reading about child development is one thing, seeing it manifested through children at play has a certain magic that can never be appreciated until experienced. If your experience is like mine, you will marvel at certain universal human traits that unify children around the world.[14]

Play futurist Yesim Kunter suggests:

Facilitating an environment that is distant but has some indications of the design topic of the project is the least biased method to get genuine feedback. We ask kids to play, naturally and make a mess, use materials that will inspire them to tell their stories. During these stories, we could pick the real motivations that inspire qualities that we need to know about in designing our project. As a result designers of observations, we are better able to engage children in more detailed conversations in adjacent design research processes including interviews, participatory and co-design.[15]

Not all research needs to be formal, as long as proper ethical practices are followed. Patricio Fuentes, principal of Gel Comm, a youth and family design agency, explains: "We observe and talk to kids on the street, and we subscribe to pulse and trend magazines – we photograph them and things that are happening. We look for outliers. We use pop-up exercises. We talk to them in a freeform manner."[16] With the depth and breadth of these investigations, designers are better equipped with insights to enhance the effectiveness of our creations. Research existing work and studies on your topic and what is already known. Then determine what needs to be added to the existing knowledge in your observations. Design researcher and strategist Katherine Bennett stated: "The goal of the observations is to see what the kids do naturally. Try out different methods for observations such as shadowing, static observations, and framed activities."[17]

▶ Figure 1.5

Non-intrusive observations: By watching kids in their natural setting, you can better understand how they play. Source: (author photo).

Observing kids:

- Prepare an observation worksheet to keep track of what you saw in the session. Summarize the findings. For continuous observations, use the same tools.
- Get an understanding of their lives. Watch them in the process of playing, creating, learning and socializing. Shadow them and watch kids in a natural context.
- Ask questions to yourself as you are observing and take notes on your findings.
- Look at their routines, activities, behavior and interest level, and look for breaks in the norm.
- Look for the visible and hidden emotions. What is the root of the feeling?
- Isolate elements of the experience and break down into small parts to analyze separately and then as a whole.
- Look for apparent behaviors and rare occurrences. Keep in mind that frequency does not mean importance.

Designers Can Learn from Educators

Designers can learn from observational methods used by teachers. Educators spend much of their day watching a diverse group of kids and focusing on the similarities and differences between children. Kids roughly go through the same developmental stages within a "normal" timeframe, but personalities, temperaments and interests create substantial differences for each child. As designers for kids, we usually start off with a generalized framework (developmental stage) and then generate depth and variations based on personalities, interests and skills in the same way teachers do.

What to look for:

- Developmental level: By observing children, we can assess their physical, cognitive and social developmental levels.
- Personalities: When observing kids at play and listening to them speak, we learn a lot about their personalities and temperaments.
- Interests: While watching children play in a natural setting and engaging in various activities, we discover their interests and preferences.
- Strategies: We discover different approaches used by various children to achieve a goal or create a desired effect.
- Skills: What skills have they mastered or do they need to practice? What are their accomplishments?[18]

Techniques used by teachers when observing children:

- Narrative Recording: Anecdotal records describe in a written narrative what happened, how it happened, when, where, and what was said and done.

- Samples: Written samples of specific behaviors to discover how often, how long, or when a particular action occurs.
- Rating Scales: These indicate the degree to which a child possesses a specific trait or behavior. Each behavior is rated on a continuum from the lowest to the highest level. For example: One: not at all like the child; five: very much like the child. A rating scale using adjectives with opposite meanings at either end, such as easy to use and hard to use.
- Checklists: Specific traits or behaviors arranged in logical order. They are useful for types of behavior or traits that can be quickly and precisely specified.
- Media Technique: Capture moments in photos or video and then write notes about what happened.[19]

For our research, we can create templates specific to what we are observing to take notes so that the process of recording is faster and easier. The method used is determined by what you are looking for. For example, if you are looking for risk behaviors, a rating scale or event sampling may be suitable methods to use. If looking for frequency in different play patterns in a group of children, checklists are good. Use descriptive words such as enjoyable, complex, technical, in-control, engaging, monotonous, tricky and held attention. In the testing, avoid words such as unique, exciting and fun. Those are too vague.

Finding Kids in the Target Demographic

If you don't have your own kids, ask people you know who have kids or if they know people who have kids who may be willing to participate in the experience. Creative director Alena St. James explains: "I talk to the kids I know. I have been talking to these kids about what they like and why since they were 5 years old, some are now in their teens. They can articulate what they and their friends like and why."[20] Always ask the parents first if they want their kids to participate, and then ask the kids directly if they would like to.

Observe kids in a non-invasive manner in family-friendly locations or at children's events. Visit a park or a children's museum or arrange with a school. Ask local colleges and universities and their child development programs if you can observe children there. When you are connected with a trusted institution, usually parents have more trust in you. When you approach caregivers, explain what you are doing and what you hope to get out of the observations. Explain how the children benefit. What will you offer the parents and children in return for their input?

Documentation

Pictures and videos are invaluable tools to revisit the experience, to document our process, and to use to explain the story behind our solutions to others. Little details in pictures and videos may show up later that were not so obvious during the observational session. Design researcher and strategist Katherine Bennett recommends:

> Collect observations without infringing. Take notes, sketch and don't move. Sometimes taking pictures is intrusive. After getting permission set up a video camera and leave it alone and have someone else check that it is running. Explain to them why the camera is there. For example tell them that you are recording a video to show other people who couldn't be there. Think ahead to what photos and videos you are going to need in the future for your design inspiration and presentations. Collect pictures of yourself actually doing the work in action. Put together a highlight reel of your research. Make a list of photos that you will need throughout the process. Do you need photos of the team, mug shots of people you interviewed, and overall shots of the experience? If you are working on a team assign roles of the researchers: a note taker, a photographer, and an interviewer.[21]

Obtain permission from caregivers to take recordings and tell them exactly how they will be used. If the plan is to publish the pictures for commercial purposes, online, or in presentations, awards competitions or printed publications, have the caregiver sign a quote/photo/model release up front. Use common sense in regard to "reasonable expectation of privacy." If someone doesn't want their child's picture taken, don't do it. Use a model release form written for guardians of minors from your institution, your company or a lawyer.

Preparation and Recording Behavioral Observations

Develop focused questions based on what are you trying to find out. What do you hope to learn from the observations?

Include on recording worksheet:

- Name, setting (location), date and time, age, gender
- Behavior: Behaviors can fit into categories such as the child's physical presence, disposition and temperament, connections with others, interests and preferences, and modes of thinking and learning.
- Visible behaviors/Invisible behaviors: Make each behavior discrete – behavior 1, behavior 2.

Additional topics to include as determined by the project:

- Object/service used
- Interactions and adjustments
- Feelings of the observed
- Time duration of the activity
- Frequency
- Intensity

- Behavioral flow (sequence of events to complete a task)
- Latency (the time between the start and end of a behavior)

The Session – Observations

When observing children, ask yourself:

- Is the activity age and developmentally appropriate?
- What does the child do naturally?
- What challenges do they face? How do they work around them?
- Observe them alone and interactions they have with others. Is there a group dynamic?
- What clues do we get from the environment or context?
- Is there an unmet need?
- Ask questions that are unexpected and might provoke new perspectives.
- Notice the steps they take to interact with an environment, object or software.
- Is the experience fulfilling?
- Identify gaps in experience. What tools does the child need to accomplish the goals?
- Discover how they really use it compared with the intended use.
- Look at analogous experiences – what are they doing already?
- What are their repeated actions?
- What are the different cognitive levels and skill levels of children at different ages?
- What happens before the experience, at the experience and after the experience?

Analysis of Observations

When you have completed the observations, ask yourself …

- What is your impression of what happened during the time that you observed?
- What did you actually observe? What are the behaviors that anyone watching would see? What behaviors did you find that led to your subjective impression?
- Analyze the notes, worksheets, photos and videos from the observations and focus on the actions, thoughts, motivators and underlying emotions.
- Based on the observations and impressions, what conclusions did you reach? Are any further assessments needed?
- Choose the insights that are most reasonable based on your experience. Compose a list of goals, strategies and theories from the observations.
- Given what you know, what would be an appropriate plan of action? What should be done next? What would be helpful?
- Brainstorm ways the observations can influence directions for your project.
- Use insights from the observations to develop deeper conversations in co-design research and interviews with children.

- If possible, have a discussion with the children about the results of your observations and give them the opportunity to expand on the topic.
- Revisit photos and videos throughout the project to see if they have new meanings.

When summarizing the research, record what was learned and what is not included in the analysis. Later on in the project, these details that seemed unimportant at the time might become your greatest inspiration.[22]

Interviews, Surveys and Focus Groups with Kids

Interviews with Kids

Kids' perspectives as users are especially helpful because many are not yet tainted by the expected solutions. Illustrator Terry Kasuba suggests: "Step into their shoes and talk to them about their real experience not just what you think their experience is. What do they think about, what makes them happy, frustrated, mad, afraid …? They give great feedback." They often speak their mind without holding back, and because they are just learning to put things together in expected ways, they make connections that are often completely unexpected and that could help us think about things in new ways. Patricio Fuentes, principal at Gel Comm, suggests: "Listen to them – let them speak – don't filter their ideas – they know more than you do. As we grow, we put on sunglasses and have rules."[23] Design researcher and strategist Katherine Bennett stated:

> The goal for interviews is hearing what they say about what they do. Interviews are helpful after the observations or in a follow-up to talk about what you saw. Begin by just talking. Avoid leading questions. When you ask them specific questions, you can't find the unexpected things they would have said with a casual discussion about the topic. It's better to ask them to explain what they are doing (or have done) and ask them questions about the issues they raise in their answers, probing to gain a deeper understanding.[24]

Interviews can be conducted individually or in a group or both. Individual discussions emphasize a personal experience. In a group, you hear everyone's voice and get diverse opinions that can be built upon. The goal is for your questions to lead to further discussion. Personal interviews include unstructured, open-ended questions. Katherine Bennett suggests:

> Think about the wording of your prompts. Don't ask people what they want in a future design. You will get an answer but it is a question they are not capable of answering because they only can base it on past experience. You are trying to find what people haven't thought about or what they can't or won't say.[25]

For younger kids, 10–15 minutes is long enough, and for teens you may be able to work with them for a half hour. Spend time on the interview questions to get them right. If you asked a kid what was their favorite or to further describe something in detail, most likely they will talk about the last thing they saw. It is also common for them to say that everything was their favorite. Katherine Bennett added: "Don't ask direct questions. Use a laddering technique. When asking questions try only asking further questions fueled by something they say." It is important to pick up clues other then what kids say in body language, facial expressions or the tone of their voice. It is equally important to think about your body language, Katherine Bennett suggested: "Keep a sense of fun. Get into a play state. Be honest and explain that you want to know what they think and why. Sitting at 90 degrees is better than knee-to-knee."[26] Author and designer Fridolin Beisert suggests: "When designing for kids we need to get into their mental world. In order to connect and be with them on their level we should speak with them at their height (get on your knees) let them finish their sentences and encourage them to open up by asking them to tell you more."[27] Advice for Interviews:

- Start with broad questions asking them about their life and interests, and then move into more specific questions about the project topic fueled by the answers.
- Speak their language. Understand the right language to use based on their developmental level so that they feel comfortable and interpretations are minimal.
- Ask the right question. Closed-ended questions get to the point quickly. Open-ended questions begin the flow of a narrative.
- Watch their body language and take notes. Are they fidgety or excited when you ask a particular question? Analyze why later on.
- One-on-one or two to three kids is enough. Do not overwhelm them with a large group.
- Write down exactly what they said, not what you think they meant, and analyze it later. Direct quotes that inspire you are helpful in validating a direction.
- Create engaging stories and have them build on them.
- Reverse roles. Have the kids interview you and see what kind of questions they ask.

Surveys with Kids

Using a questionnaire, you can analyze a sample user group to collect insights and data. Since this is a quantitative method, the larger the group, the more reliable your results will be. In-person (as compared with online) is the best way to gather opinions and preferences for kids. Surveys can be used with children aged 7 and above, but extra care needs to be taken to ensure that questions are understandable and answerable. Children younger than seven generally don't have the full cognitive abilities developed to complete surveys. If you are working with younger children, use short, simple questions with the parents present. Peer researchers can be used with older children.

- Create a comfortable environment where children can be themselves.
- Keep questions literal, short and clear. Focus on actions and feelings.
- Use multimedia for the questions (images, video, audio, tablets, computers).
- Use the minimum number of response options, including midpoints on scales (two or three).
- Use vocabulary that is relevant to the age group.
- Questions about behavior should be asked of the parents, not the children.
- Avoid ambiguity, negatively phrased statements, and retrospective behavior questions. They are hard for kids to answer.

Focus Groups with Kids

In a focus group, a moderator uses a scripted series of questions or topics to lead a discussion among a group of kids. The goal and strength is to promote the exchange of thoughts, ideas, messages and information. A focus group for kids could last between 15 minutes and an hour. It takes at least three groups to get balanced results. The developmental differences among kids highlight the importance of identifying focus group strategies that are specifically catered to children's communication. It is important that the children understand the questions, have the opportunity to reflect on their own experiences, and can effectively communicate their thoughts and feelings. In my interview with Bill Goodwin, founder and creative director of Goodwin Design, he noted: "Kids are different one-on-one than in a group. You see the social influences in the group. There is always an alpha kid who sways the thinking and conversation of the entire group."[28] Consider that kids communicate differently. Some like to look; others interact and other want to talk. Make sure to plan exercises with tools to engage all types of children. Use drawing, sorting pictures, making collages, writing stories, paired tasks and role-play exercises.

Preparing for the focus group:

- Be clear about the issues that we need to be focused on.
- Recruit between four and six kids with common characteristics. Be mindful of group composition. Are they friends or strangers? Don't include kids who are more than 2 years apart in age.
- Prepare a questionnaire and materials to interact with for up to five open-ended questions. Align the questions and exercises with the research objectives. Avoid "Why?" questions. Be prepared with back-up matter to expand on what the kids say.

Tips on focus groups with kids:

- Establish ground rules for the kids (be a good listener, respectful of others, etc.).
- Engage them. Build trust and rapport and create a relaxed environment.
- Get them thinking about the topic you are designing for.

- Teach them the terminology of the project and explain why what you are doing is important.
- Start off with simple one-word-answer questions and move into items that are more complex.
- Use phrases that react and show interest and support what they are saying, such as "Tell me more about …" or "That's so interesting …"
- Prepare a show and tell. Show them similar things (to what you are developing) they use every day and have them explain them to you.
- Gather quotes to influence your thinking about the project.
- Summarize what is happening throughout the process by writing down insights for the group to see (on a whiteboard or large pad).
- End the session with each kid talking about what you and they learned.

Experiential: Act and Do

Spend time in the field immersing yourself in, participating in and experiencing the project topic activity exactly as kids do. Play with kids in a park, at a children's museum, at a science center. Be with kids on their level, relate to things that they know, make it a game. Eric Poesch, senior vice president, Design and Development, at Uncle Milton Toys, recommends:

> In addition to observing play, participate in play, actively engaging a child, meeting them right where they're at. Before my wife and I had children of our own, we offered to babysit for our friends who had preschool age children. It was in these instances that I was able to become a playmate, following the lead of the child, adhering to the rules of their imagined world. From these opportunities, I distinctly remember fully appreciating how children learn through repetition.[29]

Play researcher Karen Feder suggests:

> Another way to ensure that you are actually designing for the user, is to take on a child-centered approach in the early phase of the design process. If you want to get a deeper insight into children and their everyday life, you can swap the co-process around. So, instead of inviting children into your design process, you, as an adult, should let yourself be invited into the life of children – on their terms. Imagining you are taking on an internship as a child, and spend a whole day together with children on their conditions, following their daily routine, in their everyday environments. You are then learning by experience, from the inside of the real life of children. It is not easy, because you have to let go of the control, and let the children be in charge. All of a sudden, they are facilitating the process, instead of you as a designer. They are the experts – experts of being children, and you are the outsider. You are now in

the hands of the children and you are not the one with the answers anymore – they are. You are co-creating knowledge in the moment. And you have to trust that you are going to get something out of this, even though you don't know what – yet. It is a very explorative approach, and precisely because of that, you are creating the opportunity to derive rich insights you couldn't have envisioned beforehand. This process can give you the answers to the questions you didn't know that you were supposed to ask – which could be some of the most valuable, precious and innovative knowledge in the design process.[30]

Empathy

Empathy is the term most often used when we discuss feeling and understanding the perspectives of people who are different from ourselves. We gain insight by observing, listening, learning and experiencing. Increasing empathy is one of the first aspects of research and understanding in the design thinking process. Mariana Prieto, a designer for social impact, pointed out how getting in touch with universal feelings is an important aspect in understanding users:

> With the speed of globalization the similarities in the lives of people around the world will become more and more apparent in the future. People's day-to-day lives may be drastically different, but who we are inside is the same. Our dreams of wanting a better life for our children are the same everywhere you go. If you strip people down to their core you'll see that their dreams, their fear of seeing someone they love get hurt, caring about what others think, the aspirations for their children's education, are strikingly similar everywhere but just expressed in different forms.

She explained methods used to create empathy by the team at IDEO.org on the Divine Diva Project:

> You can do many interviews and ask teens about what they like to do and what their lives are like … OR you can join them and see it first hand, which is a much more complete and rich experience. So that's what we did. We became Zambian teenagers for a couple of weeks making friends and going to the park with them, going shopping to the local flea market, going clubbing at night and doing whatever it was that they would do if we weren't there. The more we got to know them the more we could feed our intuition and gain an understanding about who they are and how to design for their world.[31]

To better empathize with kids:

- Plan a homestay and follow the kid throughout 24 hours of their day or at least a few hours.

- Role-play in character for a day. Hang out.
- Dress like a kid for a day. How does it feel?
- Create a play state for your design process.
- Use self-reflection. What is play to you? Write it down, draw it.
- Play in the mindset of a kid.

Although we often think of play as something that kids do, the benefits of play can enrich us throughout our lifetime. What changes is how we play. Evolutionary psychologist Peter Gray says: "We tend to talk about children 'playing' while in adults, play is commonly blended with other motives, having to do with adult responsibilities. Adults bringing a 'playful attitude' or 'playful spirit' to their activities."[32] Adults might partake in playful experiences in the physical play of sports such as kayaking, or a creative hobby in the maker culture such as hacking, knitting or creating online user-created content. When I ask professionals what their favorite type of play was as a child, there is always a strong connection with where it has led them personally and professionally as an adult. Play is currently being accepted as a business method for innovation. Companies such as LEGO have design methods such as Serious Play[33] for business professionals and designers to use in all forms of innovation and business performance. Here, tools and technologies to stimulate alternative realities or a play state are seen as opening up opportunities that could not be found in other ways. This is why I highly encourage designers to act like a child, a beginner, and play throughout the design process. When I spoke with author and designer Fridolin Beisert about the role of play in creativity, he commented: "I recently realized that my book *Creative Strategies* is about how to be a kid. It takes away the rules we have accumulated as adults and is about designing for the purpose of experiencing play."[34]

Play helps us learn through developing hypotheses about the world and testing them over and over until we become immersed and more self-directed and come up with theories and mastery. In my courses, throughout the term, we engage in playful exercises such as building play scenarios out of loose parts and hacking the playground to enhance the play experiences; we have tea parties with our favorite snacks from childhood, students come to class dressed like kids, and homework involves going out to play. These exercises offer the same cognitive, social, emotional and physical benefits for us as designers as they do for kids. They also allow us not to take ourselves too seriously and to get into a play state when designing for kids.

Teaching Research

While conducting teaching research, we as designers serve as facilitators and teach the kids about our topic. We can teach them a lot about what we do as designers and a process that they can use in many aspects of their lives. We can inspire them to create. Plan a workshop to teach kids about the topic we are designing for without being too specific about requesting feedback on a particular project. If learning outcomes for the sessions are developed and the

Create a sock puppet of your user to better understand their wants and needs.
Source: author photo.

learning assessed, then we and the children are able to benefit as a shared experience.

Architect Alice Fung explained her experience with teaching research that was used in the process of the design of a school environment:

> It is key to involve kids at appropriate levels. In our case, we collaborated with a wonderful teacher and involved the 3rd and 4th graders in activities to engage the topic that augmented what they were learning: the metric system and water. Looking for connections to the curriculum and slipping in our activity almost seamlessly reinforced our objectives and as well the curriculum's. The collaboration began at the programming stage. We printed out a giant metric-scale campus map into letter-sized segments that were distributed to the students. In teams, they located their assigned area of campus, made observations and mapped their visual discoveries. They documented what they heard: conversations, the freeway, birds, and sirens. We taped the map back together for an overview of the site conditions and uses. Students came up with wild dreams of what they imagined the campus could be. The work that we did with the students travelled home to the parents. As the design developed, we presented models to the teachers, students

and families. Their responses at those events became valuable feedback.[35]

Many students and professionals who design for kids also work with kids, teaching classes or volunteering at a local church or school, because they enjoy being around them. It is also a great way to get to understand kids better to influence your own work. I asked entertainment designer and creative director Andy Sklar what he learned from teaching a kids' class to 4th through 8th graders on theme park design. He noted:

> I learned that the act of play has not changed that much since I was a kid. When designing do what you would do as a kid such as composing with paper and blocks and building models. To set up the project I created a template. We created a park and each kid was given a land, and a ride to design. The kids went outside of the template quite quickly, which informed me about the importance of not having such a rigid template in the beginning. Let the ideas and imagination flow. By having the kids working on a shared big picture my intent was to introduce to them how collaborative the environmental design experience is. They individually had their ride and land to work on but they also had to work with the template of park and how their attraction and land adhered and worked with the neighboring area. Upon review of everyone's ride and attraction we were able to come up with a name of the proposed park. Needless to say there were a lot of roller coasters and thrill ride experiences.[36]

Some children and teens can actually take on the full role of designer themselves. There are many examples of children and teens who have been very successful in launching their own brands, apps, products or services. I developed coursework for teens and taught product and environmental design for many years at Los Angeles County High School for the Arts along with several non-profits in LA County as well as volunteering to develop creative workshops for my son's elementary school. I was always impressed with the outcomes. Where designers can help is in using our skills, connections and insights for kids to bring their own ideas into reality.

User Research with Kids

The best way to create a project that is relevant and user-centric is to have continual work with kids for testing and retesting. If we don't test our concepts and designs with kids continually, we can only imagine what they would think or do or feel. Dan Winger, senior innovation designer at the LEGO group, explains an insight that came from a study:

> Since our products cater to children, no one at LEGO is our core user. Therefore it's essential that we test often to make decisions based

on validated learnings, rather than assumptions. Several years ago, when conducting a foundation study across various ages we noticed a gap in our users at age 4 we noticed a reoccurring pattern in which Duplo products (larger bricks for preschoolers) were perceived as too "babyish," while our existing System products were too challenging to build. This insight was the direct inspiration for our LEGO Junior's product line.[37]

For the most inclusive research, putting together a random selection of kids, ages, genders and backgrounds often offers stronger insights because of the diversity of the kids involved. Include a range of users, from extreme users to casual users and those who currently don't use it. When I was working on testing a girls' toy, the company I was working with only wanted to test it with girls in the target age group. I pushed back, explaining the importance of testing it with boys and older and younger kids. Not only did this give us more information about the differences between the target and other children, but it could also lead to expanding the product to new users. If the parents are available, include them in the process, especially with younger kids, but don't let them speak for the kids. Their opinions and reactions matter and offer another point-of-view than the kids'.

When we are working directly with users, we hope to uncover:

- The kids' motivators and drivers and behaviors that surprise us
- How they define and voice their priorities and values
- The differences between what they say and what they do
- Needs that kids don't even know they have

Methods for User Testing

Concept Testing

Usually in the early stages of project development, concept testing is to get assurance that the concept is on track, or helps to choose the best concept out of many. It is about the idea, not the execution of that idea. We don't look for refined solutions in concept testing, but general thoughts about the topic or approaches to potential solutions. Eric Poesch, design director at Uncle Milton Toys, holds focus groups for concept testing. He explained: "We bring in kids and their parents, and always find it hugely valuable. We never know what feedback we're going to get, but it's guaranteed there will always be something we didn't expect."[38]

Concept testing is more difficult to do with younger kids because they can't visualize something that does not exist yet, so use as many drawings and models as you can to show them the concept. If you are testing prototypes, remember that they will be aggressive with your prototypes, so make them rigid. Sometimes parents are better at getting the answers (especially with younger kids) and transferring the insights to you. Bill Goodwin, creative director of Goodwin Design, explained:

Kids look at the world naturally through our design process. They explore, and discover. We try to capture key moments with mom and kids. We show them visuals and style boards. We ask them about the work we create. What do you like? Why they like it? They say we (as designers) overthink too much. They understand a lot of social norms – such as specific colors or tones, and their meanings such as light blue feeling pure. Kids pick up these things.[39]

Playtesting

This type of testing is best in the later concept phase through the design phase. Kids of all ages can participate. It includes open-ended play with observations of kids learning with tools or prototypes we created for testing. I find playtesting to be particularly helpful when looking at ergonomics, motor skill capabilities and thought processes. These are all things that are very difficult to interpret from secondary sources and from other products.

Play researcher Karen Feder explains the value for designers:

> The designer can understand how their thoughts, ideas, concepts, prototypes and designs relate to children and their everyday life. Not only by testing the function of a specific product, but also by taking their time to dwell on how the children actually engage in the process, the decisions they make and why. This knowledge is not only valuable for this one specific design project, but for the many more to come.[40]

Tom Mott, senior producer and interactive learning designer, explains:

> You're trying to design fun, frustration-free, engaging experiences, so you need to know what the capability of your target audience is. You MUST PLAYTEST. Playtest! Playtest! Playtest! Watch kids play! Interact with kids! See what crazy, unexpected things they say and do. You'll always be amazed.[41]

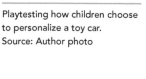
▶ Figure 1.7

Playtesting how children choose to personalize a toy car.
Source: Author photo

Usability Testing

This type of testing usually takes place during the later design phases of development, when we have some solutions defined to see if it is working and what needs to be changed. If you cannot test the real thing, design researcher and strategist Katherine Bennett recommends:

> Choose something similar to what you are trying to do to get a baseline test. Make sure it is simple and you are specific about the usability you are testing. Watch them use it and take notes and video. If testing human factors try to identify the problems with the current products.[42]

Play designer Cas Holman explains her process:

> We put cardboard or plywood parts in front of kids to see if the concept will fly – if they are engaged, interested and want to figure it out. The Providence Children's Museum has been a great partner for this, and we bring new ideas there to see if our designs make sense to kids, are too easy, or not intuitive, or sometimes just don't work. We want our designs to be challenging but not frustrating. Easy is boring. Then when the pieces are designed we test them again to make sure the design does what the concept set out to do (did they learn about how gears work or did they just learn how to make the toy work?). The last round of user testing is done with the first article (or samples from the factory) to make sure the materials we chose will hold up and work how we'd intended. By then we know the product really well. Once teachers and kids are using the products I'm constantly learning. I have skype dates with a few classrooms around the US and the kids tell me what they want. It's kind of a Career Day meets focus group.[43]

Design Prompts for Testing:

Does the product or interface work how you thought it would?
Is it easy to understand and use?
What do they do first? Next?
What do they do that is unexpected that can influence the design?
Do they do what you thought they would do?
What was easy? Difficult? Frustrating?
Do they follow the rules or instructions? Do they make their own rules?
How much time is absorbed by the experiences?

Preparing for the Test

Recruiting Kids

Think about a place where you have a personal connection with the target age group. Your former grammar school, a local science museum where you have a membership, a friend with kids who has friends or neighbors, your church. If you don't have a connection, the best places to start asking to do testing are private schools, kids' museums, colleges, parks and libraries.

Questions to ask yourself when recruiting kids:

- Consider the type of testing you want to do: Concept, Play, or Usability Testing.
- Know exactly what you are going to test.
- Are you looking for small groups or individuals or both?
- Who are you going to talk to? Who qualifies? Who doesn't?
- Make sure you have kids who are the appropriate ages. Will you be testing kids of different ages to see how kids of different developmental stages respond? Sometimes your users are different than you think they are. Are all the kids going to be the same age, and are you looking for personality differences or skill abilities within one age group?
- Do the kids represent the market you are you testing for? For mass-market projects, include kids of different demographics, or do testing at different locations that serve different demographics. For lifestyle products, include kids in the specific demographic in the testing as well as other children. For educational, include an education expert in preparing and moderating your test and assess specific learning outcomes.
- Include an equal number of boys and girls unless you are testing for gender-specific issues. Include kids from different genders to see how they respond.
- Ensure cultural diversity.

Compensation

Due to child labor laws, compensation for children is often a challenge so make sure you know what the current laws are for your context. In many research studies, adults are paid for their participation. When children and youth are compensated for their time, insights and contributions, they learn the value of being an expert. If adult members of a co-design group receive a royalty payment as a design partner, children should be included too. If you will be compensating children, understand what the going rate is for adults by asking professionals. Give the same amount (given to the parents). Additionally, a small gift is usually an immediate positive take-away for the child.

Plan the goals of the session:

- What do you want to learn? Are you information gathering, trying to see if your hypothesis is correct? Are you looking to expand your ideas with their interpretations? Are you observing to get insights into something you didn't know? Set up one or two simple goals for each testing session. Align the research goals with the exercises.
- When asking questions, keep them light or make them funny. Keep them open-ended.

- Understand the developmental stage of the age you are designing for and choose age-appropriate exercises. Create a balanced teaching/learning session as a give-and-take incentive where the kids are educated as much as they educate you, or better yet, more.
- Develop an agenda, but let the kids guide the sessions, and go with the flow of their thoughts and actions. Create multiple ways and questions to try to get results – if one doesn't work, quickly move on to the next, and the next.
- Do you want kid-only sessions? Keep in mind that younger kids may feel more comfortable with an adult they know around. Would a kid/parent session work best? Is there something you need to learn from the parent too? Don't let the parent take over – explain ahead of time – give them something very specific to do to help. Should a separate session with the parent be arranged?
- When doing testing, learn the personalities of the kids. Are they shy or out-spoken? What are their likes and dislikes? This will help you to understand different types of users.
- Are you going to do group testing? What is the group makeup? Friends? Kids who don't know each other? How will this influence the test?
- Develop simple questions to use during testing to get them to talk.
- Use hands-on interactive activities so that they can show you what they think instead of telling you. Drawing and model building are great tools to see their imagination and thinking. Ask them to explain what they made. Learn to read the meanings behind the drawings and models. Act out a scene, tell a story or create a scenario. With digital products, test out an analog version – or vice versa.

The Environment

Natural Observations
Testing is often better in a context familiar to the children than in your environment. A casual, playful space works best for natural observations, because the flow of behavior in a natural setting is more valid. The total situation can be assessed, leading to avenues you had not thought of exploring before. Often conducted on a micro-scale, the results are harder to generalize to a wider group. Participant observation is a variation of natural observation in which you join the individual or group. When in their environment, conduct a contextual inquiry to learn more about them. Ask them to show you around. How is their room decorated with characters? Themes? Which toys do they go to first? Creative tools? Pretend play? Free play or structured play? Look at what they are wearing – what does it say about them? The more you know about them, the more informed you will be on testing the right kids with the right concepts.

Controlled Observations
These are usually conducted in a private room, often with a control group and a variable group. Participants know they are being watched; therefore, they may

act differently. These are easiest to replicate and test for reliability. If they are in a testing room, don't make them feel as if they are in a cage at the zoo being observed. In a testing room, it may take a little longer to get acquainted – have plenty of crayons, markers, model-building materials and toys for them to interact with.

Introductions

- Be a friend. This may seem obvious, but introduce yourself and tell them a funny story about when you were a kid. Tell them about what you do and what you are working on. Your story can be inspiring.
- Make them the experts and let them lead. Think of yourself as one who needs to learn and the kids as the teachers.
- Tell them what you will be doing throughout the session and why.
- Questioning should be a give and take. Ask them about themselves. Tell them about you.
- Make sure that it is clear that they can engage in activities or stop the session at any time.

Working with Different Ages

According to design researcher and strategist Katherine Bennett,

> Different methods work well for different ages of kids. For example with little kids there is a lot of observation but with middle schoolers you can ask how well they perceive something. Kids are chaotic by nature. The younger they are the harder it is. Get younger children in dyads with a friend. Triads work well for older kids. Kids in middle school are thoughtful although by the time they are in high school they are cynical. Going back to the same kids for several observations and interviews at different points in the process is usually helpful. For example work with them on insights, testing mock-ups, validation, ergonomics and safety.[44]

Preschoolers (2–5-Year-Olds)

- These kids are super high energy and scattered.
- Try to limit group testing with 2–5-year-olds and focus mostly on individual or small group testing.
- They have a harder time visualizing concepts and communicating their thoughts.
- Keep your eyes on their engagement and body language. Are they smiling or uninterested? Are they wiggling around or running on to the next activity?
- If you must do group testing, try to make it purely observational and unstructured. Limit to 10 minutes or less per task.

- Watch them interact with objects and environments through natural play.
- Create tasks that reflect a diverse range of fine and gross motor skill development.

Early Grade Schoolers (6–9-Year-Olds)

- They are usually happy to share what they think if they are with a group of comfortable friends.
- They can communicate their thoughts effectively except for explaining why.
- Get them to tell stories about themselves, their friends and family. You can learn a lot by reading through the lines of these stories.
- Find out about rules with toys and computers in their house, and make sure they don't inhibit your testing.
- Limit to 15 minutes or less per task.
- Have them participate in open exploration of your project.
- Encourage them to explain what they are doing as they are doing it.
- If they are having a hard time articulating it, prompt them. How would they describe it to a friend?
- Keep in mind that some kids are afraid of doing or saying the wrong thing.
- Get them thinking – have them design – use big sheets of paper – all working together.
- Give them a template to constrain their thoughts.
- Create different stations with different activities to produce various ways of looking at the solution.

Later Grade and Middle School (10–13-Year-Olds)

- This group are the most articulate and are swayed by peers, so often they are better on their own or in a small group.
- These kids start to hold back, so make them comfortable.
- If you need to use a testing lab, they are usually the group that is most comfortable in this environment.
- They have a better idea of why they do things and a broader cultural context, so develop deeper questions.

High Schoolers (Teens)

- You can communicate with teens pretty much as you would with an adult, although they are a bit more self-conscious and cynical.
- They can be strategic, conceptual and critical, and use reasoning.
- These kids travel and work in packs. Use the nature of the group to your advantage.
- When you get them into a topic of interest, they have very strong opinions.
- Identity plays a strong role in their perspectives. Use this to understand the differences between individuals within a tight demographic.
- Older kids are not as available, so you need to meet them in their world/space.

Finishing up the Session

Give them a warning that the testing will be done in five minutes and give them enough time to have closure and complete what you are asking them to do. Let them finish whatever they started. Ask children what they learned from the session and what they liked the best, and what they didn't like at all. Tell them what you learned as well. Thank them, and follow up once the project is released to show them and congratulate them on their input and the results personalized with their specific contributions.

Follow-up

Since the research can be a learning experience for the child and parent as well, whenever possible, it is important to follow through with information on the results. The child and adult can receive an official certificate of thanks as a contributor, a free product or vouchers, and a report on the results of the testing.

Design Prompt: Testing Plan Worksheet

Prepare a one-sheet that includes your plan for the test with the following topics:

Title, Sample picture of the project, Ages, Time, Safety Requirements, Testing Goals, Procedure, Investigations, Notes, Materials.

Secondary Research Sources: Learn and Expand

Use secondary research as a platform to guide any subsequent primary research you may conduct. It also may be the only available source of specific pieces of information for you to use. Learn about kids through the news profiles, magazine articles, blogs and social networks, and case studies. Analyze behaviors and the effect of the cultural context. Look into opportunities for innovations. Read academic journals, books and scientific reports. These provide the facts and figures of your challenge with data and most up-to-date theories. Look beyond that page and interpret. When using data from secondary research, compare the sources. Are they consistent? If not, dive deeper. Then make sure to verify the credibility of the information by checking the appendix and references. If it is credible, analyze the data to seek actionable findings to move the project forward.

- What are the implications of the research?
- What will this research study lead to in the future?
- Are there common stories people tell?
- What does the data suggest?

Child Development Theories

Many theories of child development have been created over the past 100+ years that try to help us better understand kids' behavior. Piaget's stages of cognitive development, Erikson's psychosocial stages, Bronfenbrenner's ecological model and Vygotsky's sociocultural theory are some of the most respected classic theories in child development. For designers, understanding the basic concepts behind these theories helps us to understand a little more about what children may be experiencing, the way children act and why. Researchers are continually reshaping these classic theories with new studies, so be sure to keep on top of the latest studies. In what we create, designers can also challenge them with what we make for kids to bring about new areas of thinking and learning about development.

Theories usually fit with a specific genre of thinking:

Biology-based approaches: Emphasis on inherited biological factors and processes

Psychoanalytic theories: Emphasis on unconscious drives and interactions with others and the forces that develop within the mind

Environment-based theories: Emphasis on the role of the external world

Cognition-based theories: Emphasis on the role of cognition and processing for information

Contextual theories: Emphasis on the interaction between people and their environments (historical, social and cultural contexts)

Theories often conflict with one another, so sometimes people choose one approach that resonates with them and matches their beliefs, and others pick what seems to make sense from different theories based on the context we are designing for. For example, if we are working on a project that involves problem-solving, understanding theories that emphasize the role of cognition and processing for information might inspire concepts or help us figure out how to best execute a design. If we are looking at the design of a space, a greater understanding of environment-based theories may influence our perspective.

Cultural Research

Culture influences how we learn, live and behave. Context affects any design; therefore, making a project timely and culturally relevant is vital for developing solutions that are accepted, new and up to date. Trend research, behaviors and attitudes of subcultures, specific cultures within a global context, children's culture, pop culture, cross-cultural perspectives and design culture can all be considered cultural research.

With cultural research, we can find out topics such as

- How people in a specific culture or subculture behave and what they value
- How they will use, view and accept what we are creating
- What kids and parents care about today
- How cross-cultural attitudes can affect the design to fit into a given cultural context

Changes in social norms are continually occurring, and cultural studies are frequently updated. Large design offices and marketing firms hire experts in sociology, cultural anthropology and fields of social science studies to bring their point-of-view to the project team. Many colleges and universities are associated with professionals who work in this field. Meet with them if you can. University and public libraries subscribe to online resources and access to insider information, such as trend forecasting companies and academic journal publications, that you can't find in general search engines. Learn about the sources that can help with your project and see if you can obtain access for your research.

Children's and Youth Culture

This includes diverse areas in the field of childhood, from history to current pop culture. Researchers study and develop insights into questions such as what we can expect in the future based on emerging demographic trends. Reading scientific research reports, books and articles about children's culture, play and future visions gives a project a culturally relevant context. Interview experts on the cultural landscape and get input on your project topic at the beginning and the final design towards the end. In most cultures, the adult world, children's culture and youth culture interact and influence one another, but they are very distinct from each other. As designers, we often think of these as target markets that we need to understand. We get up to speed on their lifestyles, megatrends, and their wants and needs.

Today, businesses recognize that children, youth and families are a dominant market, so many products, experiences and services are designed specifically for them. Parents spend much of their free time with children engaging in their culture. They visit kids' themed restaurants, plan kids' vacations and join in with kid-specific activities.

Popular Culture

Pop culture permeates everyday lives in our society. It is what makes the times we are living in different. These phenomena, cultural activities or commercial products are often influenced by the mass media and consumerism and are created to appeal to the mass market. These shared interests often help kids connect with one another. Popular culture categories for kids are entertainment (movies, music, TV), toys, video games, fashion and books. Many designers are involved in the field of creating character and entertainment properties, comic books and products. In fact, so much of childhood today is shaped by the pop culture that it has reshaped much of their experience. How can you use kids' excitement about pop culture to motivate and inspire them? Can the characters in pop culture media help them to understand other topics? Plan a debate on pop culture topics. In testing sessions, have them write a story using characters from their favorite TV show, book or movie.

Cross-Cultural Perspectives

Views of childhood, education and play vary widely across different cultures. Different cultures emphasize specific values and skills. For example, cultures view the role of play and learning in children's lives differently. The treatment of children may depend on economic factors and family and community structures. Children working in some regions is considered unacceptable child labor in others. In some cultures, children are innocent beings, in others inconveniences, and in others possessions. Organizations such as the UN and UNICEF engage in standardizing fundamental rights for all children. Western families tend to have a hands-on approach to nurturing and teaching children until they are self-sufficient, while other cultures might allow children to explore freely. In the US, we see many variations in how parents from different heritages and backgrounds raise children. Generational differences also affect the raising of children. Is your design for a specific culture or subculture? In an increasingly globalized world, what varies when we look at the project from cross-cultural perspectives? When a toy or medium from one culture is brought to another with no cultural context, what are the implications? Taking these values and framing the project around them helps to tell the story.

Design Culture

This field studies design objects, services, experiences, the work of designers, design production (marketing, advertising, distribution) and consumption. It includes what these objects look like, how they are used, and the meanings and functions they perform. Visual, product, lifestyle and consumer trends are a large part of design culture. Questions that may arise in this type of research could be about how design culture will change through technology or consumer socialization and the stages at which young people develop consumer skills. It is essential to see other design solutions, but those designs are already solutions, not the source. How can you understand the context but in your work contribute something new?

Historical Research

Surprisingly, the history of the modern child as we know it began as recently as the late 1800s due to the changing views of children in society. Before that time, they were treated as mini-versions of adults. Today we know they are not. Through studying the history of childhood, design, and design for children, we see connections and consistencies. We can learn from looking back to look forward. How does one idea or context lead to the next? What factors in the social environment influenced the evolution? How did the cultural context affect styling? Are there designers of the past with exciting points of view that we can learn from?

Historical research:

- Gives us context and an understanding of patterns
- Helps us see what worked and what didn't and why
- Helps us invent the future from what we learned from the past
- Shows us what remains constant and what changes
- Shows us how our solution fits into what has existed before and what it may lead to
- Stops us from reinventing and keeps us moving forward

Research the history of your project topic and look at the toys, media, books, characters, stories and environments that surround it. Understand why various projects were appropriate at specific times. Develop a view of the cultural relevancy of your project and understand why it reflects our times today.

Designers and content creators can look towards nostalgic media for inspiration. Eric Poesch, senior vice president, Design and Development, at Uncle Milton Toys, loves watching movies for visual inspiration and embedding story into products: "Sci-Fi and Action Adventure are my favorite genres. I often derive inspiration from movies and will reference certain scenes when brainstorming and developing product concepts for key themes like space, exploration, and dinosaurs."[45]

Suggestions for designers when studying the history of design for children:

- What was the child's place in the overall society?
- Look at general design themes in the design of children's projects. How did they fit into a broader context?
- Study the history of the design discipline you are designing in (graphic, product, interface, environmental design). What can we do today that we could not do in the past?
- Look at products for kids as design artifacts. How do they reflect a time and social structure?
- What role did the branding, marketing and advertising to kids play from a historical perspective?
- What was the business success of the project you are looking at? How long did it last? Is it a fad or timeless?
- How do characters in media reflect the times? Look at Peter Rabbit, Buck Rogers, Mickey Mouse. Why are some still relevant and others not? Have they evolved for today's audience?
- Watch historical media, cartoons, TV shows, movies; read vintage kids' and comic books. How do they represent the times?
- The visual history of childhood/design (an exhibit that is inspiring to look at is MoMA's Century of the Child exhibit).[46]

Charles and Ray Eames

Interview with Eames Demetrios, director, Eames Office

Eames Demetrios is the grandson of Charles and Ray Eames, a creative in his own right and director, Eames Office. We spoke about [the Eameses'] work specifically for kids and what we can learn from them when designing for children today.

What drew Charles and Ray's interest in designing for kids?

With everything they did, they saw a need. Happiness is a need; pleasure is a need. They felt everyone learned best from direct experiences and wanted to give some of those direct experiences to kids. They understood the value of spectacle. The journey of scale in *Powers of Ten* is a spectacle. The Probability Machine had 16,000 marbles and is reliable and hypnotic.

[Charles and Ray] had a lot of respect for kids. [The Eameses] thought of the role of the designer was to be a good host. They also saw the needs of the guest can be different for children. They saw and heard them.

The Eameses celebrated that toys were not models or perfect replicas – toys get to the essence of the idea. A toy train holds the spirit of a train. The toys of *Toccata for Toy Trains* and *Tops* look at the core of topness, the essence, and connections between different sets of objects.

How did their attitude towards play affect their designs for children?

The rich complexity of the Solar Do Nothing Machine says a lot about the Eames attitude towards play. It was part of how they approached any problem or challenge. People ask what would they do if they were alive today? First off, they would have played with new technology.

They didn't compartmentalize play to kids – they saw it on a continuum. They loved the super ball for its extreme expression of form and function. They devised *Mathematica* the exhibit so it should work equally to intrigue a 9-year-old yet not insult the professional. When is the last time you saw a math or science exhibit that you are sure will still be fresh 50 years later?

What could designers today learn from a design history perspective in studying Charles and Ray Eames' work for children?

Here are some ideas:

Think about the honesty of materials: A kite is an excellent pass/fail test. The answer is easy – it either flies or not. THAT shows the success or failure of the design.

Think about who your guests are. Kids can tell whether a toy is good or not. They know more about some things

▲ Figure 1.8

Eames elephant 1945.
Source: Eames office (used with permission).

at that age than we give them credit for. Any 5-year-old can tell you if a toy elephant is a good toy elephant or not. The Eameses showed real mastery of the materials.

The most important thing to them was the designers' understanding of the idea and to work with that. When working on *Mathematica* they didn't look for a high educator as a math consultant, they looked for an active mathematician (in that case, Ray Redheffer). Why? They wanted to work with broad and unfiltered information. In their approach the Eameses never delegated understanding, they wanted the design to flow from the learning. Too often designers are in the situation, whether intentionally or not, where they are asked add aesthetics to content rather than find the aesthetics OF the material.

And, as a small aside, while they liked to use humor, that humor was always on point – it always advanced the message. The most entertaining graphics were still all in the service of the idea.

Can you choose projects from Charles and Ray that represent each of these and explain?

Playing: Think about the importance of open-ended play: A great example is the House of Cards. Kids who play with the House of Cards can play for a long time. The Eameses were inviting folks to hold and arrange images in a time way before today where we all do this all the time. It is a great social activity too.

Learning: The *Mathematica* exhibit's objective was simple: to let people in on the fun. We are given a dreary image of math and science, but people *in* mathematics are having fun. The Bell Curve created by the Probability machine, the Minimal Surfaces display or the utterly compelling vortex of the Celestial Mechanics exhibit promote close observation that leads to learning.

Growing: Emotionally – Physically – Intellectually. There are all kinds of growth. *Powers of Ten*, the movie, is emotional. At about 10–12 years old you see where you fit in the context of the universe. Many people have told me the impact of that movie on their young selves.

Collections of toys, travels, memories and self-reflection of childhood affected the subject of Charles and Ray's work.

They collected toys that were examples of good design. Value and cost were not the same – some things are expensive for a reason: good quality. And some things are probably still worth not much more than what they paid for them.

They loved the circus and clowns – The circus is filled with Eamesian lessons: Take your pleasure seriously. What better examples than the high wire or the trapeze? It came down to how well you prepared the stunt and effect, and yet those people were having fun.

Can you explain how their work in multiple disciplines informed their approach?

Charles and Ray said: How do you design for use by another person? You create for yourself, BUT you aim for the *universal* part of yourself, NOT the little areas where you are different, but the 98% where we are all the same, and it will connect as well. Use yourself as a laboratory.

Value the personal story – How do we design for the commonalities? Their friends had kids and grandkids. So the Eameses asked: What would be of value to these kids? Well, building, playing with images, and so on. They wanted to have a good present to give.

Storytelling was a valuable tool and kid driven – puppet shows – Mathematica – world fair projects. Convey an idea through the story of the product.

Advice for young designers they once gave was this: You have a limited amount of time that is genuinely yours, don't waste it. Use it to explore, to see if your idea is good, invest yourself in it, take that risk – who else will? Don't wait for permission to start that design process, do what you can.

Powers of Ten would be considered a STEAM film today. What do you think translates well in these films today to communicate STEM and STEAM concepts?

To me, scale is the new geography, and understanding scale is a form of literacy. And *Powers of Ten* is an expression of that. At the Eames Office of today, that is a focus of our efforts. I have done workshops with 3rd graders, but also corporate seminars. In all these cases, groups make their powers of 10 as we put each participant's images on the wall to illuminate, discuss and understand. There are so many ways to explore these ideas around scale. So many ways to take your powers of 10 journey, but they all stem from the visceral experience created by Charles and Ray Eames so many years ago.[47]

Sustainability and Safety

As children are our target group, we have a responsibility to design safe product offerings, production methods and marketing messages that support positive values. In many cases, safety and sustainability are considered together. I interviewed sustainability expert Professor Heidrun Mumper-Drumm, who offered:

> Designers are constantly going in circles to invent the newer and better. How can we sell more of this cereal? We need to come to design from a different place. Designing for sustainability helps to tackle environmental challenges and also provides a host of opportunities for true innovation.[48]

When we design for play, children have a right to safe play with safe products in safe environments. Caregivers should feel confident that children are not exposed to products and experiences and inappropriate messages that influence or affect them negatively or may cause emotional or physical harm. Designers are responsible for thinking creatively about sustainability and safety issues and testing products thoroughly. We must design within safety constraints and make sure the project is safe for physical, cognitive, and social and emotional growth. We must be aware of false claims about our project and make sure it does what we say it is going to do. A holistic approach and asking the right questions about sustainability and safety in designing for children is essential. Heidrun Mumper-Drumm alleged:

> Designers can preempt problems during their process by looking forward to the potential outcomes of what they are making. Look at the "best" practices around health and safety and consider: how can we make them "better." What are the current standards in children's products, packaging, marketing, and production practices? Where are they lacking and how can they be improved?[49]

There are many design challenges around sustainability when designing for children:

- Safety: Safety is of highest importance. Today, many more sustainable materials rival the unsustainable choice in safety performance.
- Lifespan: Children's items have a short lifespan. They outgrow quickly. How do we keep products that are designed for short-term use safe or, better yet, design for long-term value?
- Cost: Eco-materials currently are more expensive (bamboo, recycled plastics, organic cotton). How do we bring down and fill the social equity gap around cost?
- Marketing: Marketing sustainability is often not a priority for big brands to model the change. Smaller companies know this is their advantage. How do we influence large companies to shift their priorities and help small companies grow?

Design Prompts: Safety and Sustainability

- Focus on areas where we can make the most positive change. How do we support children's playing, learning and growing with the things we produce, the messages we communicate, the environments we build? What should we design? Why should it exist? Do we want to be a part of what this project stands for?
- What are the positive and negative effects of our design decisions? How will this creation affect a child today and their future?
- How do we teach children the difference between want and need and encourage them to be conscious consumers and eliminate unnecessary purchases and make the most of what they already have? At the same time, how do we support them to make a positive footprint on the Earth?
- Does the project we create bring the child closer to a positive goal or further away? Can we include a positive impact to counteract an unavoidable negative one?

- What is the real effect? Physically on the environment? Socially on the people? Economically on the community and the business?
- How can a designer take on the responsibility to model and guide parents, companies and children to make better choices?

Design for Long-Term Value

- Design products, spaces and communications to optimize safety, materials and energy. Eliminate harmful substances at all levels, including chemical, mechanical and electrical safety. Minimize the size, shape, weight, number of parts, and materials and processes used to save energy and simplify for recyclability.
- Consider the end of life. Although we can't eliminate the end of life of a product, environment, packaging or printed communications, the longer we encourage the use and reuse of what we make, the fewer new materials are introduced into the system. Support a closed loop system at the end of life. Consider new business models, take-back programs, disposal and reuse.
- Design for physical, cognitive, social and emotional growth, including short- and long-term benefits. Carefully consider the latest and greatest and the value it offers.
- Because of the quick user turn-around for children's products, clothing and toys, promote reuse, multi-use, levels of use, short-term use and long-term use.
- Build in emotional connection and attachment to objects as a path to longevity. Consider customization and DIY as a method to produce attachment.
- Include a resale market or barter system recycling program, products or services that grow with the child in children's goods.
- Heirloom-quality materials expand the lifespan. Never use more than is needed to achieve the goal.

Sustainability in the Design Process

- Create ethical products/service offerings. Design for age appropriateness to avoid misuse or underuse of a product. Consider the ethical and environmental impact at each stage of use.
- Model sustainability, safety, inclusion and ethical practices to children.
- Prioritize safety, user experience, performance and then cost. Is it beneficial, safe and healthy for individuals and communities throughout its life cycle? Always strive for higher safety standards than those required by law.
- Include a life cycle analysis as an integral part of the design process. Work closely with team members who are engineers, product planners and production coordinators to avoid mistakes that create waste.
- Design for a minimal effect on the environment at all levels: materials, manufacturing, products, packaging, environments and communications. Ask: What about the physical product or packaging can be eliminated? What potential harm can occur to the child?
- Proudly showcase the style created from eco-design, materials and safety in the design language.

- Avoid introducing new materials and technologies into the system whenever possible. Choose technology for longevity. Allow room for the redesign of mistakes to reduce wasted inventory.
- Consider methods to reduce shipping emissions and miles: Ready to assemble, knockdown, direct to consumer, palette configurations, DIY.
- Advise the user how to use safely and how to recycle if possible.

Life Cycle Assessment and LEED

When conducting a life cycle assessment (LCA), design and product engineers have precise requirements on how to select sustainable raw materials and processes and to build in recyclability. Clear requirements during the design stage can potentially aid in qualifying for sustainable product standard certification, giving assurance to the customer. There are different methods to use.

More information about the process is available at www.okala.net/.

Leadership in Energy and Environmental Design (LEED) is the most widely used third-party verification for green buildings. LEED-certified buildings are resource efficient. They use less water and energy and reduce greenhouse gas emissions (https://new.usgbc.org/leed).

Sustainability in Children's Products

Interview with Heidrun Mumper-Drumm
Professor Heidrun Mumper-Drumm is the director of Sustainability Initiatives, ArtCenter College of Design, and vice president, International Council of Design. We discussed the challenges in making sustainable products for children.

What do you see as the most significant challenges in making children's sustainable products?

- Designers need to know what the current standards for sustainability are. CE – European community safety, Lions Mark British standard, Green Blue sustainable packaging, and Sustainable Apparel Coalition are all excellent places to start to understand current standards. There is a lot to be done. We can continuously be improving in this area. Nevertheless, there are significant gaps in global standards and accountability. Even trusted brands outsource and sometimes don't know their supply chains, effectively "outsourcing" their responsibility for issues such as toxicity. Glow in the dark, fluorescent colors are fun, but what about trace radiation? Look carefully at flame-retardants, electronics, battery usage and energy efficiency. An excellent way to manage is to create a list

of things never to use, and a secondary list of things to use only in trace amounts if necessary.
- Design for responsibility is rarely done. Why don't we design children's products for repair? Kids can repair them, or learn to fix them from those more experienced. The notion of the reusable, hand-me-down has been forgotten. We no longer make things with the quality to last or hand off to other children and grandchildren.
- We remember toys that we "invested" the most in. When I was a kid, I went to Laguna Beach with my family to an art show and got an easel. It was a big purchase and many hours were spent being an "artist." I cut and sewed clothes for my traditional Danish Troll doll and built shoe box environments. Now many kids want what is new.

What do you think designers need to understand about designing for children?

- Always keep the experimental, exploratory value of play at the forefront. It is important to base a project on an educational goal for the child. Learn about child development and play to influence your decision-making process.

- Develop a rational and thoughtful approach to designing for sustainability. Expose children to skills that prepare them for sustainable behaviors. Play and toys can model sustainable practices in the real world. For example, using play food that includes food waste for kids that are playing "store" or with a kitchen set. Why isn't there always a compost bin built into a play kitchen? What kind of play can model these behaviors?
- Promote independent thinking. A baby does not wear a Disney bib because they like it? Rather than promoting a play experience, parents are reflecting themselves in the child. What is appropriate and what kinds of values do we want to instill in children?[50]

Ergonomics/Human Factors/Inclusive Design

Human factors relate to how individuals behave physically and psychologically with products, environments or services. When designing for children, there are many considerations connected with physical and cognitive growth and their relationship to one another that designers need to know to keep them safe. We collect data on averages to see the benchmarks on sizes, heights, weights, sensory inputs, attention and other attributes to make design decisions. Shri Jambhekar, user experience designer and design educator, suggests: "When looking for information see if the research has been converted into a standard. There are many websites and books with information on ergonomics that list the standards."[51]

Three Areas of Ergonomics

Physical Ergonomics

This includes the physical load on the human body. A designer must consider the size of a child, or the result may be an object too large or small for its purpose. Designers use anthropometric data to make sure products and environments are usable, body movements and reach zones are appropriate, and the designs are safe. Physical properties of the projects we design, including forms, buttons, dimensions, placement and materials, can help improve the product environment or interface. For example, proper positioning when sitting at a desk in a classroom can prevent back pain. A backpack with the right size of materials and placement of straps could help with posture. A spoon that is easy to hold could prevent food from falling all over the floor, and more nutrients would be consumed.

Cognitive Ergonomics

This area is concerned with mental processes, such as perception, memory, reasoning and motor response, as they affect interactions among humans and products, environments and experiences. Most children follow the same stages of development, but each child differs individually in the rate, frequency, length of time and strength with which they utilize the cognitive skills. They also have different motivations, fears and interests, which affects interaction and decision-making. For example, a child with well-developed spatial skills will understand and have an easier time navigating a virtual environment. A child who is afraid of a character will not proceed with an experience. A laundry detergent pod or a soap bottle can look like candy, and a child may try to eat it.

Testing ergonomics for sizing for this bug eye toy on different kids' faces made it so that it fits comfortably.
Source: Eric Poesch (used with permission).

Organizational Ergonomics

For children, it relates to school and other institutional settings and is used for topics such as computer use and furniture design. Product and interface and interaction designers are challenged by ergonomic issues related to the increased use of digital devices in schools and at home, and how to tackle what has typically been considered an adult work-related injury. For example, some schools are starting to increase movement in the classroom through the curriculum, layout and furniture.

One type of exercise user experience and design educator Brian Boyl uses, which combines both physical and cognitive ergonomics, is posture studies:

> Look at the touch points throughout the experience – what are kids physically doing at each interaction. What cognitive resources do they have available? If someone is sitting and looking directly at a computer, you have more focused attention than if someone is walking down the street with a cell phone. The interaction design needs to be different both physically and cognitively for these contexts. Considering design interactions such as button sizes, type hierarchy, and visual contrast for different contexts can make the information easier to absorb for users.[52]

When developing specifically for kids, we look at a range of kids within a specific age group. A taxonomy Laughery teaches is:

1. Design out the danger in the first place (designer's responsibility)
2. Design to guard against potential problems (that might cause harm)
3. Warn against it in a way a kid might understand (this is the last resort)[53]

- Children in their learning mode are vulnerable and inadvertently do dangerous things that might harm them – hence design with utmost care
- Anticipate all kinds of issues ahead of time, because kids are capable of very creative approaches to play and problem-solving (creating some unanticipated matters)
- Design with their context in mind and not as adults observing and designing for them (go beyond observation)
- Prevent visual design that might send confusing messages to kids – especially smaller kids who still don't have language comprehension
- Look at all activities throughout the day and take out any possibilities of danger in the first place
- Check the design in all the three modes – learn, entertain, play (look at the solution from each of these perspectives separately)[54]

Inclusive Design

When designing for inclusion, we consider accessibility and the variations among people, including children and those with special needs based on growth, experience and abilities. Over the past several decades, there has been progress in including children with physical differences; however, we are just starting to design around those with cognitive and sensory differences.

Just being smaller and having fewer learned experiences in an adult world are already putting typical developing kids at a disadvantage. Most of their environment is supersized for kids, and they learn to adapt to it. How can designers interpret these objects and spaces with universal design and transform them into creative, kid-friendly experiences? Designing for adaptability and preventing injuries for both able-bodied and physically challenged children can be integrated throughout the process. Interaction designer Brian Boyl suggests that designers consider: "What are the extremes of which you are going to design? Don't design down. If well considered it will not just be a better design it will work for everyone."[55]

The benefits of inclusive design support not only the child but also the community of children that surrounds them. Children should be able to choose how they want to engage in the play space or with an object with developmentally appropriate challenges, building comfortably on the skills they already possess to help them gain new skills. Contact among children with different abilities is often not enough to get children playing together or to create genuinely inclusive play experiences. Educators and programmers who use intentional strategies to promote disability awareness promote quality inclusive play opportunities while providing all children with the tools to ask questions, get accurate information, explore their feelings and learn how to interact with their peers positively.[56]

Content creators for stories and characters can consider including all types of children in their stories. Art director and illustrator creative director Douglas Day suggests:

A strong character is a role model for the child to build their confidence. I recommend designing characters with differences in mind. Different personalities and abilities help kids take on the roles of various characters. Kids will see even with diversity they can do things together and be accepted by the group for their differences. This also creates more interesting scenarios about overcoming conflict in the stories.[57]

Sometimes kids with special needs are not invited to participate in experiences, and for whatever reason, it is not possible to include them. When children with special needs are shut out of an experience in the physical world, it is important to offer an analogous experience. In this way, they can benefit too.

Principles of Universal Design

These principles were initially created by the Center for Universal Design at North Carolina State University in 1997 by a working group of architects, product designers, engineers and environmental design researchers. The purpose of the Principles is to guide the design of environments, products and communications. They include

- Equitable Use
- Flexibility in Use
- Simple and Intuitive Use
- Perceptible Information
- Tolerance for Error
- Low Physical Effort
- Size and Space for Approach and Use[58]

Safety

We are lucky to have many health and safety standards that are outlined by agencies such as the Consumer Product Safety Commission (for toys and juvenile products) (www.cpsc.gov/), ASTM (for playground safety) (www.astm.org/) and the Food and Drug Administration (FDA) (for food and drugs) (www.fda.gov/), to name a few. These guidelines allow us to prioritize safety in our designs. Also, there is plenty of data connected to ergonomics for children that helps us to design within a kid-friendly framework (www.humanics-es.com/recc-children.htm).

Inclusive Design Tips from a Creative Director and Mom

Interview with Valerie Poliakoff Struski, executive creative director with an MS in Industrial Design. She is Mom of a remarkable child who happens to have some special needs.

How do you define inclusive design?
Inclusive design is a big, broad topic. It takes into consideration more than just your typical kid. The intention is for all kids with all abilities.

Keep in mind that many special needs children have the same needs as any other child, but the conversations between the parent and child are very different. For example, if a child wants their independence but they have to say, "Can you help me walk?" and ask permission. They want to play dress-up, but physically they can't do it by themselves. When the stacked blocks keep falling in our family, we say, "What do you do when you fall? We keep getting up."

An excellent place to start developing inclusive designs is to look at the most common issues. CP is the most common motor disability in children. It's an umbrella term that refers to a group of disorders affecting a person's ability to move. Physical impairments can range from mild to severe, additionally visual, learning, hearing, speech, epilepsy and cognitive disabilities are possible.

What do you think are the most essential topics designers should be working on to create more inclusive designs for children? Why?

Many environments are difficult to navigate for special needs kids. There are playgrounds, children's museums and spaces in nature everywhere where designers didn't consider all of the details. Grass, sand and mulch are uneven surfaces and are difficult to walk on and make it impossible for these kids to enter or use the space. They say, "You are not welcome here." The children are left out of the complete experience.

Design spaces that enable independence. Children with special needs need freedom. A wheelchair fosters independence better than a stroller because many kids can move themselves around and not have to be pushed by a parent. Playgrounds are terrible. If you can't get to or on the slide or down the slide what is the point? We pick spaces to go that our daughter can maneuver and wander off on her own in her wheelchair and we can watch her from a distance. Science and Natural History Museums are big spaces with good lines of sight, and she can wander away.

Do you have suggestions for approaches to the research process?

Talk to Physical and Occupational Therapists. All kids develop fine and gross motor skills at varying rates, not just kids with special needs. With only a little more information from these experts, you can design a product to go younger and older to reach a broader range of typical kids and kids with special needs. When you talk to the PT or OT then tell them about your idea and what you are thinking of making and ask them how you can make it more accessible? Ask them questions like: Would bigger be better? Would handles help? How about the material if it is harder or softer?

Products designed specifically for kids with special needs are costly for most families. You can add a zero or two to end of the price tag of a comparable product for the mass-market audience. Designing for kids with special needs doesn't need to be a massive hurdle or much more expensive. Everyday products for kids can be modified with a few details. I have this great tricycle which has an extra strap intended for a younger kid to give them trunk stability. The designer was probably not thinking "universal design," but the details make it work for my daughter.

Are there specific issues related to each of these areas that you think need work? Why?

Products: When products are not designed for kids with special needs parents can curate other offerings that work for their kids. We look for products that won't frustrate her. Any kid gets frustrated when they can't make something do what they want to do. We offer our daughter challenges but don't introduce products where she will get too frustrated. I don't know if she will get to play with regular Legos but Duplo the large blocks are fantastic.

Environments: Design surfaces and movement through space that include all children. Nature and disabilities don't go together. Even designed nature is hard to access. Kids with special needs also need to be in and experience nature.

Technology: Consider the sensitivity levels when making an app or electronic devices. Today they are very exact and reactive. The sensitivity can be modified to adjust to have a broader range for varying fine motor development. Again many kids with delayed fine motor development can't use a tablet or learn from the content on it. They are missing out.

Media: Create inclusive content. Shows like Daniel Tiger on PBS that has a lot of life lessons about trying again and a song about falling down. That includes all kids and is particularly meaningful for a kid with extra challenges.

How do you see inclusive design benefits the community of children?

When kids play together with the same toys or in spaces that are used by all children, a commonality is formed. All of the kids in our daughter's preschool are learning at an early age about inclusivity. It teaches the community about the similarities, not the differences. Our daughter loves surfing, horseback riding and skating at the skate park just like the other kids – she just has her own way of doing it.[59]

Children with special needs have the same needs as other kids. They want freedom and independence.
Source: Valerie Poliakoff Struski (used with permission).

Business Factors

Design does not stand in isolation. Often a design is part of a system and an overall business strategy. Successes and failures don't just rely on the quality of an idea. Successful projects also rely on making the right design and business decisions. Mari Nakano, former design & interaction lead for UNICEF's Office of Innovation, explained to me the critical role of designers in business in the coming years:

> Designers are going to have to know how to talk business while still holding true to our foundational goal to create things that will truly and positively impact children's lives. We'll have to think more deeply about things like the scalability of our work and whom we partner with to develop and design products and services. We'll need to know how to work in interdisciplinary teams comprised of both designers and non-designers alike and with people with varying personalities. I'm not sure yet if we should prepare ourselves to be jack-of-all-trade-type designers, which is sort of the trend now, but I do feel we'll need to consider developing expertise in a separate skillset that complements where we want to be (e.g. Policy or international development if you want to work for a place like UNICEF). We have to be prepared to position ourselves in boardrooms and high-level meetings, so communication is quite key to our success in this environment. We've got a lot to share, we provide valuable observation, and we're good at untangling complex issues, but we have to learn how to articulate how our creativity will add value. Basically, we have to think about how we take on more leadership roles, which ultimately means that we need to beef up our skills in business and advocacy.[60]

Designers are trained with deep skills that we can use in the business world to help children grow. In his TED talk *The 4 Superpowers of Design*, Kevin Bethune, vice president of Strategic Design at BCG Digital Ventures, explains how designers have four significant superpowers:

1. X-ray vision: We can tap into implicit behaviors, attitudes, motivations of people
2. Shapeshifting: We can emulate the behavior of those we are designing for – turning feelings into products and services
3. Extrasensory perception (ESP): We can think of traits that people have. The hopes, fears, and multiple senses
4. Ability to make others superhuman: We can develop products and services that enhance human potential.[61]

In an interview with Kevin Bethune, he explained:

> Children's early-life experiences involve many emotional triggers that are present between their friends, their parents and how they feel as they develop their own voice, personality and self-esteem. Designers can investigate these triggers and find ways to inspire kids through their natural inflection points as they figure out life.

With the power that business has in children's lives and their development, how do we use this to best support them in a positive way? Kevin suggests: "Instead of treating children as an afterthought (e.g. 'taking down' adult product to present a kid's version) in their product lines root products and service at these key inflections, to begin an authentic conversation."[62] Many companies have come to realize the importance of including products and services specifically for kids in their overall strategy. Some companies focus on kids and families, but today many are new to the youth and family market. Patricio Fuentes, principal at Gel Comm, explains:

> Every company needs a youth initiative since they will be a customer some day. Kids are growing up faster; brands need to take the market seriously. By getting into this market, it helps them understand this audience when they grow up and how they will evolve and build loyalty over time.[63]

> Due to technology, kids today have a more significant sense of speaking for themselves in the adult world and creating their own innovative products and services for themselves and their peers. So instead of just designing for kids, companies are now looking at and trying to figure out where they fit within the context of innovation by kids.[64]

Positioning

Knowing what is currently on the market and positioning your project in comparison with others is an essential driver in design differentiation. It helps buyers and users understand where a product "fits" in terms of what it offers, and how it compares with competing products. Ania Borysiewicz, creative director and graphic designer of Ladder Design, recommends:

Know everything about the project and company you are designing for. You can't just research it and read about it. You need to experience it and feel it emotionally. Ask others about their feelings as well. Sometimes it is anecdotal and not scientific, but the stories can be powerful.[65]

Eric Poesch, director of design at Uncle Milton Toys, suggests: "Retail audits are critical, to see what's out there, what's selling, and often more importantly, what's being marked down. I buy a lot of samples to play with and test." [66] Graphic designer and illustrator Melinda Beck suggests:

It is important to look at the environment in which the graphics will sit. I spend time looking at what others have done with similar design problems. What makes this client different from their competition? What type of visual best communicates their unique point-of-view?[67]

To position a project:

- Segment your market and understand the overall industry
- Pick your target: Identify which segment and user the project appeals to the most and why. Do you need to create a new segment?
- Position your project: Highlight the unique advantages of your project.
- Develop a positioning statement: Write a short explanation of why your plan is the best solution to the target segment's requirements.

Case Studies

Case studies tell the story from start to finish of people, products, environments or services. They show us how the industry works and critique related approaches, solutions and systems. If there are case studies publicly available on your project topic, read and review them to see what you can learn. We can also do our own informal case study of closely related projects in the children's industry we are designing in. Choose a plan for which you can get an interview with the designers, engineers, and production and business people who worked on it. Interview the experts on how the project unfolded and why design and business decisions were made. Prepare a summary of your interviews along with the learnings that you can move forward with as inspiration.

Some key business topics to investigate:

- Business models, costs and revenue models associated with doing business in children's industries
- Trends in kids' and youth industries and market sectors

- Marketing methods, distribution channels for children's products, including specialty and mass market, brick and mortar, digital distribution
- Branding, competitive analysis and positioning of existing offerings in the design space
- Manufacturing processes and labor issues
- Intellectual property opportunities

Your new business insights will help you to structure a business strategy for your project. That business strategy can evolve into a business plan and pitch to potential partners and investors. This could help to bring it to life in the real world.

Ethical/Sustainable Business

As consumers' priorities have shifted and companies have learned that doing good is good for business, more and more companies have developed sustainability initiatives to help better the planet. As designers, we can persuade companies, parents and children to use better materials and production processes, use positive business models, teach children that less is more, and use design thinking to solve social issues for children. Heidrun Mumper-Drumm explained: "The application of design strategies that address sustainability is a skill set designers must be capable of. If you don't have it, you will not be prepared. Companies aim for best practices, and designers should be able to do even better."[68]

- Consider children's' rights throughout the design process.
- Use ethical business practices and create ethical product offerings.
- In addition to qualifications, hire employees for diversity.
- Hire from, engage, support and give back to the community.
- Ensure safe and responsible working conditions throughout the supply chain.
- For distribution facilities and retail stores, set up a reduced energy and emissions program, and reduce product packaging, waste and transportation.
- Reduce waste and energy consumption in the office through recycling and work from home programs.
- Avoid products that promote media addiction, gun violence, gambling and tobacco.
- Include a recall program on all products. Work in close collaboration with suppliers, licensors and relevant authorities to alert them and arrange for necessary actions.
- Support initiatives in underserved communities.
- Support fair trade practices, mentoring and children's education programs.

Intellectual Property

Intellectual property (IP) is the legal rights over artistic and commercial creations of the mind. Under IP law, owners are granted exclusive rights to intangible assets, such as ideas, discoveries and inventions. The most common forms of IP include copyrights, trademarks, patents and trade secrets.

When designing for children, there are specific focus areas to consider, depending on the project you are designing and the industry you are creating within. A few examples are:

- When developing a medical device for children, you may focus on areas within the design that could obtain a utility patent. The same could be true for a toy invention. This may also help to get investor funding, secure a licensing deal from a toy company, and help deter competitors from infringing.
- When developing a story and character property in the form of a children's book, television show, film or game, you will want to obtain copyright protection of the written and visual content. The story and characters can be licensed to various media formats as well as for product merchandise.
- When developing a brand, you will want to trademark your company name, product line names, product names and logos. You will also want to trademark the visual language that supports it. This will help consumers identify the source of the goods as being authentic.
- If you have a project that has a smart twist, a secret formula or method of making it, you might want to keep this a trade secret and not disclose it to anyone.

Michelle Katz, an intellectual property and start-up attorney, offered some suggestions for designers:

1. Create original characters and storylines. Don't base yours on current trends or existing characters. Keep checking to see if you are consciously or subconsciously changing it or altering something that already exists.
2. Don't brand your product or company too early. Wait until you have your project figured out. Stay away from descriptive terminology. For example, naming your company a children's toy company.
3. Make sure a product is new and novel. The innovation can be small, but it should be there.
4. Make sure that when you promote your product to any third parties that you don't make representations about the impact of your product. For example, don't say "educational product/improves learning" if you don't have a factual, research basis to prove it.

Researching

Start the process of learning about the different areas of IP and the laws in the countries where you intend to obtain it. In the US, the Patent and Trademark Office (uspto.gov) and the Copyright Office (copyright.gov) websites both have information on the requirements to obtain IP, search databases with existing IP and online application processes. Early on in your project, study all of the steps from beginning to end, so you know what is involved in the process.

The second step is to research and uncover the IP that already exists that is close to what you would like to accomplish with your project. Getting a general idea of the type of IP in your project categories that others have obtained will give you a context of where others have been successful in acquiring IP, and it may inspire areas of development in your project. For example, in your search, you may discover that what you have in mind to develop may already exist and may be legally owned by someone else. Knowing this early on in the process allows you to build on existing work, not repeat it. You don't want to invent a new technology that is already owned by someone else or design a logo around a brand name that is already taken in the classification you will need.

Developing and Obtaining

Once you know the type of IP you will try to acquire, you will include that thinking in your design development. For example, if you have identified that one of your designs has a novel, new and non-obvious approach to the structure that can be further developed into a patent, you may want to pursue that design. If you have a unique ornamental aesthetic to your product design, you may consider a design patent. When brainstorming brand names, if you have identified a name that is perfect and currently not owned within the classification that will be required for your product or service, then you can feel more comfortable designing logos, packaging and graphics with that name and may begin the process of obtaining it. You may also find that you need to buy or license IP rights from another creator or owner. For example, using a children's character from a movie or using someone else's patent in your design will require permissions and most likely fees, given that they want to sell or license them to you.

Maximizing

Do you think J.K. Rowling knew that the copyright and the trademark for the Harry Potter brand would turn into one of the most popular global brands, including books, movies, theme parks and real-world housing developments? Publishers, toy companies and media companies in the children's market rely heavily on brand awareness through licensing collaborations. Stories and characters are made into books, films, television shows, games, clothing and a variety of other products. Many big companies have brand licensing departments.

Artists could also license the copyright on their surface designs or illustrations to reproduce, sell and distribute the work on products in the children's market, such as children's clothing, backpacks and notebooks.

If the patents on the products you designed can be used in many other products or may have value to others in different regions of the world for production, the patent can be sold or licensed to others to use. Enforcement of IP right could also be a strong defense against potential infringers or, hopefully, an incentive for companies to pay you for your ideas instead of using them without permission.

- Licensing out IP rights to other companies can be lucrative through license fees and royalties.
- Businesses can enter new product categories through licensing or co-branding with minimal risk and cost.
- Licensing brands can increase their exposure and can lead to new markets. For example, kids may feel more connected to a cereal or toy and proud to wear a cap and shirt with the logo to show their allegiance.
- Character and personality merchandising (real person) is a marketing technique that is used to increase the appeal of products or services to potential customers who have an affinity with that character or personality.

Types of IP

- **Patents.** In the US, patents give the owner the right to try to prevent others from making, using, selling, importing or distributing a patented invention without permission.

Utility patents, also known as invention patents, are given to the creation of a new (or improved), useful and non-obvious product, process or machine. The most common utility patents are assigned to:

- Machines (something composed of moving parts)
- Articles of manufacture (products)
- Processes (business processes, software)
- Compositions of matter (formulations)

Design patents are for the visual elements, the distinct configuration, distinct surface ornamentation or both. Design patents that represent the formal language or brand aesthetic can prove to be valuable to companies that have a particular look and feel to their products. They cover only what the consumer sees (i.e., not ergonomics). According to Michelle Katz,

Even small variations on a design patent can be outside the protection of the original inventor's patent. For example, a children's dollhouse that has a hexagonal domed and candied roof would not keep others from making a single dome roof with points and differently shaped candies.

- **Copyrights.** Literary works, visual art, musical compositions, performances and software are all copyrights that the creators can own. Michelle Katz recommended:

Remember it is not the concept that is protected; it's how it is described and expressed in its tangible form. Create your best version of a style guide and register for copyrights early and often. Copyright the drawings, characters, research, user manuals, marketing materials, charts, graphics and interface design. Label your work with copyright notices. Registering gives you a record of when the work was created.

- **Trademark.** A trademark (TM) is a recognizable insignia, phrase or other symbol that denotes a specific product or service and legally differentiates it from all other products. TM law protects consumers so that they will get the product or service they expect. TM law also helps the brand owner for the time and money they spent on making consumers familiar with and approving of their brand. The names of products and brands, logos and slogans are trademarked. Some brands in the children's market, such as Popsicle, or Band-Aid are so prominent that they have almost replaced the common words "ice pop" and "bandage." IP lawyer Michelle Katz suggests:

In choosing a brand, you want to create a unique identity which consumers will recognize by sight and sound and associate with your products or services.

Brand

A brand represents the long-term strategy and personality of a company. The goal of a brand is to connect with customers who align with its values, products and services. Patricio Fuentes explains: "As parents, we teach consistency, sense of connection, comfort, and warmth. It is similar with brands. The user can be sure that they are getting authentic products, services, and experiences while supporting companies they believe in."[69] Ania Borysiewicz, creative director/graphic designer, explains: "It is important for consumers to know about a company and what they stand for. Do they trust the owners and their vision? Do they want to be part of this community?"[70]

Every business has a company brand that they design around, and also individual product and service experiences that they create around, which support the company brand. Art director Scott Allen points out:

"Branding" is a term that comes from the world of marketing. "The Brand" is one of many elements a designer should consider. Branding is important to (some) adults / parents. The actual experience you are designing is more important for the child. Understand the DNA of your project, of which brand identity is just one element.

He continues to point out:

It is essential to distinguish between: DESIGN OF THE EXPERIENCE and DESIGN OF PROMOTIONS AND ADVERTISING

DESIGN OF THE EXPERIENCE
The most important things are: story, message, fun, look & feel. While designing the experience, brand plays a role but is in the background.
Examples include:

- A picture book
- A theme park attraction
- Online game
- Toy
- Interactive (real world)
- Interactive app
- Museum exhibit
- Augmented-reality experience

DESIGN OF PROMOTIONS AND ADVERTISING
Commerce – Designed to sell
Brand is essential to the adult and (to some extent) the child.

Examples include:

- Advertisement
- Billboard

- Catalog
- Poster
- Print Ad
- Banner Ad
- Animated giff
- Packaging[71]

Patricio Fuentes, principal at Gel Comm, explains:

> The process of designing for kids is not different than any other branding project. We give our clients a market overview – trend boards, market, competition, colorways, shape and form, and stylistic direction. Style guides are the backbone of the brand. Mature brand managers appreciate thoughtfully designed style guides. In the licensing space, it defines the execution of the partners. They are also used in organizational leadership to get everyone on the same page with things like a value proposition.[72]

The younger the kid, the more the brand portrays its personality to the adult's values. As kids get older, they influence the adults and make their own purchasing decisions.

Jini Zopf, senior manager of Hot Wheels Packaging Design at Mattel, revealed: "It is challenging designing brands for kids because you need to think about multiple audiences: The user (child) is different than the consumer (parents). We are also designing for the retailers (customer)." Jini Zopf added: "Parents are always making sure it is 'on brand' yet want to see 'What is new?' We need to show what is different from their current toys at home yet still connected with the brand they are familiar with. We purposely design around the transitions in our product offering."

It is essential to understand the age demographic and the sub-demographics within an age group. For example, Jini Zopf said: "For Hot Wheels, our entertainment licenses are for a specific boy type versus our classic Hot Wheels boy. We purposely design around their interests even within sub-categories."

Jini Zopf discussed her research process:

> For our early research we do competitive shopping. We look at other products in the same category or age demographic. We try to understand what is selling and why? We work with kids directly and want to hear exactly what the kids are thinking. We then brainstorm ideas and begin to concept. After testing our ideas with kids within the demographic, we refine and move into production. Once the product/packaging is in the market we keep tabs on sales and talk to the buyers at retail. We use these findings to determine if it the product/packaging is still relevant. We can adjust and make improvements based on the pulse of the market.[73]

Kids today are very brand aware. Ania Borysiewicz explains:

Branding stays with you from an early age. All the products in our life are (modeling) behaviors. For example my 2-year-old daughter knows the logos of foods that we eat are all naturals or gluten-free from Whole Foods we get samples for Costco and what town we are in by the Boba shop logo. She loves Happy Baby Puffs because the container is easy hold and the way the puffs dispense is fun for her.[74]

The average 3-year-old can identify dozens of logos. Psychologist Alison Gopnik says: "Children are extremely good at this, and it's important to them to ask, 'How do I divvy things up in my culture?'" Kids need to understand brands, and they will. What we owe them isn't protection. Instead, it's brands that promote decent, healthy products, and an education that supports a robust understanding of why and how those brands are working so hard to sell us on their stuff.[75]

Bill Goodwin, creative director and founder of Goodwin Design, stated:

Brand is a big deal in our world. We lose sight that this has been around before branding was even a term. When kids are developing it is a way to communicate with each other. A way for them to discover and express who they are. Their social personality. Brands show what you are into.

Brands market values to kids. Bill Goodwin explained: "This is the world we live in. They are being prepared for the real world economy and social model. When we were kids we were to be seen and not heard at the table – then kids got their own tables – now they have their own food."[76]

Some brands for kids focus on consumerism and tell children, "You are what you buy."[77] Others focus on integrity and authenticity, which are very important in the kids' market today. Jini Zopf said: "Brands stand for certain values and kids have emotional and social affiliations with them. You call out things they care about and why it should matter."[78]

Aspects of a Brand

The brand promise: The tangible and intangible (products, services, feelings) the company tells the consumer they will deliver on.

Brand perceptions: How consumers perceive a company, products, services.

Brand expectations: Based on the promise that what customers expect must be met in every interaction.

Design and business: Brand persona (identity) – The appearance and personality of the brand.

Brand elements: The tangible elements (brand logo, messaging and packaging) that are consistent to communicate brand promise, shape brand perceptions, meet brand expectations and define your brand persona.

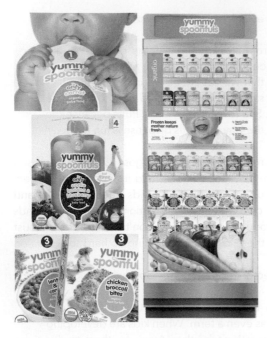

Organic food packaging and branding system designed to offer healthy and independent eating.
Source: Goodwin Design (used with permission).

Brand design: The unique colors, typefaces, characters, form language, words, slogans and visual styles created to represent the brand. How do all these elements interact with one another? Consideration of varied uses, sizes, interactions across media.

Smart Brands

Express the consumer experience and target concepts and feelings that already exist. Ania Borysiewicz explains:

> Smart branding relies on knowing detailed demographics about the target users in this case they know the exact parents and children that you are going to sell to. What do they value? A design may be a nice design but will it speak to that specific person or group of people?[79]

Smart brands:

Are clever: The design communicates something emotive to a viewer.
Are flexible: The design should not be rigid and should able to adapt to a multitude of uses with ease.

Portray individuality: The design should convey character by understanding what it represents, its history, its function and the ethos behind it.

Are simple: Don't overcomplicate. The simpler the system, the clearer the cohesion and uniformity of the communication.

Packing Design Considerations from Jini Zopf at Mattel

- Designs for younger kids have more of a pop, bubble gum, whimsical, brighter, lighter and playful feel. As kids get older they are aspirational, sleeker, fresher, more tech feeling.
- We know that the entire experience of opening the box and building the toy is part of the play. We try to think how playful we can make the process. We cannot count on the fact that the child will read the instructions, so we need to make them visual enough that they can complete the building process with ease. We need to consider their reading level so they can do it themselves and not have to depend on a parent. We also know that our color choices and fonts will help to relate to the kid.
- The copy on the package communicates the benefits and details of that particular toy. We need to show more than the play patterns. We need to present it in a way that appeals to this child's interests and maturity level. We also take into consideration what mom or dad is looking for. Are they concerned about the educational value of this toy? How can we help to alleviate those concerns?
- Design and storytelling comes from the "show, don't tell" mentality. You never want to SAY it's fresh or fun. You need to SHOW it visually and find ways to bake it into the imagery and tone of the copy. We need to guide the consumer through the out-of-the-box experience: How do we tease a kid's imagination? We want to make them excited about the prospect of playing with it. As they engage in the package by opening and assembling, we want to deliver on that promise.
- With online shopping, the purchase decision is made before they touch and feel the package. In the store, the box has the role of informing the purchase decision. We think of it like this: What do we want to communicate at 10 feet away in the aisle, then 5 feet away? When they pick it up, or turn it over, how will this give them the assurance that this product is what they need for their child?[80]

Design Prompts: Branding

Research: Who is your target customer/s? How is this project/company positioned in the current market? What is being offered that is unique? How do other companies visually represent what they stand for? How can you communicate your point-of-difference? Create a positioning statement.

Define personality: What values will you promote in the brand? How will you deliver them to the customer? Create a list of keywords that build an attitude that supports the brand. Everything that you do reflects back to these values. How can these be communicated through naming and visuals?

Strategize: How will you reflect the needs of your target customer? What strategy will best maximize the brand? What formats and outlets are best to promote it? Will you need a local, national, regional or global brand strategy? How might they differ?

Execute: What type of IP (trademarks and copyrights) will support the brand? Developing brand visuals, for the project experience and marketing materials.

Developing Visual Communication for a Brand

Interview with Ania Borysiewicz. Ania teaches brand development and packaging design. She is a creative director and graphic designer of Ladder Design.

What process do you use to teach brand development?
Designers need to bring out the emotion of the brand. Think about: What people will connect to? What story does the company want to tell?

- First define the key brand attributes, the words that are chosen to represent the brand values.
- Develop a challenge to come up with words (adjectives) that may work. For example: Can you brainstorm 200 words in 5 minutes?
- Live with a thesaurus. Ask: What is this company now? What do they want to be?
- Pick six keywords. Use super detailed words that are always adjectives. Never use nouns. Don't choose synonyms, colors or words like unique or innovative. They don't add anything.
- Place the words on a circle. Does it feel balanced? Are the words equally spaced? Play with the words and swapping them with new ones until they all make sense. Often times you think you have it and then you have to redo the process again since a word might be too close to another word's meaning and therefore creating a domino effect.
- Do you feel something? Are there holes in the story that need to be filled?
- Adjectives for the key attributes easily make a description and the visuals (the photos, the illustration, the font and the colors) should easily come to your head.
- It is easy to determine if the visual works. Does it match the words you choose in the key attributes?

How do you go about choosing specific fonts, colors, graphics, illustrations, packaging materials?
People ask, "Which typeface is best for this?" You can't answer that question. It depends on everything else. If you pick a really sweet typeface you need to choose colors that compensate for it by balancing, contrasting and complementing. Once one aspect of the design is chosen it does not live by itself. For example the typeface may not fulfill all attributes but with the colors, illustration and other pieces they all balance out to create a feeling. You need to know a few steps ahead about the other elements to make a decision because everything you do creates a domino effect. For example the logo may set the foundation for other parts of the project. A complicated logo may steer you to make all other parts minimal. If the color is loud and type is loud they cancel each other out and don't work.

As a mom who is also a designer, what are essential things to think about related to design for kids?
Get in the mindset of the parent and child. What are their emotions? We both might be having a meltdown or only slept a short time. I might be negotiating with my kid and I want them to meet me halfway on the purchase.

- Less is more. Don't overwhelm people with visual choices. Too much choice brings about indecision. Too much information makes it hard to understand.
- I want the products to work intuitively. I don't have time to study directions.
- Be honest and authentic. Don't make promises or glowing claims. Even if it is proven I don't trust it.
- Don't patronize parents or children. I don't like packaging that tries to educate the parent.
- Don't fill all the space. I need space.
- I don't like reading too many words. For example five words or less on packaging is enough. I can't process all of the benefits. If I want to know more I can visit a website or read more info later for product if I am interested.
- Make everything fast. I don't have time. I have a toddler pulling at me in store and when I am online. I have a critical amount of time to make a decision.
- I think a lot about how much space a product will take up in the house. It can't be too big.
- Think about sustainability. As a parent I definitely choose less plastic.[81]

Trend Research

Trend research and analysis helps us to predict and create for the future. We can incorporate the research on trends and forecasts into our projects to develop a better products and services experience. In my interview with Tessa Mansfield, content and creative director at Stylus, an innovation research and trends firm that I use and value, she explained:

Trends provide businesses with compelling narratives, visual inspiration and social context from which they can plan and create products or services, and against which they can measure their brand. Regardless of the size or type of business or industry, everyone can benefit from gaining a broader perspective and understanding of trends and their movement. And for businesses to innovate or disrupt the convention, they should look outside of their own industry to cross-industry trends, benchmarks and best practice in parallel and neighboring worlds to consider how these could translate to their own.[82]

Many cultural and design trends for kids tend to be rooted in nostalgia, since often adults are also the consumers. Others start with the adaptation of a youth audience. When I asked Tessa Mansfield why she thinks many trends begin with kids and teens, she stated:

> We look at a wide range of consumer groups to inform our trend research, but the younger generations are particularly interesting because they are the connected consumers of the future. They're the fastest technology adopters, are super social, and are not constrained by the limitations of the past. With this in mind, younger consumers are experimental, imaginative, and build networks quickly.
>
> While they don't hold the family purse strings, they often hold purchase influence in the household. Partially due to universal connectivity, these generations are in some ways growing up faster than previous generations. And, in terms of personal and social diversity, they appear to be the most open and accepting consumer groups that we've seen.[83]

It makes sense, since each generation is trying to create their own world that is different from the adults around them. They are explorers, open to new ideas, are experimenting and trying new things out. They are not set on their "brands" yet. They are looking for what they can make their own. New trends start as a reaction to what is mainstream. What is mainstream that needs to change?

When researching trends, we look at the host of different things that come together economically, politically, and in terms of consumer attitudes, psychology, behavior, communication, technology and media, to name just a few. Tessa Mansfield noted:

> Trends don't exist in a vacuum – they evolve over time, and one needs to recognize the whole lifecycle of a trend. We track the evolution of trends because we believe it's critical to understand their development in order to chart their future trajectory. So for each trend we identify, we consider its past, present and future iterations.

For designers, trend forecasting also refers to the change in aesthetic, the use of materials, processes and colors, shapes, technology and anything that we might have in our palette to work with. Tessa Mansfield explained:

> It's essential for designers to work through visual trends, understanding the aesthetic references of the past to apply them to future design directions. This is true across disciplines – from fashion to product design – and makes for an informed and enlightened design process.[84]

There are many companies that provide services for trend forecasting, or we can do it on our own. Creative director Alena St. James, who does trend research for her personal project development, explained:

> I use trend services when developing colorways, textures, patterns and materials. I am also looking for the ability to take advantage of new manufacturing processes. However, I believe designers need to rely on multiple sources as one trend supplier could negate another's trend. Additionally, you need to see what trends are happening in specific geographic areas (the US vs. Europe vs. Asia).[85]

Types of Trends

Fads: A fad is fleeting and affects people in a short period of time. They are essential if you are designing in a fad-driven market such as clothing or toys.

Micro and macro trends: A micro trend typically lasts a few years. A macro trend lasts in the 5 to 10-year range. Different kinds and a high number of trendsetters adopt the trend. For example gender neutrality in children's clothing has evolved from a micro to a macro trend.

Megatrends: Megatrends take years to develop and are long lasting; they shift the world. Megatrends of today are globalization, increased urbanization, the movement towards sustainability and the increase of technology usage.[86]

Fast culture includes fads, memes, fashion, viral content, natural language, trending topics, news events and micro trends. Slow culture includes macro trends, consumer behaviors and movements that have a cultural impact over many years.[87] Looking at fast culture trends helps us to understand how or why something caught on. Looking at slow culture trends gives us a greater understanding of how and why this will impact society in the long term.

Once we learn about a trend, we define patterns to think about why the direction is relevant. What made this take off as a trend? Once the patterns are predicted, then we can use that information to propose the future and make design decisions. We can envision how our project fits into what is happening in the world. For example, Erin Rechner, senior editor, Kidswear at WGSN, explained to me how sustainability, gender neutrality and technology are affecting the kids' and youth clothing market:

We see a big shift towards more seasonless and versatile pieces that create a timeless wardrobe. Less is more, is a true shift in consumer culture, moving away from fast-fashion and heavy environmental impact. We still see a push towards gender neutrality for kidswear, especially with the continued focus on Athleisure trends. Even high-end brands that used to do just occasionwear are breaking into more casual clothing designs to pair with their more traditional pieces. As instant information is more prevalent than ever on Instagram, Snapchat, Instagram Stories and Facebook live, everyone is looking to the Gen Z and millennials for inspiration. Fashion trends never start with young kids, but definitely juniors and young men, especially celebrities. Everyone thirsts for newness and these are the fastest platforms that can provide this trend-led info.[88]

I spoke with Liliana Becerra, a design and trends strategist and author of the book *CMF Design: The Fundamental Principles of Colour, Material and Finish Design*, about her recommendations for designers who are starting trend research for a project that is designed for kids:

It is important for the research to be grounded in the current and future context of science, technology, and culture. Since kids are now born with a digital chip in their heads, start by looking into emerging trends in the area of "Technology and Kids." I would also make this a global search since technology is not equally available around the world to all kids.

Although looking into design for kids isn't the only requirement. She also recommended looking at trends in the broader culture: "Also include a number of variants that are important in the context of today like co-creativity, co-living, inter-generational, etc."[89] This will show how kids and teens intersect with other aspects of society.

This will lead to trends that are apparent from a global perspective or a regional perspective. We can also learn a lot about how trends in other parts of the world might affect the territories we are creating our product service or system for. Liliana gives an example of this: "If researching education search Finnish education which is considered the best in the world and kids there officially begin learning at school in first grade, not before so that they have time to play and grow a bit more before starting the formal education process." This can then inform us on best practices and a goal to strive for.[90]

What to look for in kids'/youth trends:

- Cultural trends (kids' culture, broader culture, education trends)
- Market-specific trends (toy trends, clothing trends, home décor, play trends, street culture trends, action sports trends)
- Territory-specific trends (global trends, national, regional)
- Demographic trends (ethnic populations, target markets, Gen Z, Gen Alpha)
- Technology trends (hardware, software, media)

- Trends in design (form, color, font, aesthetic, visual, new and future materials)

Where to look:

- Observe the trendsetters (Cool hunt – trend spot)
- Look at major trend-setting cities (NYC, LA, London and Tokyo). What is new there?
- Review the media that appeal to the trendsetters, blogs, magazines, and entertainment
- Read progressive companies' annual reports
- Visit trade shows in related or tangential fields
- Conduct focus groups with trendsetters

Cool Hunting and Trend Spotting

Cool hunters will often seek out individuals who are leaders and early adopters from within their target demographic, observing and making predictions of what is and will be next. In Malcolm Gladwell's book *The Tipping Point: How Little Things Can Make a Big Difference*, "Ideas and products and messages and behaviors spread as viruses do." "The Law of the Few" is, as Gladwell states, "The success of any kind of social epidemic is heavily dependent on the involvement of people with a particular and rare set of social gifts."[91] These are the people to watch.

Interview with Ali Otto, Director of Design for Baby Boy, Old Navy

Ali Otto designs trend-right baby boy apparel for the Old Navy customer. We spoke about many issues around child development and children's culture, materials and processes that she considers when putting together a clothing collection.

What do you like about designing for kids versus a general user?

I enjoy creating an entire collection primarily driven out of emotion. We can be more creative, more playful, more colorful and address the fashion trends faster than our adult counterparts. We can tell a more cohesive story because of the curated collections we design.

Do you ever work with kids in your research process? What types of research do you find to be the most helpful?

Yes, I think the most helpful research comes from our live models and in-depth look at customer lifestyle. We have live fit models that come in at least twice a year to wear our product and to address any questions we have. We travel to different countries looking for trend-right clothing around the world for a broad perspective. We listen to our consumer for insights into the things that they like/dislike. Several designers flew out to meet our customers one on one. Old Navy has conducted in-depth ethnographies that look at who our customer is, his/her habits, practices and lifestyle.

How can clothes support the developing child in each of these areas?

Cognitive development

Texture, fabrication and 3-D elements are both aesthetic and can support cognitive development. We have tees that have dual play functionality with pockets, zippers and Velcro. Graphics/print and pattern content can provide for more academic enrichment through print, pattern and graphics applications. Positive concepts/collections can teach the child about his/her environment.

Physiological changes

I prioritize quality and comfortable fabrications. Hand feel is one of the most important aspects when our team determines fabrication. We discuss functionality as it relates to the parents and the child. Ease/restrictions/fit and functionality go hand-in-hand. Designing details such as adjustable elastic waistbands or reinforced knees for durability. At Old Navy, we design styles with rolled cuffs to support the baby's growth. We use technological advances such as knit-like denim for maximum comfort, cool-max for temperature regulation, and stay bright technology to address colorfastness.

Social and emotional development

I think designers can empower kids through positive concepts, prints, patterns and graphics. I noticed my son at an early age gravitated towards "princess dresses." He was attracted to the color and fabrication. The clothes made him feel special. As he has aged, he is more discriminating about the color, style and characters he wears.

Play

Kids are inspired by the clothes they wear. The clothes allow them to create and participate in their own imaginative environment. Costumes are a great example of this, as kids personify characters. Their clothes, from a physical aspect, also influence kids; clothing can also allow for an increase in movement like running or jumping. We consider durability, aesthetic/concept and functionality as it relates to technology.

Media and pop culture influences such as characters, music, video games, co-branding, and phrases are all used in product lines by the Gap/Old Navy. These images have a positive impact such as socializing kids, give them power, and present their identity, although some schools ban them? What do you see as the pros and cons of their role?

Licensed products/pop culture influences depend on the user. I think licensed brands are popular because they are inclusive and people/kids can relate to them, but these images can also be polarizing and generalize/amplify the characteristics of a race or gender. At Old Navy there is an excellent awareness of these generalizations, for examples, we now sell superheroes across both genders. We promote characters that are "positive" and fit our brand filters.

How do trends in children's culture influence your designs?

Our team travels the world for inspiration. We recently visited Korea, Japan, LA, New York and Europe to track the trends that are occurring in the different marketplaces. We have an architecture built to support these growing trends. As a designer, you must also keep a pulse on adult trends, as there is a clear connection between children's trends and adult ones. For example, this "playful personality" trend we see in children's wear (through patchwork, fun icons, etc.) can be directly related to the early 90s trend that we're seeing in the adult fashion world. We capture trends in color, print and

pattern (scale and content) that influence the way we design. We also make sure to identify silhouettes that are trending.

Are there specific issues related to kids and sustainability that you have achieved or are working on?

The foundation of our company is built on strict policies and high standards of production. From a design community, we are very conscious about our carbon footprint and do things in-house. We can't always affect the production side (another team). We recycle and donate sample garments and are currently working with our mills on denim processing that uses recycled water to rinse our denim. We have innovations that we are working on, mostly related to fabrication, that address the issue of sustainability. For example, we are looking to develop eco-friendly/organic fabrications for our Layette line.

What do you see as the future of kids' fashion over the next 10 years as it relates to new materials and technologies?

Smart clothes for kids: Technology and how it functions with kids' clothing is growing more critical for the tween age.

Technology: As it relates to how our children move/create/play. Fabrications designed to address their individual needs, cool-max, stain resistant, stretch, etc.

Mass-customization: This is a niche movement. There is a trend, but more for tween/teen ages as the child seeks further independence and individuality.

Specifics for kids' clothing

Age considerations

Babies: Primarily focused on their different personalities, soft/sweet/sassy, and based on emotion and the need for functionality.

Toddler: Slightly older, more focus on activities and the child's lifestyle.

Kids: This age ranges from 5 to 18 years old, a wide age gap and points of view based on the company.

Functionality and performance

Durability: Quality of material and fabric. We have wash test restrictions that our garments need to pass because we expect children's garments to be washed more than adult garments.

Flexibility and function: Old Navy has the Karate Jean that is on the market; it specifically targets how boys move as well as how they like to play. It has bi-stretch as well as reinforced knees.

Safety: We have a list of Gap Inc. PI restrictions that must be met to be compliant with Gap Inc. standards. Some of the top safety issues are: Saliva tests, crocking (rubbing off of color), pull test (can an infant pull off a button or 3-D element), length of ties/pulls to prevent children from getting their fingers stuck or pulled. Other safety considerations range from fur length, to trim height, zipper pulls, as well as additional choking hazards. For example, graphic techniques such as sequins or beads cannot be used on our baby/toddler garments because they do not meet our safety guidelines.

Design decisions unique to the kids market.

Materials: Cozy, soft hand feel, technology

Print and pattern: Colorful, emotional, conversational icons
Colors: Bright, optimistic and playful
Graphics text: Empowering uplifting graphics with a playful wink/humor
Illustrations: Emotional hand-drawn illustrations that evoke whimsical/playful humor
Style: High-fashion inspirations, developing and growing more fashion-forward along with more basic everyday silhouettes. Styles for baby/toddler include Critter details (bear ears), Prints/Patterns, Layette 0–12months (footed styles), Rompers/one pieces
Branding: Emotional and aspirational. Built around different lifestyles: Millennial Mom, social media and online components are the first points of entry.[92]

Color

Color shapes the personality, marketability and usability of our designs. Colors or color combinations can be chosen to elicit behavior, promote action, set a mood, and enhance functionality and attachment. Color preferences are a combination between the hard-wired (the relationship between our eyes' biological capabilities and our brains) and what is learned through experience. I sat down with Richard Keyes, a long-time colleague of mine, associate professor at ArtCenter College of Design and color consultant, to discuss color. He usually leans in the direction of hard-wired. He explained:

> A terrific book called *The Righteous Mind* by Jonathan Haidt proposes that we are born with a starter operating system, that is adapted through experience. We are hard-wired to notice yellow and black stripes on an animal since the creatures that display them are potentially dangerous to us. Yellow is the first color we see as infants. We have many more red cones in the eye than either green or blue cones, so we see more variations in red than in other colors. We also have infinite responses to subtleties, like changes in skin color through capillary action. Females have a greater ability to process colors in that color range. 50% of women can see more colors than 90% of men. In the mammal kingdom, only primates can only see red. There may be something to girls liking pink.[93]

It is in the learned aspect of color that many differences between adults' and kids' preferences are most prevalent. Richard Keyes continues:

> The biggest tool in the learned color toolbox is language. I can say that young children prefer saturated warm colors that fit easy

to understand verbal categories (BLUE is easy, INDIGO is hard). They need to be able to describe what you designed for them and name it. Learned color experiences also come from cultural associations and pop culture influences. Don't give them colors they don't understand. They will say, "That's not for me."[94]

What Are You Designing?

The most crucial factor in choosing color is that it must be appropriate to reinforce the project concepts and usability. Richard Keyes suggests: "Ask specific questions about the project. How do we best use color in the designs to achieve its goals?" When working on a website or app, what colors will read the best to move the eye around to drive focus to an action? When designing a product, how could color be used as clues to move the interaction forward? When creating a color palette for an animated kids' show, which colors reinforce the storytelling? Illustrator and animator Hsinping Pan explains the importance of color to reinforce the story in the Land of Nod Little Golden Books animation: "In the storyline books bring colors to children's' worlds. Before the book bus arrives, the world is black and white and it turns colorful when kids get the books. I designed the color bubbles to spread the colors. So people can understand the impact of the book right away."[95]

Environments

Color and light are used in relationship with one another and can set a mood in an environment. Shorter wavelengths (blue) tend to make us calm, whereas longer wavelengths make us more alert.[96] Richard Keyes suggests:

> As a designer considering an environment, my first consideration about color is "does it increase clarity? Then you need to look at is it indoors or outdoors?" People say blue is soothing in an environment. When people started painting waiting rooms blue, they were

▶ Figure 1.12

Land of Nod Little Golden Books promotion.
Source: Hsinping Pan (used with permission).

soothed at first, but then kids got antsy. So there is a lot more to think about such as how much time they will be there. Green tends to be a sheltering color and nature-focused. To convey something is grand and make it feel more expensive use sophisticated colors.

Digital (Screen-based)

Richard Keyes stated: "Color on a screen is much more vivid than color in reflected light, and the value contrast is stronger. We expect a highly saturated world in a digital environment." Color can be used as a hierarchy to show the level of importance, to motivate a child to press a button or engage in an activity.

Graphics, Branding, Packaging

Although many branding experts believe that color offers an instantaneous method for conveying meaning and messages, Richard Keyes doesn't entirely agree. He says:

> I don't think the color in branding is anywhere near as important as some experts tell us. If there is a field in which the big dogs tend to use certain colors (blue, in telecommunications, for instance), then a new company can use color to align themselves with the current color trend, or conversely, use color to show that they are not like the others.

> Terry Montimore, partner and creative director at Goodwin Design, stated: "Nickelodeon has pretty much made orange and bright green the go-to kids' colors, but blue also has a good girl/boy crossover appeal." The more serious the business, the darker the colors; the more casual and light-hearted the business, the brighter and lighter the colors. Richard Keyes said: "When communicating a message, for example, working on a public service announcement, it may make sense to use saturated warm colors pointing towards urgency. There are some basic things like using red messages to say stop, be aware, or it is an emergency."

Product

Color can be used to promote direction of movement, functionality and emotional attachment. Richard Keyes says:

> Bright saturated colors tend to cheapen high-end objects, so most of our costly possessions tend toward neutral. Stereotypes sometimes work because of the speed and clarity of message. Raymond Loewy's *The MAYA Principle* is a good example of how to introduce something new to users for adoption.[97]

Illustration

Richard Keyes explains: "Movie posters, editorial work, video game boxes, visual development for a film will all have different approaches." For story and characters, Richard Keyes suggests:

> If you are trying to create a feeling of magic or make something mysterious use greens and purples, and limited value contrast helps, too. Historically purple was hard to manufacture and rare, so it was used for royalty and is still considered a royal color. There are also cultural stereotypes to consider. Disney replaced traditional girly colors with aqua and purple for *Frozen*.

Influence of Science on Color

What do scientific studies say about color perception in children? Color can impact learning, memory, mood and emotion.[98] Some researchers will tell you that there is substantial scientific evidence for color preferences and why we are drawn to specific colors and not others.[99] Children prefer more colors at once and have more intense reactions to colors than adults do.[100]

Functionality Affects Color

School buses are yellow so we can see them better. They send out a strong warning. Yellow can be seen in the peripheral vision: "Lateral peripheral vision for detecting yellows is 1.24 times greater than for red."[101] Social and cultural factors influence white for baby clothing and offer a feeling of purity. Does white baby clothing make sense from a functional perspective? A baby will spit up and have diaper leaks, while a toddler will spill food all over their clothes and play in the mud. A black glossy toy probably doesn't make sense for a product that a baby will put in their mouth continually.

Richard Keyes recommends: "When a user is interacting with a product or an interface the colors used guide the eye through a path. It is important to make sure all buttons or interactions within a path follow the same colors, so they don't get confused." He suggests, use a group of functions that are the same color so they relate: "Saturated colors point towards vitality. Red points toward 'notice me' and stands out. Overall if you are trying to make something stand out do something with color the user doesn't expect."[102]

Markets Affect Colors

Many designers research and choose "In colors," trending from trend forecasting companies, for their palettes. From a sales and marketing perspective, these colors fit in with what is going on in a broader cultural aesthetic context. The fact that many designers follow these predictions and include them in their designs is what makes them popular. Richard Keyes states:

> Color companies like Pantone decide on the color of the year and everyone wants to use it. If something is associated with cool, then we adopt it. Becoming socialized is a big part of childhood. Kids want to be socially accepted, and awareness of trend and what is cool helps with that.[103]

Creative director Alena St. James explained: "I recognize and utilize trends. But, I also use colors that I feel deserve to make a comeback or that can be a disruptor in the marketplace, that will deliver a subtle to the loud excitement. I like everyone to feel like they won't get lost in the crowd."[104]

The market that we design for affects the colors we choose. When creating for the children's mass market, we tend to be less risky with color and use bright and warm primary and secondary colors of the rainbow. When designing for a niche markets, colors are often more "special," complex and sophisticated, or brand specific. Here you will see tertiary tones and shades and tints of primary and secondary colors.

Cultural Associations

Cultural associations to color are learned, so a child's relationship to the symbolism can include cultural norms, gender and identity, and peer affiliations. What are the cultural affiliations of the colors you will be using for a global audience or a specific region of the world?

A color can be used for symbolism. Perhaps the most intuitive color connection is green – the color of outdoors, eco-friendly, nature, and the environment. Orange is considered a "fun" color.

Gender-based colors are driven by markets, which in turn can drive preferences. Currently in the US, color and gender are stereotyped for children: pink for girls, blue for boys, or primary colors. This is gradually breaking down. When designers choose these colors, it invites a child to identify with the ownership of a product through the color associations they have learned. When they don't select expected colors, they are saying that this is gender neutral or that gender is not an issue.

Emotions and Associations of Color

Emotional Responses to Color

Individuals may have a particular feeling about colors; for example, green and brown tend to represent earthy and natural. An ecological valence theory (EVT) has suggested that the preference for color is determined by the average response to everything the person associates with that color. Positive emotional experiences with a particular color are likely to increase the propensity to develop a preference for that color, and vice versa.

Parental Opinions Affect Color

Adults respond to color differently than kids do. Richard Keyes thinks: "Since experience is part of the mix of life knowledge, adults have more experiences to apply to hues."[105] Parents tend to be drawn to colors that they like themselves and for their children. Blue is a color that is favored by most people, independently of culture, country, age, socioeconomic bracket or gender.

Developmental Stages Affect Color

Color is connected to visual and perceptual development and experiences, so children's developmental stage can influence their interpretation of color. Color preferences evolve, becoming more sophisticated and personal throughout childhood.

Babies

Research on visual development can influence design decisions, since their visual acuity is developing. An example of this is that very young infants respond best to high-contrast visuals.[106] For a couple of years, we saw a lot of baby products made out of black and white and red for high contrast. Many choices of colors for babies are based on parental preferences or themes.

Young Children

Young children's color preferences and reactions tend to span across cultures, suggesting that these preferences are hard-wired. Yellow tends to be a favorite among little kids. They make the connection to it being a bright, happy color. Browns and grays consistently test poorly. Children prefer more pop primary and secondary colors – red, yellow, blue, orange, green and purple.[107] Richard Keyes suggests:

> Depending on the age and sophistication of the individual child, saturated, and warm colors are more popular. ... and, 8.5% of the world's population cannot see all the colors that the rest of us do so it is good to back color up with another design cue.[108]

Simple colors like red, blue and yellow are more straightforward to describe than crimson or magenta. The color preferences of the child can influence their attachment to or rejection of a product, service or experience. It is common for a child to quickly and decisively change what their favorite color is.

Later Childhood

Colors start to get darker and represent an older demographic, pulling colors from a more mature market but at the same time making them their own. Tweens tend to prefer colors that are not childish and associated with little kids. By 9 years old, their language skills have developed to describe a broader palette of colors and tertiary hues.

Teens

They are open to experimenting with more sophisticated and complex tones. Many younger teenage girls love varying shades of purple and pink. As they reach their late teens, they often show preferences that look similar to those of adults: black, blue and then white. Peers influence color choices. Teenagers wearing black might say "I am part of the artsy crowd," or a color from a particular brand might represent lifestyle preferences. This in turn connects them to other like-minded peers.[109]

Form, Materials and Finishes

Functionality is usually the first trait that comes to mind when choosing forms and materials in any project for kids and adults alike. Our decisions around functionality affect the materials we choose. This in turn influences safety and aesthetics. Questions we may have when making these decisions might look like this:

- Which materials will perform best considering where and how the product or environment will be used?
- What are the motor skills capabilities of the children, and how does that affect your choices?
- What are the safety considerations of the item and on the materials used?
- What properties should the material have? Should it be hard, soft, flexible, elastic or dense?
- What level of strength, resilience and durability does this product or environment need to function and not break easily?
- What are the chemical properties required?
- What is the projected life of the material in relation to the products and environments of use? Will it hold up over time and withstand being a hand-me-down?
- How quickly will it break down in a sunny, rainy, or stable environment?
- What color should the material be? Should it be clear so you can see if a part inside is loose?
- Do recycled or eco-materials have any performance constraints compared with the alternatives?
- Which form factors make sense based on how it will be used? Does it need to be rounded or blocky or have subtle curves?
- What finish makes sense to enhance the durability, strength and safety of the project?

Materials Safety

Safety is a number one concern when designing for children. Kids' brains are developing rapidly, and exposure to toxic materials can influence this development. When I spoke with Dice Yamaguchi, Faculty of Visual Arts & Media Studies, at Pasadena City College, he said: "Anyone designing for children should be concerned about using any materials and finishes with hormone disruptors such as BPA and elements/chemicals that cause developmental damage and/or damage to children, mothers and their fetuses." He is also an advocate of using recycled or waste materials in new product designs. He suggests: "Consider the use of reclaimed wood, and wood waste materials such as particleboard and MDF instead of virgin wood."[110] This will curb the introduction of new materials into the system.

Sometimes, safe materials used in products for children represent better materials for the environment as well. David Tyler, University of Oregon chemistry professor, says that parents are usually more concerned about immediate impacts rather than long-range ones. "Say your kid chews on this doll, and it leaches out some of those plasticizers or PVC or whatever. That would be a much more serious concern than some far-off global-warming disaster," he said. "Even though the result is the same. I'll either die or get sick. Toxicity issues seem to outweigh water use and global-warming potential. It makes sense. It's more immediate."[111]

Although there are materials safety standards across the board for all users, there are additional safety concerns and regulations that are applied in the children's market to protect children from themselves and designers' decisions. For example, one consumer products safety commission regulation for children's sleepwear above size 9 months and up to size 14 requires that the fabric and garments must pass specific flammability tests or must be tight fitting. It is essential to know not only what the standards are early on in the design process but also why the standard is implemented: to prevent the risk of injury from fires, which happens most frequently with loose-fitting sleeve and leg cuffs. This helps the designer to make safe choices about materials selection and design. Learn the safety standards in your project category, and don't look at them as limitations but instead, as a design challenge. Also, keep in mind that safety standards change over time. Older materials might not meet the current materials safety standards, so always check the current guidelines at www.cpsc.gov/.

Designing using safe materials also means ensuring that they are safely handled and are age appropriate. Children's behaviors and cognitive abilities may also influence their risk. Kids use things differently from the intended use more regularly than adults, so we need to think about unexpected uses of a design. Babies and toddlers will put things in their mouths, kids will touch a touchscreen when their hands are covered with jam, and they will grab a hot metal chair that is out in the sun. Designers must not just think through a typical use case but extend out the possibilities of use to insane, crazy atypical use cases to ensure safety. Kenneth S. Kutska, executive director of the International Playground Safety Institute, explains how designers need to consider uses of playground equipment:

As manufacturers and designers push for new interesting and challenging play experiences they have started to get into uncharted waters with regards to how children might foreseeably use the equipment. Is this use reasonable or unreasonable? Nobody can know absolutely how a child might use a piece of equipment. As they master the intended use of the equipment they will then by their own curiosity begin to experiment in different unintended ways. With ADA [Americans with Disabilities Act] requirements we have provided easy access to many pieces of equipment that can put the youngest most vulnerable user in peril just by being there.[112]

Age appropriateness is also another important consideration. Even if a 2-year-old is brilliant, there are multiple safety reasons, based on their physical and cognitive development, why they should not be playing with an 8-year-old's toy. They may put it in their mouth or use it as a projectile, while an older child has hopefully learned not to.

Different markets provide designers with a different level of freedom. In the mass market, many parents buy on price and trusted brands. Being good for the environment isn't usually considered as the top priority in a purchasing decision. Just ask some parents: Have they considered the off-gassing of foam and rubber in the car seat they purchased? How about the carpet in the child's room? Some think that toys will only be used for a short time, so as long as their kid plays with it and it is the right price, the choice and safety of the materials is in the domain of the company making the right decisions. As designers for the mass market, we make the choices for them, although we have less freedom to explore materials and often decide on what are the safest materials and finishes we can create with to meet the legal standards within the cost structure. Since many products can be made in more than one way, designers consider the best options available and usually make tough decisions. I asked Bill Goodwin, creative director and founder of Goodwin Design, about the waste that is generated through packaging, since he has many mass-market consumer brands as clients. He stated:

> We approach our clients with the hope of being sustainable – most of the times it is out of our hands. We also rely on what is available from the manufacturers. Most packaging uses other criteria. Retail is the driver – archaic systems – shipping, distribution, and marketing. We speak to it but because of the scale large companies rely on existing solutions – there are many out of the gate constraints minimal opportunity to innovate. LCA often is not included in the design process and falls in the packaging engineers, procurement management limiting our impact in decision making. There are these slow bursts of energy to try something new and incremental changes.[113]

In the higher-end or more niche markets, progressive parents or kids have the knowledge, time and money to open up the range of choices. They can afford to be cautious about materials, products and environments that surround kids. They feed their kids organic food, purchase all-natural personal care products, and make sure the toys and the playgrounds they play in meet safety and sustainability protocols. Many of these parents choose products that protect the planet and their children's futures. Bill Goodwin explained:

> With a start-up we can build the whole model. Use the thinnest PET and biodegradable materials and choose best possible solutions with a minimal footprint. Vietnam is so far ahead because they have old stuff. The big guys can't take the risk to adapt. They start smaller brands or invest in small start-ups to acquire.[114]

Sustainability and Materials Selection

Because kids are always changing in size, attitude and capabilities, they outgrow stuff mentally and physically very quickly. This impacts our decisions on our designs, materials and the environment in many ways. Our design decisions can also affect a child's attitude towards materialism, consumption, acquisition and identity. Advances in material processing and technology mean that recycled materials are accepted as equal. Today, sustainability is being framed as a process to drive innovation. Choose quality materials that are durable. Heidrun Mumper-Drumm states:

> Today toys are not expected to last long. They are made of poor quality materials, to be replaced by other, new, short-term toys. Toys at one time were designed for durability, because they were meant to last through many hours of play, for more than one child. Parents rarely invest in durable toys as important tools for play.
>
> As designers, we should encourage parents to invest in the basics. Wooden blocks can be played with from 2 to 10 years old or more. Children play with them differently as they get older, and all ages can play together. They can use them for stacking when they are younger and then design cities. Wooden blocks are an expensive purchase, with a wide range of possible play for everyone in the family.[115]

Baseline Safety Materials for Kids

Woods

Wood toys, furniture and other products have a natural, warm, classic crafted quality to them that you can't find in any other materials. They are loved for their quality, durability and wood grains that build the character with age. Particleboard, MDF and reclaimed wood allow the use of wood waste instead of virgin wood.

◀ Figure 1.13

The choice of this pronounced grain Japanese hardwood for these blocks extends the life of the toy from a stacking toy for a younger child to a puzzle for an older child.
Source: Fridolin Beisert (used with permission).

Materials

Solid hardwoods, softwoods, bamboo and Forest Stewardship Council (FSC)-certified wood. FSC is widely considered the best forestry certification program. When choosing composite wood products, use formaldehyde-free plywood, particleboard or MDF.

Finishes

Thoroughly sand off surfaces and edges. Coat with no finish, natural wax, water-based vanishes, non-toxic finish and lead-free paint. Don't use paints that contain volatile organic compounds (VOCs).

Textiles

Safety, comfort and ease of care should always be a top priority when choosing material for children's garments, toys, bedding and other textile products. Stick to organic textiles (better cotton, wool, felt, hemp, soy, cashmere/silk) and cotton/polyester blends. Use flame-resistant fabrics.

Reduce or eliminate the use of chemical manufacturing processes or finishes that may induce allergies or cause rashes. Avoid flame retardants. Consider length and trim height for fur, flannel, minky and terry cloth for safety.

Finishes

Use non-toxic dyes and inks.

Paper, Cardboard and Packaging

Paper accounts for much of North America's solid waste. Furthermore, the pulp and paper industry is the third largest industrial buyer of elemental chlorine used to whiten paper, a process that is linked to a proven cancer-causing chemical called dioxin. Packaging, which serves as a marketing tool for children's products and is often excessive to showcase the product and grab the customer, has a massive impact on the environment. Use openings versus plastic-covered windows.

If possible, use chlorine free (PCF), recycled papers and cardboard instead of virgin materials. When this is not possible, use elemental chlorine free (ECF) papers. When using virgin paper, use FSC-certified lines.

Finishes

Use papers colored with water-based dyes or soy-based inks. Avoid metalized foils and coverings. Reduce ink coverage in use.

Plastics

Many kids' products are made of plastics. For parents who want to minimize chemical exposure for their children, there is a growing movement to eliminate the use of petroleum-based plastics and any plastics used in children's products.

Plasticizers, the additives that give PVC its flexibility, are endocrine disrupters that have been linked to cancerous tumors, birth defects and other developmental disorders. These additives could make their way from a product into a child's bloodstream. The environmental cost of manufacturing PVC is toxic and generates air and water pollution near the factories where it's produced. Lamination and co-molding of materials reduce its chance of being recycled and introduce a failure point.

Materials

Safer plastics include those made out of polyethylene polymers, polypropylene, and bio-based materials. Better plastics: BPA-free plastics, PVC-free, phthalate-free. Some scientists argue that polyester, when made from recycled products, is greener then cotton.

Bioplastic made from corn and potato starch as well as other natural materials have a lot of potential in the future from a health and safety as well as sustainability perspective. They are still expensive because of their limited use and are only appropriate in specific applications based on their strength and durability.

In 2012, the US Food and Drug Administration banned the sale of baby bottles that contain bisphenol A (BPA), a compound frequently found in plastics. Since then, store shelves have been lined with BPA-free bottles for babies and adults alike. Recent research reveals that a standard BPA replacement, bisphenol S (BPS), may be just as harmful.[116]

Additional Materials

Natural rubber: Use those processed to exclude the latex-like nitrosamine proteins.

Glass: When considering use and breakage in the design, glass can be a safe alternative to plastics for some products (bottles, tableware).

Metals: Stainless steel. Understand the composition of the metal alloy.

Food-grade silicone: Can withstand heating and freezing without leaching or off-gassing hazardous chemicals. It is also odor and stain resistant, and hypoallergenic.

Non-toxic art materials: Crayons and paints should say ASTM D-4236 on the package, which means that they've been evaluated by the American Society for Testing and Materials.

Foam, batting and loose fill: Choose materials that are certified safe for children's products. Construction strength is essential to contain the contents.

Recycled materials: Industrial and post-consumer.

Textiles: Recycled polyester, denim.

Plastics: high-density polyethylene (HDPE).

Woods: Salvaged.

Materials to Avoid

Metals: Antimony, arsenic, cadmium, cobalt, lead, mercury and molybdenum; alloys when you don't know their makeup. Avoid magnets if it is not possible to secure them so that it is impossible for them to come loose. They are dangerous if swallowed.

Plastics: Phthalates are a group of chemicals added to plastics and polyvinyl chloride (PVC).

Non-food-grade silicone: Not yet thoroughly tested.

Chemicals: Formaldehyde in all materials, flame retardant.

Excessive use of batteries: Full of toxins.

Off-gassing materials: Carpet, foam, synthetic and recycled rubbers.

Form

Form follows function, but it also follows safety, manufacturing and aesthetics. Do children have enough life experience to recognize beauty? Do they have taste? Do they care? Do their parents? Will grooves or holes on the surface attract and trap dirt and dust, making it harder to clean than a smooth surface? Is it easy to wash to avoid germs? Are there sharp points or squared corners that may be dangerous if a toddler learning to walk falls on it? Choking is a particular risk for kids aged 3 or younger, because they tend to put objects in their mouths. A balloon that changes form seems safe in one shape, but as I learned in a baby CPR class, just like peanuts and grapes, popped balloons can be deadly for young children.[117]

Choking

Toys should be large enough – at least 1¼ inches (3 centimeters) in diameter and 2¼ inches (6 centimeters) in length – so that they can't be swallowed or lodged in the windpipe. A small-parts tester, or choke tube, can determine if a toy is too small. These tubes are designed to be about the same diameter as a child's windpipe. If an object fits inside the tube, then it's too small for a young child.[118]

Children should not be given toys with cords or strings, sharp edges, small parts, loud noises or propelled objects. Avoid children's "play" jewelry and cosmetics that may be made of hazardous materials and chemicals.

Balloons, tiny magnets and button batteries are among the most dangerous objects for small children.

Metals: Avoid thin metal, sharp edges.
Fabric forms: Consider sewing and technical construction.
Wood forms: Use round corners and avoid sharp edges. Consider construction.

Finishes

Consider the texture on the surface of the product and the final layer of finish. Finishes can be used to enhance durability, performance and aesthetics. Choose finishes applied to the product wisely. Avoid chemical finishes. Use non-toxic finishes and glue.

What does it feel like when it is touched? Put in a child's mouth?
Does it have a rough finish, a smooth finish and hide scratches?
Will it chip or break off?
Is it durable?
Does it feel authentic to the material?
Are they easy-to-clean finishes?
Will they look new over years of use? When exposed to sunlight?

Technology Research

In the last decade, the media and technology environments for children and youth have evolved, offering them both benefits and challenges. Art director

Scott Allen states: "Kids will always need guidance from all adults in their life. Each kid learns and plays in his/her own way. New TECH on the horizon will benefit both educators and the kids they teach."[119] Currently, the effect of media and technology usage among kids is a big experiment that is currently being closely watched by researchers in real time. For example, a positive use of technology could be lights to communicate action or reinforce cause and effect and promote positive reinforcement. On the contrary side, the noise of electronic toys and devices can be deafening (even louder if a child holds them directly to the ears) and can contribute to hearing damage. Keeping on top of the most current research will help you to make the most informed decisions to use it to best benefit child development. Over the next decades, it is going to be crucial for designers to become more technologically savvy. Mari Nakano, former design and interaction lead for UNICEF's Office of Innovation, suggests:

> Technology, like any innovation, could affect the development space for better or for worse. It really depends on who is applying it and how it's being applied. Like any new idea, we need to create checks and balances, and I think the design practice can offer insight into how we positively implement new or improved technologies into the field. As a community, designers are already diving deep into how we can integrate more thorough and measurable creative practices into the social space. In 15 years, though, I think it's essential that we find our way into major decision-making processes and spaces to actually enforce, as opposed to suggesting, more thoughtful and non-harmful innovations into the world.[120]

The designer's role is to learn how to adapt to changing technology. Technology will always change, and it is up to us to understand the best way to work with it. Art director Scott Allen spoke about designers keeping on top of the latest technology and the promise it holds in the future and how it benefits kids. He explained:

> Technology holds the promise of keeping things fresh. I am always exploring "what's next" … fun and exciting tools to create new experiences. There are two areas to consider: TECH designers use to *create* new designs and TECH kids *interact* with to experience new designs.

1. TECH as a design tool:

New technology can provide new ways of making.

One example is Wacom's CINTIQ which has streamlined and revolutionized the way designers physically work. It makes the creation process more fluid and productive.

Another example is LAIKA's use of 3D printing. Introducing 3D printing to their stop-motion production process opened up new possibilities.

BUT KEEP IN MIND: NEVER let a new technology drive the design.

2. TECH as user interface:

New Technology can revolutionize the way we experience stories.

Today we swipe a screen with our fingers … tomorrow we will activate our environment (both virtual and real) with our voice and our body.

Virtual Reality and Augmented Reality are two TECH "shining stars" right now. There is a lot of attention and investment right now in VR and AR. I think, someday, a new "morphed" version of the two will be the next big thing. There will be a lot of successes and failures before we get there. Kids will benefit in the future from learning and play experiences that are more interactive than today, more transportive than today and more individually tailored than today. But will all kids have equal access to these experiences? I hope so.[121]

When researching technology, center it around the appropriate usage. It is easy to get sucked into the wow factor of technology, but without a lasting benefit, after the initial impact its importance will probably fade. Question what is best done from a developmental perspective in the physical world and what is best executed through the use of technology. Does this technology create a better long-lasting experience?

▶ Figure 1.14

Playing with technology and making in the design process.
Source: Georgia Siapno (used with permission).

Technology and Research

- Read about the role of technology and its effects on child development.
- Understand the media content and technology being used and age appropriateness.
- Look at technologies being offered to the general public in other areas: for example, how will clean energy technologies and autonomous vehicles affect kids' lives?
- Look into emerging trends in the field of "Technology and Kids."
- Play with technology as a new way of making.

Design: Visual Inspiration

As makers of visuals, designers often look towards visual representations for inspiration. I have seen many similar motifs used in design for kids: explosions, movement, signage-style graphics, exaggerated cartoon illustrations, hand-drawn bold text, and Bubble letters. Terry Montimore, creative director at Goodwin Design, explained what makes these approaches kid friendly: "These have movement, life and energy ... they give off a certain feeling of 'safe' rebelliousness. They think their parents' cereal doesn't look like mine." He suggested:

- Illustration style is an ever-changing target. Look at cereal box illustration and current kids cartoon styles to inspire new approaches.
- An illustration is used to draw you in. We use a hero shot that demonstrates the scale of the product.
- Kids are impulsive and immediate. Go visual as much as possible. Use fonts with some personality. Include layers of discovery. The more expensive the product the more you need to read. Do not visually complicate the message.[122]

Creative director Alena St. James explained her designs of prints and patterns:

When designing for younger kids I use less detail, so designs tend to be bigger, bolder. Older choices are based on peer mentality. I always try to create something that is fresh and ideally expresses fun, irreverence and/or individuality. In today's highly visual world kids have developed greater visual literacy. I can push the boundaries of color and pattern to reflect this. For example much of today's illustration reflects this crossover in kids and adult style. I don't use a lot of phrases for kids' items, since language style and word choices differ on geographical location.[123]

► Figure 1.15

Steampunk piggy bank for Ritzenhoff: "When I re-visit past trends or vintage ideas, I always put a twist on it and consider how it reflects today and tomorrow."
Source: Alena St. James (used with permission).

Patricio Fuentes, principal at Gel Comm, explains:

> The adult market is aging down and the kids market is aging up therefore kid trends are blurring with adult trends. When designing for kids bring energy, life, and fun to it. Make it softer, brighter, friendly and happy. Use narrative, a dynamic story that will drive the imagination. Younger kids' designs are often wilder while older kids are more sophisticated and minimal. Older kids have so much coming at them it needs to be organized.[124]

Design Prompt: Make a Trend Board

Collect the information gathered from the research and create a trend board with visual images that reflect a vision. Name your own future trend and use five keywords that define it. Use visuals of all forms, colors, materials, finishes and technologies. Include both visual and cultural trends.

Graphic Design and Illustration for Kids and Teens

Melinda Beck, illustrator, animator and graphic designer, has received two Emmy nominations for the animations she created for Nickelodeon. She has also created award-winning graphics for children as well as adults for clients such as Chicco Baby, *The New Yorker*, *The New York Times*, Random House, Scholastic, Sesame Workshop, Target and *Time* Magazine. She shared her insights on her design process from graphic design and illustration.

Can you describe your design process? What methods do you use to create your ideas and work?
My process is the same, whether I am creating work for kids or adults. I make a lot of rough sketches and list words that I think are relevant. The goal is to get as many ideas as possible out there, and not be judgmental. The great ideas are often not that different from the terrible ones. I also look at visual inspiration while I am sketching,

other design, and forms of art like photographs, movies, music or literature.

The next step is to develop the rough sketches into tighter sketches, and final designs.

For animations and books I create a storyboard; leading the viewer step-by-step through the narrative. My goal is to teach in an entertaining and fun way, being careful not to lose the viewer at any point.

How do you keep your work, style and point-of-view consistent but adjust it to the goals of the project?

I first try to entertain my own inner child. It is essential to play and have fun while doing the project or else you end up with a joyless piece of art. The balance is to be mindful of who the client is, while also being true to

my own aesthetic. I use my voice to communicate who they are.

When designing for kids, I find I can imagine more fantastical and exaggerate. Kids appreciate the unreal; push it to the point of humorous impossibility. Kids are always looking at things in new ways, and I try to do that when designing for them.

What are the most important things to think about in illustration design that will appeal to different developmental stages?

I see two groups of kids: 0 to 11 and 11 to 14. The big difference being 11–14 seems to like imagery that looks older and not too cute. The humor can be a little darker and even sarcastic. The 0–11 crowd does not seem to

▶ Figure 1.16

Song animation for Noggin Network.
Source: Melinda Beck (used with permission).

be too concerned with things looking too simplistic or babyish.

When designing for 0–11 I always try to have an education component. Children are inquisitive about the world and how it works. They are still asking questions.

Two of the children's books you have illustrated are tailored to different ages – L is for Lollygag – grades 6 and up – Lines, which is a picture book for age 2–4. Both embed visual storytelling and rely on the interaction between text and image to tell the story but in different ways. Can you explain how your approach differed for each?
For the picture book *Lines* the author Sarvinder Naberhaus created minimal text so I had a lot of freedom to interpret. This was a story of how different objects are made up of lines, circles, triangles, squares and rectangles. Kids of this age don't read yet so it was essential to have a story that could be told visually. I created a journey through a world rich with cars, trucks, trees, people, houses and buildings that are made up of these simple shapes.

L is for Lollygag is filled with visual puns based on definitions of obscure words. Here the visuals are not free standing and rely on an understanding of the text. A much more sophisticated interaction between word and visuals that is age appropriate for a tween or teenager.

How does the medium you are designing for affect the way you approach the work?
A logo has to stand up to being applied to many different surfaces and in many different sizes. To accommodate this logos should be more straightforward, capturing just the essence of the client. Illustrations do not have this constraint so they can be much more elaborate. If they are in a book where they will be looked at over and over again, it is nice to have a visually rich image that offers something new each time one looks at it. With animation one is adding the elements of sound, time and movement. These elements are used to create visual surprises, and the viewer engaged by varying the shots: close up, distance shots, cuts and timing.

What are some of the challenges you have when you are working with already existing branding/character properties?
I look at who the client is and try to capture their essence while still using my own voice. For *Sesame Street* I thought back to watching the show in the 70s. The bright acidic colors of the Muppets and their bold graphic quality inspired my design. For *Yo Gabba Gabba!* I was inspired by the quirkiness of the furry characters and used photos of fake fur and pom poms within an eclectic and odd selection of fonts.

What do you think designers should consider when designing for kids?
So many designs for kids I have seen a million times. They don't teach anything new or have any humor or visual surprise. I really appreciate it when designers make an effort to break out of the ordinary and create something that is more visually unique and unique to their own personal point-of-view.

What do designers need to know about being a kid today that is different than in the past? Why?
Kids are on the internet early and are much more aware of other places and ways of living. They are exposed to more visual experiences. I can draw from a broader range and more sophisticated pool of influences.[125]

Idea Generation

After the initial pass at design research, there will be plenty of material to ruminate over and continue into the brainstorming phase. We shape the research, give it meaning and then shift focus to developing concepts for solutions. Play researcher Karen Feder explains:

> An important part of going from the research process to the design phase, is the reflection. What did you see, hear, read, experience, or perceive? What surprised and astonished you? What went straight to your heart? Keep that in mind, and try to figure out, what is at stake. Why is it important? This part of the process is not only about an analytic approach, but also about

intuition and hunches. About being human and figuring out not only how you can find a solution, but create the opportunity to get to the right solution.[126]

Now it is time to visualize, analyze, organize and synthesize all that you have learned.

Design Prompt: Prepare for Brainstorming

- Choose visual-thinking tools
- Filter data, notice emergent themes, cluster information, synthesize information
- Interpret observations into ways to guide the project
- Identify unmet needs
- Create a list of requirements that the project must address
- Refine the scope
- Develop insights and opportunities
- Decide on targets
- Redefine goals and objectives, point-of-view, the mission, project brief based on what was learned

Brainstorming

Usually thought of in the beginning phase of the design process, brainstorming can also be used at any stage. This is where we as designers get to play and exercise our creativity. We use divergent thinking while generating ideas. We use convergent thought to select the best ideas. We get to think like a kid, act silly, wild and crazy with nobody judging us. Returning to this childlike freedom as part of our creative thinking process might be the most influential reason why we decided to become a designer to begin with. The goal is not to solve the problem or find the perfect idea but to come up with many possibilities. The best way to see the unmet need is to envision many designs to choose from. Good and bad. Immerse yourself in multiple approaches: draw, build, dance, sing and tell stories. Imagine anything is possible. Rein it in later. Encourage wild and seemingly unrealistic ideas. With unlimited freedom we tend to voice what we really want without limitations. Although brainstorming is a free-for-all, it is helpful to have a goal and some useful constraints.

Set up goals and criteria:

- What problem are we trying to solve? Consider everything.
- What parameters do you need to consider?
- Break the project down into smaller chunks, make connections, and look deeper.
- Set up the criteria that will be used to evaluate ideas later on.
- Allow for failures.

Brainstorming sessions can take place individually or in collaborations. Each has something unique to offer. When brainstorming on our own, we get into a flow within our head, making connections, and we have time to think and focus. A benefit of team brainstorming is that everyone on the team has unique life experience and perspective to offer different ideas and solutions. Brainstorming with kids is also often extremely helpful. Children are comfortable just saying what they think and have little problem accessing their own creativity. Include some individual and group and brainstorming with kids in your project development.

Consider the tools used in the session. Construction paper will create different results than index cards. Recycled cardboard from kids' cereal boxes will generate different results than foam core. Finger paints will create different results than drawing pencils. Square posts will create different results than hexagons.

Tips for Brainstorming

- Consider the timing. When are you most productive? When is the best time of the day for the team or kids?
- Get out. Hold a brainstorm session outside, off-site or at the observational research site.
- Frame the problem. Don't make it too broad or too narrow.
- Ask the right questions. How might we?
- Use humor and reserve judgment. This is about a lot of ideas, not the best idea.
- Think disruptively.
- Diverge a little outside of the scope while staying on topic. Keep pushing beyond the obvious solutions.
- Use words and images and work three-dimensionally with different materials.
- Try many things out. Make and make again. Make some work. Make some fail. Fail fast and learn.

Tips for Team Brainstorms

- Consider the makeup of the group. Embrace diversity, experience and expertise.
- Invite all stakeholders with conflicting views and interests.
- Make everyone on the team comfortable and treat as equals. You never know where a good idea is going to come from.
- Bounce back and forth between generating and sharing.
- Encourage everyone to speak freely. Encourage members to use their voice.
- Take turns leading the group. Have each person reframe the question.

- Be willing to relinquish ownership of your ideas and accept others.
- Allow one person to speak at a time. We often learn more from listening than speaking.
- Encourage cross-pollination. Build on each other's ideas. Have each member take an idea to the next level.
- Avoid moving from sharing ideas to critiquing them.
- Promote challenges. Can your team come up with 100 ideas in an hour?

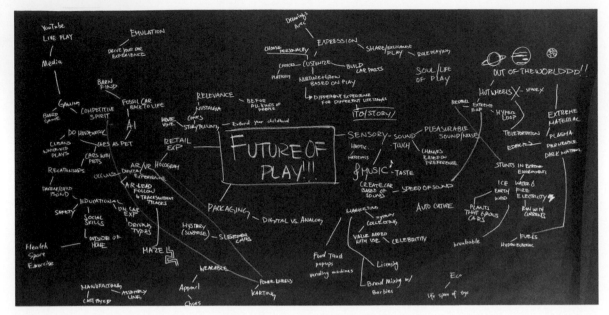

▲ Figure 1.17

Mind mapping the future of play.
Source: Jeremy Dambrosio, Harmonie Tsai, Jack Xu (used with permission).

25 Ways to Start Brainstorming When Creating for Kids

1. Mind mapping: Free-flowing visual depiction of thoughts. Graphic way to represent ideas and concepts.
2. 2D images: Make sketches big and small with diverse levels of detail. Experiment with kids' tools such as crayons and finger paint. Rip out a page from a coloring book and build on it.
3. 3D models and materials and technology experimentation: In addition to working with professional modeling materials, build ideas out of easy-to-access kids' materials including straws, recyclables, pom poms and glitter. Get some technology or a tech toy and get it to work.
4. Photos, videos: Take pictures and make videos of kids playing. Create your own experience and tape it. Build a photo journal or a clip reel.
5. Installation: Get some big sheets of cardboard and hack a space to design experience or a place for the experience to take place.
6. Storytelling and performance: Create character bios and put on a puppet show. Create a storyboard or comic strip to depict a story. Dress up as characters and put on a play.
7. Social media: Join a forum or follow a topic. Write your opinions and discover what you have to say in response to other people's views.

8. Play: On your own and with friends. Use multiple play patterns to approach a topic, from open-ended to structured, from social play to constructive play.
9. Personas: Define guiding personas and role-play. Pretend you are a kid in your target demographic. Dress like them. What would they do? Keep a daily diary of your persona.
10. Use media for inspiration: Listen to kids' music, watch kids' movies, read kids' books, use games and apps.
11. Play with kids: Play a game, a sport, interact, talk, or just kick a ball around.
12. Word play: Use words that evoke an emotion, have personality, a spirit, and send a message. Build phrases.
13. Adjacent relations: Consider before, during and after the use or experience. Consider closely related topics. What is it? What isn't it? How it is communicated?
14. Creativity cards: Make your own or use ones that are published.
15. Experience map: Plot out needs, actions or motivations of the user over time. What is the customer journey? What are the critical decision points? Storyboard it like a graphic novel.
16. Self-reflection: Create a self-portrait of yourself as a child. Write a story about your childhood. Build a diorama to tell the story.
17. Build scenarios: Change the variable and consider outcomes. Build a future scenario. What do we need today to get there?
18. Use mental models and schemas: What does a kid think of this topic based on their experience?
19. Visualize in Venn diagrams and infographics: How are your topic and others closely related? How can unrelated questions relate? What is logical? Not logical? What is the data? How can that data be represented visually?
20. Typologies: Break out the user values with different definitions. Build a strategic landscape. Determine use priorities.
21. Cognitive walkthrough: Put yourself in the frame of mind of the kid. Walk through the tasks and ask questions as a novel user.
22. Co-design workshops: Include kids in your brainstorming session. Listen to what they say.
23. Debates: Create opposing arguments on a topic and role-play.
24. Framework: Organize the motivators and trade-off decisions of the users.
25. Shift focus: Start with different words for brainstorming. How does each of these change the approach to the project?

Bonus:

25.5 Senses: highlight prioritizing each of the senses and develop solutions for each of the senses. Use the five senses in your brainstorming session.

◀ Figure 1.18

Brainstorming user scenarios with toys.
Source: author photo.

Methods Designers Use in Brainstorming

Tom Mott, senior producer and interactive learning designer:

> I'm a pencil-and-paper person; I pantomime a lot. I act things out physically and verbally. I imagine I'm the kid, playing with the thing. You can cut through a lot of theoretical garbage if you just start manipulating things in real life. If I'm designing an app, I'll draw a screen on the page and pretend I'm tapping around while visualizing what I'd want to happen. If I'm designing a board game, I'll draw a game board and get some pieces from other games and pretend I'm playing.[127]

Andy Sklar, entertainment designer, Universal Studios:

> I sketch out concepts and look at the big picture. I will compose a lookbook of inspired ideas that reflect the overall vision and visit similar examples that are already in existence. I play with layout and positioning. I like working in elevation concept forms as well to help articulate details of architecture, icons, and graphics. From there CAD designers construct it for reality.[128]

Jini Zopf, senior manager, Hot Wheels Packaging Design at Mattel:

> For brainstorming when working on younger kids' products brainstorming starts with aesthetics right off the bat. Older kids we start with what is unique about the play pattern? We start with the toy and then the elements you want to pull out. Is it STEAM? Does it have a licensor?[129]

Hsinping Pan, illustrator and animator:

> I always feel a bit of a panic when I start a new project and drawing in the sketchbook always calms me down. As I keep drawing, it helps me think of more ideas. When I look back to the sketchbook later, those drawings sometimes work but sometimes don't. But I feel it is good to have gone through the process.[130]

Cas Holman, play designer:

> SKETCH!!! I sketch constantly. Messy, quick, lots of notes to myself so I can remember what I was thinking about when I drew it.[131]

Terry Fry Kasuba, illustrator:

> I usually start with writing notes and/or putting together a board with reference images. I then begin to sketch out ideas with paint. I usually put down the most obvious stuff first and get to know the subject. Sometimes I'll listen to a podcast or watch a movie on the subject while I'm painting. When I am in the mode of a project, I am thinking about it all of the time, and then I come up with some not so obvious associations which takes the illustration to another level.[132]

Fridolin Beisert, author of *Creative Strategies*:

> When brainstorming it is about creating choices and when designing it is about making decisions. If 100 ideas develop from brainstorming I pick 20.

Methods he suggests:

- Use props: Open-ended non-descript, non-branded no one-time use. Imagination supports the play.
- Combining things: Remixing – putting together things that don't belong together – see what comes of it.
- What if: Work with partners or small groups on imaginary scenarios. Ideas will clone and expand.

- Lists: Make all different categories with loads of topics and break into subcategories. This gets obviousness out of the way.
- Change environments: New physical locations provide new stimuli; go outside or to a new place to brainstorm.

Then choose 3–5 possible design solutions. Devise a strategy. For example, mine is to pick three areas to explore

- Something that will be fast to test
- Something that is familiar (so it is easier)
- Something that is new (neither of the above)

With this formula, there will be some failures, successes, and challenges.
Then question with appropriate questions such as:
What does it need to accomplish?
What type of emotional response does it promote?
What is it teaching?
What are we solving?
Then pick ten more to develop. Transfer mediums – from drawing to prototype and physically test with peers, kids and users.[133]

Ania Borysiewicz, creative director and graphic designer of Ladder Design:

It is important to brainstorm and come up with a lot of ideas. Put down whatever comes to your mind and don't stop. Different methods work for different people. Find what works for you. Set a timer for 5 minutes or 10 minutes and go. Sketch with your whole body. Act it out. Experience it physically. Use humor and laugh at it. Use your favorite pen and large paper. Write on the cement with chalk if it works. For some the tactile experience brings best ideas forward. Again, you don't make the idea happen – allow the ideas to come to you.

Co-Design

In the true co-design process, the users take an active, hands-on role in the design of the project; working directly and collaboratively is the creative process. Input from stakeholders is usually not referred to as co-design, although currently the term is being used loosely to represent many forms of collaboration.

The goal of a co-design team is to create a learning community. In co-design/participatory design teams, everyone on the team has an equal place at the table and brings their expertise to the discussion. Kids' opinions are equal and valued. They don't have to wait to grow up to contribute. Adults don't have the answers, and everyone is working together to solve the problem. The best age range for including kids in the co-design process is 7–12 years old. Kids have their sense of wants and needs, and it is our role to help interpret their expectations for them. What can we teach children about human behavior and design so they can play a role in creating the optimal experience for themselves? I asked Terry Montimore, partner and creative director at Goodwin Design Group, for his thoughts about the benefits to kids in being part of the co-design process. He responded: "Kids are involved in the products they use, they take pride in knowing that it was made for them and somewhat by them. It becomes an empowerment tool by helping create/shape designs for them and their peers."[134]

One method to create a co-design team is to form a group of equal numbers of kids and adults. Get them both to trust that it works, get them situated and comfortable. First, set goals and paths to the goal. It is essential to be ready to pivot at any time based on what is happening. The hope is that the adult sways the idea one way and the kid swings it another way, and there is a continual back and forth. Each member needs to listen – and build on the other ideas – and tease the ideas apart.

Contributing can build self-efficacy and communication skills.

- They like that they are a respected part of a team.
- They get communication practice. Kids get comfortable communicating with a large group and learn to be an advocate for their ideas.
- They practice problem-solving throughout the process.
- There is also the content that they are learning by being a part of the process – about the topic or industry that they are designing for.

Play researcher Karen Feder explains the benefits of co-design for kids and companies:

- Children have the right to participatory democracy, on decisions regarding their own life, as stated in the Convention on the Rights of the Child. This includes designs made for children, where they can influence the result through co-design. By participating actively in the process, it offers them an opportunity to speak out for themselves, to express their needs and wishes.
- Children are experts on being children, and by including them in the design process, we acknowledge their position and the value of their perspective. This acceptance is a part of what makes us feel valuable in life, which is vital if we want children to have a great childhood and grow up to be happy and satisfied with themselves.
- Children are not only giving when they are part of a co-design process. If the process is planned well, it provides the opportunity for the children to develop as well. Being introduced to a design approach supports and recognizes children's talent for being creative, explorative, curious and all the other traits they already are experts in. Skills they will still need when they grow up in a continually changing world.[135]

◀ Figure 1.19

Co-designing characters out of food for a nutrition project.
Source: author photo.

Children's Drawings in Research

Engaging kids in drawing is a great way to communicate and understand children's motor skill level, what they understand, what they are thinking and what they are feeling. Kids' drawings have been studied for more than 100 years, and there are continually new processes for using them both educationally and therapeutically.

It is essential to allow the child to tell you what the drawing means to them. They might look like they mean something particular to you, but the child has an entirely different idea of what they meant. Drawings can omit things that are actually there, they can distort things that are there, and they can add things that are not there. They can provide insights into the imagination and conceptual depictions of reality. Kids are usually freer then someone with more life experience, and the lack of experience can bring exciting ways of looking at our projects. Collages and making models are also tools to get kids thinking about your topic. Kids' drawing skills develop over time. It is helpful to understand the different stages of development for kids when engaging with them in drawing.[136]

Analyze

First, take a break. Let it sink in – what ideas do you remember later? The experience of the brainstorming process might be enough for our brains to act unconsciously and know when something feels right. This might be a radical idea for those of us who are learning about and using processes and backing it up with research in our profession, but we do it in life all of the time – although in design, we often have to justify and validate our decisions beyond it feeling right. Play researcher Karen Feder explains the importance of a good foundation of research:

> In 2014, a group of design students and a couple of teachers from the Design School Kolding in Denmark went to Shanghai, China, to study play inside private homes of Chinese families with children. The study was very intensive and lasted for almost a month, which is a long time for a research process. What the students experienced was that all the hard and focused work of research they did, in the beginning, eased the rest of the actual design process. They were so well informed by their experiential learning of the real life of children's families in Shanghai, that they constantly could qualify their decisions and ideas in the final design phase back in Denmark. This illustrates how important a good groundwork is for the foundation of a design process when you don't have the opportunity to engage the children all the way through the design process.[137]

Hone insights and analyze brainstorm content to take your ideas and build them into reality. What are the various forms the concepts can take? Develop designs and plans to execute them. Iterate continually. How do you select the best ideas to carry through to implementation?

Develop a list of criteria and apply them rigorously to the ideas on your list. Compare the ideas against explicit criteria and make judgments as to which will succeed and which will not. Look at factors that influence design decisions.

Each project's questions will be different depending on the goals, but some questions to help you choose ideas might look like this:

- Is it innovative?
- Is it better?
- Does it solve a problem?
- Does it fill a need?
- Is it feasible?
- Does it make things easier?
- Does it fit the purpose?
- Does it do what it is supposed to?
- Is it marketable?
- Why do users like it?
- Can it be clearly communicated?
- Is it profitable?
- Is it technically feasible?
- What learnings will help to refine the story to tell others?

Prototype, Engineer, Validate, Improve

Prototypes take ideas from theory to reality. They help us to see what will be involved from a practical level. Before making anything, know what you would like to learn from different types of prototypes. It is essential to get prototypes into kids' hands as quickly as possible for user feedback. Once the goals are established, you can decide the most worthwhile prototypes to build.

Do you need prototypes to …

- inform your design decision-making process?
- test various concepts with kids to edit them down to the best ones?
- test developmental appropriateness?
- test functionality and engineering?
- test the performance of various materials, processes and technologies?
- visually test scale in relationship to children's hand and body sizes within an age range?
- test forms and size for ergonomics and small and large motor skills?
- use to conduct safety tests?
- be a communication tool to get everyone on the same page?

Sketch – (Rough) Prototypes

Sketch prototypes transform drawings into simple models. Fridolin Beisert explains the value of the process of fail fast prototyping: "Through making the next steps in the process are presented. Don't wait for the perfect idea. Focus on the evolution of it. It is better to fail along the way than in the end."[138] In the early stages, go back and forth between 2D drawings and 3D models. Make multiple variations to get kids' and caregivers' input and feedback. They are cheap, easy and made as quickly as possible. They can be actual size or scale models. They can be made out of foam, paper, cardboard, clay, wood, fabric, ad hoc materials or store-bought substitutes, or through rapid prototyping using additive or subtractive technologies. As long as they are rigid enough for hands-on interaction, they can be used for concept testing, playtesting and observations. Prepare and design the prototypes. Concepts with advanced technologies can be made with similar experiences. For example, when testing interactive designs, instead of coding software, the experience can be simulated with screenshots or Post-Its or mapping-out scenarios. The interaction of pressing buttons can be tested with drawings and analog materials such as buttons and rubber stamps. Industrial designer Nathan Allen, who teaches prototyping and various rendering software, suggested:

> You don't have to reinvent everything. You want to save time and put effort into the right areas for the project. Use parts that already exist. There are a lot of mechanical parts readily available. If you need a scale mechanism buy some scales at the store or go online. Take them apart, try them out and see how they work. Use them in your model.[139]

It is useful to design multiple prototypes for testing and design flexibility into each of them. This allows kids to manipulate and change them so that you can test how kids use and interpret them. It may also be useful to integrate the making of the prototype into the testing session to get feedback on colors, use or assembly. Nathan Allan recommended: "Let them choose the colors. They know more than you do what they like."[140]

With today's technologies, it is easy to make polished-looking prototypes but keep them loose and unfinished. Nathan Allen mentioned: "If something is too complete the kids are either going to say they don't like it or feel like they can't change it." When testing your prototypes, he recommended: "Be open to feedback. Why give prototypes to kids in the first place if you are not going to learn from it? Be willing to let go."[141]

Refined Prototypes

Refined prototypes can be working prototypes (works-like prototypes), visual prototypes (looks-like prototypes) or a combination of the two. These "pre-production" prototypes are developed with working mechanics and structure, out of the intended materials (or close to them), and to the correct dimensions and tolerances, artwork, colors and packaging. They can be used as pilot prototypes for small in-field trials to see if the model meets the requirements established in the problem. With refined details, it is easier to conduct usability, physical and ergonomics testing. Does it work the way you expected? Does it do what you want it to do? Is the idea efficiently communicated? Is it easy to assemble? Multiple prototypes can test different methods of technology.

Nathan Allen and I discussed the role of software (CAD drawings) in relation to the physical prototype, and he mentioned: "Many people lean on the software to resolve a shape but there is a huge benefit to sculpting by hand and see the shapes evolve in front of you. It is hard to make six variations on the computer. It is easy to carve six different shapes out of foam."

After the refined prototypes are tested, we can finalize the experience, technical drawings, design, engineering, materials manufacturing, colors, technology, artwork and final design details. Once the design is complete, appropriate IP can be considered and addressed.

Production Prototypes

Production prototypes come out of the tooling "set-up" that is prepared for manufacturing. Soft tooling may be used for limited runs for safety testing, to conduct usability testing with a more extensive sampling pool, to get samples into sales people's hands or to presell a product. You may choose to jump right into hard tooling for long-term production runs; however, keep in mind that when a project scales, there are often changes, so allow for flexibility. A production prototype communicates a product more efficiently than a refined prototype to business partners, investors, attorneys, or packaging or marketing experts.

In our discussion, Nathan Allen reinforced the importance of meeting with manufacturers at the beginning of a project. He explained: "If you create the engineering drawings and then meet with a manufacturer you will probably need to redo them later on. They usually like to figure out how to make it for their production methods."[142]

Prototype Suggestions for Testing with Kids

- Make it durable and safe. It is pretty apparent that you can't design something and have pieces break off while you are testing it. But it happens all the time.
- Test the ergonomics. Kids of the same age are a range of sizes.
- Whenever testing several different models make them all the same color and scale so that they are equal. If they are different they may choose them for different variables than you are testing.

- Get something in their hands as fast as possible and don't tell them what it is. You will observe how they see it. After you get some reliable feedback, then tell them what it is and see if they respond differently. Consider what you learned from that.
- Testing prototypes at the beginning and later on in the process. If they are not as excited in the later stages, you know you have lost something in translation. Sometimes as we refine things we try to be serious and what we are making is no longer as playful.[143]

Production

There are usually two phases of production: preproduction and production/distribution. During preproduction, the research for manufacturing, sourcing materials vendors and tooling for manufacturing takes place. During production/distribution, the item is produced, inventoried and sent to distributors. Play designer Cas Holman shared some insights on choosing vendors:

It's a long slow arduous process to manufacture in the US. In every product I design we start sourcing locally and broaden as needed. In my mind, manufacturing partners are critical and I work hard to make sure the project is mutually beneficial. There is a saying used when working with vendors and contractors and sub-contractors which is "beat them up" over pricing. We have to be competitive but if they don't make money from manufacturing my products then I'll be looking for a new factory in no time. Or I'll be dealing with quality control issues. I prefer to make a smaller profit and enjoy my relationships.

Production Issues Related to Kids

- **Child labor and exploitive working conditions:** Child labor is work performed by a child that is likely to interfere with his or her education, or to be harmful to his or her health or physical, mental, spiritual, moral or social development (Convention on the Rights of the Child, Article 32.1);[144] International Labour Organization (ILO). Many companies outsource production and are not entirely in control of how items are made. As an advocate for children everywhere, know how and where the items that you design are made.[145]
- **Safety and ecological production:** The materials and manufacturing processes used in the end product and during the production should be safe for the environment. The use of sustainable resources, energy consumption toxicity, and recycling are essential in every project, but some are held under stricter guidelines and are in more demand in children's items. For example, art materials intended for use by children may have banned substances and must be tested and certified and include tracking labels.[146]
- **DIY, post-consumer manufacturing, customization:** In these processes kids are not just users but makers. The designer's purpose is to configure a product, service or environment into a freeform malleable format. We must understand what kids are capable of, what variations they could create, and how to design a system to satisfy those demands.

Phases of Production

Determining the best plan for production usually relies on business research into industry specifications of materials, processes, costs and distribution. We ask, "How do we best achieve the design's objectives with the safest materials and methods at a reasonable cost?"

Determine Scale

Since there are many ways to produce projects, understanding the scale of production helps us to decide the best way as it relates to cost. How many will we be making in the first run? Will there be subsequent runs? What are the sales goals? Researching the sales of similar projects gives us an overview of the commercial viability. Studying the market size or developing a distribution plan will help pinpoint some sales goals.

Production Research

Understand the industry. Why are specific modes of making used in a particular industry? How will your approach be similar? Different? Who will make it and where will it be made? How will it be made? What types of operations resources will it take to execute it? What will the makeup of operations team be? Will you set up production in-house or outsource production?

Production Planning

With the knowledge from production research, we can now develop a production plan with a timeline for each. A project might only have a short-term plan; for example, printing a poster for an event that will happen once. But many projects for extended use also have a long-term plan. How will a mini-run then roll out at a larger scale? Will the manufacturing be on demand, stocked inventory, just-in-time, or post-consumer production?

Production Stage

This stage includes small runs to get out to the customer: making a single run or multiple runs or redesigns, rigid tooling (long-term). Allow for software technology design updates, full force making, implementation, inventory and distribution. Projections of scale and timing are essential. Every year there are shortages of the top Xmas toys because companies had underestimated projections and could not turn around product quickly enough to meet the demand. Apps often miss their release dates, and designed environments are notorious for missing opening day deadlines.

Different Ways of Making

Artisanal: These projects are handmade in small quantities and usually integrated into the slow and local movement. Many are produced using sustainable methods for the conscious consumer.

Limited production: A finite amount of a product or service is made, and that is it. For example, in the fashion industry, a finite number of items may be made for a season, and when they are sold, they do not make another batch.

Mass-production: These items are usually made with high-speed equipment for large volumes and low costs. Many products on the mass-market are made through mass-production processes.

On demand: The product or printed item is made to order.

Customized: These can either be made on demand, stocked and then configured to meet the customized specifications, or customized in a factory by a company for the customer.

Post-consumer: Items made by the consumer after they are sold. This could include assembly or the making of a complete item.

Sustainable Production

- Reduce the number of product samples and shipping whenever possible.
- Consider production runs and tooling set-up for minimal waste.
- Work with companies, services and clients that mandate sustainability.
- Question when the value of working locally is higher than the savings from shipping costs.
- Use just-in-time inventory versus stocked inventory to prevent overstock.
- Manufacture in countries with strong environmental and ethical protections.

- Design hands-off efficient manufacturing.
- Produce in small batches and only what is needed in limited runs to ensure safety and salability.
- Use production facilities that are free from child labor and have safe and responsible conditions for factory workers.
- Limit airfreight and transportation energy use and emissions.
- Choose factories that have been regularly audited in accordance with social compliance standards.
- Avoid companies with a history of piracy and infringing on patents, copyrights and trademarks.

Designing, Manufacturing and Marketing Children's Footwear for Play

Interview with Jason Mayden, chief executive officer and co-founder of children's footwear start-up Super Heroic.
He is armed with the knowledge of advanced materials and processes that can unleash the potential of every child in the world through play.

Can you elaborate with specific details of how you inspire kids through the details in the product design?
Traditional footwear for children is generally focused on rigidity and durability and less on "performance play"

▲ Figure 1.20

Super Heroic footwear is designed for the spontaneity of play that keeps up with kids' movements.
Source: Super Heroic (used with permission).

features such as cushioning and traction. Children move from carpet, to gravel, to grass and wood chips all in the same day. We provide the proper support to handle their spontaneous changes in direction, environment and play.

Can you explain the advanced materials and manufacturing process that are used in your product design solutions?

Materials upper/sole
The innovation lies in the configuration of the cushioning system that is inspired by the hoof of mountain goats. The harder outer perimeter of the midsole promotes stability during multi-directional movement on uneven surfaces. The softer, more responsive inner core promotes proper heel-to-toe transition during linear motion and comfort during a resting or standing position.

Form
We created a wider forefoot with less material and an EVA [ethylene-vinyl acetate] strobel board, allowing for the child to be closer to the ground, spread their toes and provide additional interior width for their toes to spread during jumping and landing motions. The reduction of internal layers allowed us to have a sleek silhouette without compromising fit.

Closure
The "Utility Strap" is designed to provide a comfortable level of containment across the midfoot, which is the best location for proper support. This also plays a

significant factor in how the shoe moves with the child's body instead of against it.

How did you address functional issues specifically related to kids' footwear that you might not have to address to the same level of rigor with adult footwear in your designs?
We explored several ways to provide performance for a child during the 90–120 days that they are in their current size. We focused on zonal protection through laminated skins in areas of high abrasion, such as the toe box. We protected the heel counter, which is a critical fit component that is generally damaged during rear entry of the shoe, by adding a natural ledge or "heel bumper" for the child to use during removing his or her shoe. We also measured and benchmarked several lightweight performances that were popular among parents and children, all the while keeping a very strict focus on creating a responsive product that enhanced a child's ability to feel fast and safe in our product.

What kind of testing with kids did you do in product development?
Before designing the first shoe we conducted focus groups with parents and children that helped us to identify our pricing, positioning and learn about the current state of play. For play/usability testing, we captured data during a series of bio-mechanically based motion analysis tests that highlighted areas of potential innovation that have been traditionally overlooked. We have a process that allows us to gather feedback from our audience on a quarterly basis.

How did you know when you were successful in getting your product right?
In our research we saw our unboxing experience was the catalyst for parents to reminisce about their most delightful memories of play. Everything from the shape, the signature sound that plays when you open it and

the included cape were all designed to signal to the parent and the child that they have transitioned from the ordinary world to the "special world."

What is your approach to marketing and brand building for these different audiences?

Parents
We focus on discussing the importance of the preservation of play. Furthermore, we discuss and visibly display the core benefits product that is "Built for Kids." Our social content encourages and inspires various forms of play and self-expression. This approach has resonated with parents who desire to have their children to remain children for as long as possible.

Kids
We reference the "Possible Selves Theory." Simply put, if you present a child with an image of someone similar to them in a position of importance or strength that child is more likely to believe that they have the ability to achieve that outcome. With every campaign, social post or graphic element we present a child with an image of someone similar to them in a position of importance or strength. We choose positive role models to showcase young girls and children of color.

How does your product address other play patterns outside of active play? If so how?
We used insights from the "Batman Effect"[147] study where children were found to have a greater sense of perseverance while in character. Every element of our experience from the unboxing, to the placement of the stickers in the "quick start guide," to the bag that transforms to a cape, is intended to create an immersive multisensory reveal for the child. The detailing of the packaging and the need for a two-handed removal of the shoe itself are all in place to spark and enhance imaginative, object-based and fantasy play among families.[148]

Marketing and Distribution

Marketing is how a product, service or experience is communicated to the users and customers. Dan Winger, senior innovation designer of the LEGO group, states: "Communication is key. How do you communicate the product, story, and experience, to kids and parents? This is even more critical if you have a new and innovative experience that is unfamiliar to users." Distribution is how it is moved through channels to get there. Distribution is an integral part of the marketing plan.

Many fields of design overlap with marketing and distribution, including branding, promotional materials, advertising campaigns, packaging design and the purchasing experience.

Designers must learn specifics about marketing and distribution related to the industry and the project they are creating to make valid design decisions; questions such as: What is the developmental level of the target market? What value does a brand have for them? What are appropriate and ethical marketing messages to use? For example, Patricio Fuentes, principal at Gel Comm, explains:

> Designers need to understand the buyer psychology of a young person. When speaking to teens you are not speaking to an individual you are speaking to a cluster – they live in packs. Companies can't just throw money at marketing anymore. Branding is all about trust and kids today trust the small guy. It is like the David and Goliath story. They want love – not big money.

Learning the current systems in place opens up opportunities for creative inspiration for new methods of marketing and distribution that disrupt the current methods. For example, crowdfunding (raising funds and preselling to customers before your project is made) has become an alternative option for start-ups in some children's markets that have previously been challenging to access customers and distribution that has formerly been limited to large companies. In addition, user-generated content such as the unboxing videos and children and teens with their own YouTube channels have shaken up traditional advertising models.

Most of the marketing of products and services for small children is through the moms. When I asked Bill Goodwin, creative director of the Goodwin Design Group, why even today people always talk about marketing to moms and not parents in general, he replied: "The truth is the majority of decision-making is the mom. She influences the dad. Mom is more restrained." After the age of 12 years old, when teens have more control over how to spend their money, they are often marketed to directly. Bill Goodwin calls it "bimodal branding." First, you are marketing to the mom; then to the mom first and the kid; and as they get older, it is to the kid first and mom second. He explained:

> When mom is buying the product it is about her relationship with the brand. By 7 to 9 years old the kid is in control of a lot of product categories. Parents enjoy kids coming into their own and picking out their own things. They learn more about them by seeing what they like. Kids get one message we get another. From 0–14 years old you can influence kids. Beyond 14 they have already made up their minds – it is hard to influence them."[149]

Marketing Phases

Phase 1: Market Research

Market researchers pinpoint, collect and analyze target customer data, the competition and the target market environment to aid in making marketing, branding, messaging, positioning and pricing decisions. Market research is done before and after a product or service is launched. Market research can be beneficial to designers when developing a project to help make design decisions. The more we know about our users and the business landscape, the more specific we can be in our designs.

Valid market research can help determine whether and how a company should:

- Enter a new market and determine price points
- Launch a new product or service
- Promote brand awareness
- Get the most out of a marketing campaign
- Approach messaging, perception of the product or service, and customer service
- Develop a product, experience, packaging or delivery system

Phase 2: Marketing Strategy and Messages

With the research, marketers develop messages for the different audiences that they need to communicate to: kids, parents, retailers and stakeholders in the project. These messages are conveyed through design in branding, promotional materials, packaging design and advertising. As Bill Goodwin described it:

> In the early years you are selling the licenses (characters or stories). In the later years you are selling individuality. Kids want to say, "I have this and nobody else does." Kids today are praised for their uniqueness compared to 30 years ago when it was about conformity. It is the difference between the Breakfast Club (being part of the group) and Glee (be who you want to be). We live in a time of acceptance. Young kids are empowered by blurring the lines of what is acceptable. Jocks and musicians are all friends. Retail has different sectors, general and niche. In niche markets we can do customizing and tailoring.[150]

Independent illustrator Hsinping Pan explained what she considers in creating marketing materials:

> When designing a poster, first I try to understand what is the message that client wants to convey. Then I like to create an imaginative world with their messages in it. Something more abstract and can catch people's attention right away. Small details in the drawings can draw people in to look closer and have fun with it.[151]

"In this environmental art poster, the client wanted to show the community, an art paradise, and the love to the environment. I drew a soft mother like creature hugging the art village expressing the harmony of people and environment."
Source: Hsinping Pan (used with permission).

Phase 3: The Tactics

They then develop a plan with tactics for disseminating the messages, advertising, print, social media, web, TV or packaging. A distribution plan is built. Where will a company access this information? How do sustainability issues related to shipping, packaging and disposal fit in? Sometimes marketers will also be the sales voice, but often expertise in sales, especially in business-to-business relationships, will negotiate a sales transaction.

Questionable Marketing Practices Currently Being Implemented towards Children

There are regulations in place for marketing and advertising to children, and many companies follow the guidelines and believe that children's rights and truth are very important. Others find ways around the regulations. All marketing targets people's emotions, and what makes children different is their level of awareness of the truth behind the messages and their ability to make educated value judgments about what they spend their money on. Sustainability expert Heidrun Mumper-Drumm believes:

> Many companies have subversive values and approaches towards marketing to children. The approach is, How can we get into a kid's head and get them to desire what we have? They are taking advantage of their lack of experience. You can't even buy a toothbrush for a kid without a character on it. It affects how they play and relate to others. They are promoting particular values, such as how much better it is to be a princess, to children who are just beginning to develop their values.[152]

Kids under the age of 8 years old don't yet understand the persuasive intent of advertising and marketing. This put them in a vulnerable position. Along with kids' natural ability to nag over and over again to get what they want, this is proving to be hugely profitable for companies.

Child psychology and knowledge of how the brain works and the awareness of children's emotional needs are used to sell products and services to children. Just one example that I find irresponsible is when reputable companies tempt children with contests or free gifts that they are "sure to win" or when they collect data through surveys with children on existing purchases for product feedback to pull them into the next purchase. Some companies shape children's brand preferences by telling them what they need to have a meaningful life and shaping a child's entire experience. In my conversation with Bill Goodwin and Terry Montimore of the Goodwin Design Group, Bill Goodwin stated:

> A lot of companies try to work around guidelines. They market to 13-year-olds by law but they get to the younger kids through other means like social networks. Say you target a junk food ad to teens and hope it goes viral so every kid sees it. We won't lie to kids – we don't take clients that ask us to.

Terry Montimore explained:

> When I first got into this I felt like the devil marketing to kids, I didn't want to be part of it. I am always in check. The more I thought about it I knew to be honest and unapologetic and don't deceive. I hated being oversold to as a kid – you opened up the box and it was a piece of crap.[153]

Some practices currently being used are:

- "Buzz marketing," a tactic where the "cool kids" are paid to give status and spread the word.
- Commercialization of education is when schools facing budget shortfalls allow corporations access to students to market their products or services in exchange for funding. Some kids are subjected to school bus radio, a station of commercials they listen to on their way to and from school.
- Viral ads that are designed to be passed along to friends so that they look like endorsers.
- Pester power, using kids' natural ability to pester to bug their parents to make a purchase.[154]

Ania Borysiewicz, creative director/graphic designer, explains:

> Friends' use of brands influences my 2-year-old daughter. She recognizes the Mickey Mouse logo, is attracted to *Frozen* and Minion characters through her friends. At first I rejected *Frozen* but then I learned to give into it and use the characters to help her do something she needs to do. For example when she won't brush her hair I say. "Don't you want to make it long and beautiful like Elsa's?" When we need to get out the door I say "Let's

pretend to get on the horse and ride like Elsa." It helps my life. Ironically, she has not seen the movie. There is enough advertisement from her friends.[155]

The Better Business Bureau's Children's Advertising Review Unit (CARU)

CARU issues guidelines that play a significant role in ensuring responsible advertising to children under 12 in the United States. The guidelines are posted at www.caru.org. The Children's Online Privacy Protection Act (COPPA) gives parents control over what information websites can collect from their kids. Bill Goodwin, creative director of Goodwin Design, explained: "At the end of the day more kids have phones at 7 years old. At 9 to 10 years old they are on Facebook and Snapchat – no matter what the guidelines are."[156]

Marketing can be influential in promoting beneficial messages. If you are marketing a project intended to influence kids to eat more vegetables and less candy or to reduce drug use, smoking or gun violence, that is very different from girls being told that they need to be pretty and sexy and boys being told that violence, aggression and toughness are required to be a man. A designer's point-of-view and value system play into what kids experience as users.

Ethical marketing suggestions for designers:

- Design and use age-appropriate marketing messages.
- Message to educate, not preach.
- Offer positive social and emotional development; provide empowerment, not discouragement.
- Create quality media content (educational entertainment (edutainment)).
- Ensure a positive shopping experience.
- Avoid stereotyping and negative messages such as gun violence, gender and over-sexualization.
- Do not expose kids to marketing that could affect them negatively.
- Promote safe play. For example, design and show personal protection equipment when promoting athletic products.
- Give realistic impressions of product features and service experiences.
- Use gender-neutral and inclusive visual communication. Show children playing together with products across categories. The same is true when presenting roles of moms and dads.
- Use multi-racial children, special needs children, appropriate languages and culturally sensitive marketing.
- Group products and services into categories such as interactive play, creativity and learning, classic toys, baby and toddler products, not by girls or boys.
- Provide positive role models within diverse character traits, backgrounds and body types.
- Do not use characters or child-friendly advertising that is intended for older audiences to attract younger children.
- Do not market inappropriate products or services in children's environments, including parks and schools.

- Prescreen all advertisements (TV, print, online) directed to children to ensure proper messaging/content prior to distribution of these advertisements.
- Reduce the digital divide by promoting integration, not segregation.
- Advertise within or surpass federal guidelines appropriately. Choose appropriate placement of advertising online, e-mail marketing, product placement, social sharing and collecting data when marketing to children.
- Consider the impact of AI, machine learning and surveillance technology.

Design Prompt: Marketing

Develop a marketing plan based on following these three phases using ethical marketing practices:

- Market research
- Marketing strategy and messages
- The tactics

Distribution Channels

Distribution includes that invisible network that is built on relationships that will get your product or service to the users and the customer. It also includes inventory management and the shipping logistics of moving something by boat, ship, truck, car and digital download. How will you access users and physically get what you are creating to them – through parents, schools, camps, online?

The distribution model is decided based on the level of penetration that the company wants to achieve, and may include multiple channels. The mode of distribution affects many design and manufacturing design decisions. For example, a piece of children's furniture may be designed to be distributed online and through catalogs. That piece of furniture may need to be shipped and assembled by the parents. In that case, the piece is designed to fit within shipping weight and size constraints to keep the shipping costs low and reflect the shipping method, pallet sizes, container sizes, UPS, Post Office and other delivery methods.

Modes of Distribution

- **Business to consumer (B2C):** An item is sold directly to consumers. What makes children's items unique is that often the end user is the child but the consumer is the caregiver.
- **Business to business (B2B):** A business sells to another business. For example, a medical device for children is sold to doctors to use with children.
- **Indirect distribution:** The product or service reaches the end customer through a middleman (a wholesaler, a distributor, a sales agent or a manufacturer's rep).

- **Direct distribution:** The company directly sends the product or service to the end customer or through a website, direct internet, direct catalog, direct sales team.
- **Intensive distribution:** Mass-marketing product covers as much of the market as possible through numerous channels.
- **Selective distribution:** A branded company will have selective distribution. These companies are likely to have only limited outlets. Examples are Pottery Barn Kids or exclusive product offerings for Target.
- **Exclusive distribution:** Markets may include an entire industry, and territories might be appointed for a complete country.
- **Digital distribution:** The distribution of audio, video, software, apps and video games content over the web – on demand. Content distributed online may be streamed or downloaded.

A few current design challenges in distributing to kids:

- **Global distribution:** Large companies are moving to broader international markets. Some products or services are made with a global strategy that is generic throughout the world. Others are made with different approaches for different territories. How do you speak to multiple audiences? Sustainability expert Heidrun Mumper-Drumm states:

  > With globalization we are stripping away authentic cultures. We have to consider the impact. Communities, outside the reach of mass toy manufacturers, have been making the same handmade toys for their children to play with for generations. Companies go in and say, "This branded toy is better." What is going to be more appealing to a kid? The cornhusk doll or the soft, plush, goggly-eyed, stuffed toy? What does the loss of traditional crafts do to family, community and culture?[157]

- **Third world children's needs:** Getting products and services to children in need throughout the world is not easy. Many children are living in environments where they do not have access to transportation, food, medicine and education. Since mobile has become more widespread, digital distribution is starting to make a dent in education, but physical assets are still hard to distribute.
- **Reuse, recycling, resale:** Since kids' products have a short life span, there is a broad opportunity in the resale market; however, most of the time it is more expensive and uses more energy to return an item or donate it than to throw it away.
- **Small companies and big corporations:** Many large corporations have access to distribution networks that smaller companies do not. The higher-end and lifestyle markets provide more opportunities for smaller companies that produce quality products, services and experiences. This gives access only to those who can afford the best options.

- **Licensing for distribution:** Companies can expand their distribution through licensing products, brands, characters and experiences for distribution rights to other companies nationally and internationally. Companies must be sure not to lose oversight of the brand and decision-making.

Design Prompts: Distribution

To create a distribution plan:

Study the current distribution methods within your industry.

Where is your target customer currently accessing the alternatives to your product/service/experience?

How can you get the product or service from the point of origin to the end customer?

How can you control costs and save time, material resources and energy while executing the distribution strategy?

How can you build a competitive advantage through distribution?

How can you distribute to the most in need of the product or service you are offering?

Can you disrupt the current distribution models?

Post-Distribution Research and Brand Engagement

Long-term studies can be helpful in determining patterns. Evaluating a project over time will also help with redesigns of new and improved versions, and can influence how to implement updates. One benefit from following a project over time is being able to test a design in the context of scale. In the original testing phases before a project was implemented in the real world, only a few individuals tested our assumptions.

Cause and effect relationships are more noticeable, and connections can be made more clearly. We can also look at the impact the project has on the children as they grow. These studies require time and a large sampling pool but can be particularly helpful when developing projects for long-term impact, such as designing medical devices, treatment methods, educational tools or programs, or the impact an environment has on a community. Long-term engagement might be critical for children's safety. Kenneth S. Kutska, executive director of the International Playground Safety Institute, explains:

> Nobody seems to care about what happens after the ribbon cutting celebration of a new play area. It has been estimated that almost 40% of all playground injuries were caused by a lack of maintenance or improper maintenance practices.[158]

In a case like this, and often in the case of digital products that form addictive habits in children, follow-through from the creators is crucial. The collection of information embedded into our creations needs to be done in an ethical way.

There are many aspects to a project that we cannot foresee until an audience uses it. Children's book author and illustrator Marla Frazee explains the surprise she encountered while observing how teachers used her book *Roller Coaster*:

> *Roller Coaster* and it is all about fear. Nothing about a roller coaster is NOT about fear, in fact. It has been interesting for me to see how this book has been used with children as a teaching tool for feelings, for creative word choices, and to study the physics of motion and energy. I leave it to the experts to find ways to teach with my books. I am blown away by how teachers use my books with their students.[159]

When a project is released and living in the world with children interacting with it, the designers are usually deep into their next project. This time lag and the lack of allocation of resources for follow-through create missed opportunities for designers to learn from and improve upon the design. One of the most important areas for learning to measure is the success or failure of a project. Play designer Cas Holman explains:

> I gauge success from how kids play with it! If they make things I wouldn't have thought of – it's working. If they are engaged for more than 20 minutes, it's working. If they work together, communicate, assemble and disassemble, then it's working! If the teachers engage in questions and feel confident letting the kids direct themselves and show the teacher how to do things, that is a success.[160]

Methods of Follow-up

Market Research

Once a project is released, we can look at and test customer response under real-life conditions. This can give insights into product/service modifications, adjust prices or improve packaging. It can uncover users' issues related to the sales, products or service performance. Developing a plan for collecting information that will be most relevant to make future designs before a project release can then quickly be promoted on the web, included in the packaging, collected through a product feedback system or implemented as a research study. There is also a lot of user-generated content out there about products we create that can be collected from sales websites, product reviews and customer input. Eric Poesch, senior vice president, Design and Development, at Uncle Milton Toys, explained: "After the product is released you can learn a lot from online reviews and people calling in to tell you what is working, not or how it is being used."[161]

Co-creation and Brand Engagement

The days of launching a product and then forgetting about it are a missed opportunity to keep connected to the users. We have just gotten a point where it is technologically feasible and cost-effective for individuals in the mass-market to participate in the creation of the products/services they purchase. With many co-creation initiatives, companies are now brand managers and service providers, and continue relationships with customers after the purchase of a product. Pottery Barn Kids holds recreational decorating and DIY workshops for children and adults. Home Depot has classes for children to learn how to use tools and build.

Location Studies

These can be used to observe kids with the project in the natural environment. This can give a more accurate picture of customers' usage habits, shopping patterns and what is needed for maintenance and programming. The researcher visits the location, watching kids interact with the project and determining whether they use it as intended or in unexpected ways. This could be an educational tool for the preschool environment, an in-home study of a children's furniture design, watching kids play with an app, or a children's entertainment space.

Design Prompts: Follow-up

- Did the decisions or the claims hold true at scale now that we have hundreds, thousands, hundreds of thousands of products in use?
- How does the actual implementation compare with the reactions to the original drawing and models?
- Can a built environment be visited over and over to see how kids really use the space? Is it being used in the intended ways? What can be learned from the unintended ways?
- Can research observations and data be collected to really understand a product's use?
- Can a longitudinal study tell us something about how to improve upon our project?
- Are the claims that were tested during the design research phase holding to be accurate or better or worse than initially thought?
- How does the design hold up over time?
- What are the required technology updates based on data that was collected through the back end of an app or website?
- How long on average did people interact with a product, service environment or technology?
- Is your customer the same one that you thought it would be?
- If it was designed for multiple users, are they using it in all of the intended ways?

- What is the sustainability impact of the project? Are people recycling it as planned?
- What is the behavior around hand-me-down, take-back program, resale and use by multiple children? What changes can be made if it is not working?

Notes

1   Allen, Scott 2017
2   Berger, Warren 2016
3   Brucken, Carolyn; Cooper, Alban; Wilson, Sarah 2017
4   Mallory, Michael 2017
5   Beisert, Fridolin 2017
6   Bennett, Katherine 2017
7   Allen, Scott 2017
8   Feder, Karen 2017
9   Bennett, Katherine 2017
10   Prieto, Mariana 2016
11   Amatullo, Dr. Mariana 2016
12   Feder, Karen 2017
13   Prieto, Mariana 2016
14   Poesch, Eric 2017
15   Kunter, Yesim 2017
16   Fuentes, Patricio 2017
17   Bennett, Katherine 2017
18   Observing Recording, and Reporting Children's Development, CRI Preschool Assessment Instrument, retrieved from http://laffranchinid.faculty.mjc.edu/Ch5.pdf
19   Goodheart-Willcox Publisher n.d.
20   St. James, Alena 2017
21   Bennett, Katherine 2017
22   McLeod, S.A. 2015
23   Fuentes, Patricio 2017
24   Bennett, Katherine 2017
25   Bennett, Katherine 2017
26   Bennett, Katherine 2017
27   Beisert, Fridolin 2017
28   Goodwin, Bill 2017
29   Poesch, Eric 2017
30   Feder, Karen 2017
31   Prieto, Mariana 2016
32   Gray, Peter 2008 *The value of play II*
33   LEGO Serious Play
34   Beisert, Fridolin 2017
35   Fung, Alice 2017
36   Sklar, Andy 2017
37   Winger, Dan 2017
38   Poesch, Eric 2017
39   Goodwin, Bill 2017
40   Feder, Karen 2017

41   Mott, Tom 2017
42   Bennett, Katherine 2017
43   Holman, Cas 2017
44   Bennett, Katherine 2017
45   Poesch, Eric 2017
46   MoMA Century of the Child
47   Demetrios, Eames 2017
48   Mumper-Drumm, Heidrun 2016
49   Mumper-Drumm, Heidrun 2016
50   Mumper-Drumm, Heidrun 2016
51   Jambhekar, Shri 2017
52   Boyl, Brian 2017
53   Laughery, Kenneth 1992
54   Jambhekar, Shri 2017
55   Boyl, Brian 2017
56   Inclusiveplaygrounds.org
57   Day, Douglas 2017
58   National Disability Authority Centre for Excellence in Universal Design
59   Poliakoff Struski, Valerie 2017
60   Nakano, Mari 2017
61   Bethune, Kevin 2017
62   Bethune, Kevin 2018
63   Fuentes, Patricio 2017
64   *New York Times*
65   Borysiewicz, Ania 2017
66   Poesch, Eric 2017
67   Beck, Melinda 2017
68   Mumper-Drumm, Heidrun 2016
69   Fuentes, Patricio 2017
70   Borysiewicz, Ania 2017
71   Allen, Scott 2017
72   Fuentes, Patricio 2017
73   Zopf, Jini 2016
74   Borysiewicz, Ania 2017
75   Gopnik, Alison 2017
76   Goodwin, Bill 2017
77   IMDb 2008
78   Zopf, Jini 2016
79   Borysiewicz, Ania 2017
80   Zopf, Jini 2016
81   Borysiewicz, Ania 2017
82   Mansfield, Tessa 2017
83   Mansfield, Tessa 2017
84   Mansfield, Tessa 2017
85   St. James, Alena 2017
86   Haberman, Mike 2015
87   Sparks and Honey 2017
88   Rechner, Erin 2017
89   Becerra, Liliana 2017
90   Becerra, Liliana 2017
91   Gladwell, Malcolm 2002

92 Otto, Ali 2017
93 Dam, Rikke 2017
94 Keyes, Richard 2017
95 Pan, Hsinping 2017
96 Wolchover, Natalie 2012
97 Keyes, Richard 2017
98 Adawaiah Dzulkifi, Mariam; Faiz Mustafar, Muhammad 2013
99 Fields, R. Douglas 2011
100 The GiggleBellies 2013
101 Color Matters
102 Keyes, Richard 2017
103 Keyes, Richard 2017
104 St. James, Alena 2017
105 Keyes, Richard 2017
106 Huggamind.com
107 Tarkett *Children's Perceptions of Color and Space*
108 Keyes, Richard 2017
109 Empowered by Color
110 Yamaguchi, Dice 2017
111 Passaro, Jamie 2015
112 Kutska, Ken 2017
113 Goodwin, Bill 2017
114 Goodwin, Bill 2017
115 Mumper-Drumm, Heidrun 2016
116 www.scientificamerican.com/article/bpa-freeplastic-containers-may-be-just-as-hazardous/
117 *New York Times, Balloons Made of Latex Pose Choking Hazard*
118 Explanation by author
119 Allen, Scott 2017
120 Nakano, Mari 2017
121 Allen, Scott 2017
122 Montimore, Terry 2017
123 St. James, Alena 2017
124 Fuentes, Patricio 2017
125 Beck, Melinda 2017
126 Feder, Karen 2017
127 Mott, Tom 2017
128 Sklar, Andy 2017
129 Zopf, Jini 2016
130 Pan, Hsinping 2017
131 Holman, Cas 2017
132 Fry Kasuba, Terri 2017
133 Beisert, Fridolin 2017
134 Montimore, Terry 2017
135 Feder, Karen 2017
136 Learning Design
137 Feder, Karen 2017
138 Beisert, Fridolin 2017
139 Allen, Nathan 2017
140 Allen, Nathan 2017
141 Allen, Nathan 2017

142 Allen, Nathan 2017
143 Allen, Nathan 2017
144 United Nations Human Rights Office of the High Commissioner
145 International Labour Organization
146 Consumer Products Safety Commission, *Art and Craft Safety Guide*
147 Young, Sarah 2017
148 Mayden, Jason 2018
149 Goodwin, Bill 2017
150 Goodwin, Bill 2017
151 Pan, Hsinping 2017
152 Mumper-Drumm, Heidrun 2016
153 Goodwin, Bill 2017
154 MediaSmarts
155 Borysiewicz, Ania 2017
156 Goodwin, Bill 2017
157 Mumper-Drumm, Heidrun 2016
158 Kutska, Ken 2017
159 Interview September 28 2017
160 Holman, Cas 2017
161 Poesch, Eric 2017

Part II | **Child Development**

Chapter 2

Development Stages

Child Development for Designers

Child development refers to a child's individual progress from dependency to increasing autonomy. It includes biological, cognitive, psychological and emotional changes that occur between birth and the end of adolescence. All areas of development and learning are essential.

Developmental milestones are a set of functional skills or age-specific tasks that most children can do within a specific age range. For example, most children learn how to walk shortly before or after their first birthday and learn to read between the ages of 5 and 7. Our goal as designers is to best understand what is taking place in children's brains, bodies and environments during the diverse phases of development to make informed decisions that suit the needs of our users in the demographic. At every age and developmental stage, products and services must meet the basic needs and wants of kids and their caregivers. From a business perspectives designers think of stages of development as our target market/s. Through creating products and experiences that aid in learning, playing and growing, we provide support, encouragement, structure and interventions to enable children to progress through each stage as smoothly and successfully as possible. Integrating the most current research in the field of developmental psychology can serve as a driver for our design approach and decisions.

Parents usually focus on the needs of their children as individuals; however, since most designers are creating for many children as opposed to an individual child, we take a broader perspective on development. We ask: What are the general norms? We recognize the widely diverse range of behaviors and abilities at different stages, influenced by nature and nurture. We consider variations based on a variety of differences, including children's unique set of personality traits, interests and environment. We then ask: How can we best support the differences in children?

Influences on Development

Humans have an extended childhood. Babies and children are dependent on adults for longer than any other species. In her TED talk *What Do Babies Think?* psychologist Alison Gopnik refers to

◄ Figure 2.1

Parents can identify unique traits, likes and dislikes of
their child from the day they are born.
Source: author photo.

babies and children as the research & development stage of human
species. They are protected, learn through trial and error, they think
blue sky and have good ideas. As adults we put those ideas to use,
do things, and put things out there in the world – more like production and marketing.[1]

Development changes are genetically determined; however, the environment, social and economic circumstances, parenting and peer relationships
all influence how children benefit or experience a disadvantage in childhood.
This includes physical, cognitive capabilities, temperament, and social and emotional makeup, to name only a few.

- Environmental factors that can influence development include housing, community, learning environment and exposure to different types of situations.
- Relationship factors that can affect growth include family life, attachment,
parenting styles, extended family relationships, peer relationships, and social
support networks.
- Biological factors that can influence development include gender, physical
health and mental health.

Nature versus Nurture

Plato and Descartes suggested that certain traits
are inborn; nativists believe most behaviors and
characteristics are the results of inheritance. John Locke
suggested that the mind begins as a blank slate.[2] We
can learn from studying this history, but the reality is that
there is not an easy way to separate the genetic, environmental, social experiences and cultural influences
that form a person. Instead, today, many researchers
are interested in seeing how genes and environment
impact each other. A 50-year study on identical twins by
Dr. Beben Benyamin from the Queensland Brain Institute
and researchers at the VU University of Amsterdam
revealed: "About 50 percent of individual differences are
genetic and 50 percent are environmental."[3] Designers
can support the child, family, community and society
through nurturing those differences in the products,
experiences and environments we create.

▶ Figure 2.2

Observing identical twins highlights how children with the same genetic makeup who are raised in the same environment can be very different from one another.
Source: Molly Boyl (used with permission).

What Is Universal and Different for All Children?

There are many factors that are rooted in biology and universal for all children around the world. In general, healthy children go through developmental stages at roughly the same time. All kids want their needs met, to be loved, safe, heard and taken care of. Where substantial differences lie is related to personalities, interests and culture.

Stages of Development

In our design projects, first, we determine the developmental factors or traits that we are designing for. Are they physical, cognitive, social and emotional, or a combination? Are you designing for what would be considered healthy progress or slow or advanced development? What is a child experiencing within each of these ranges? For example, a learning activity might be targeted to a specific learning outcome for a target age group. The activity and environment might include activities for students that support the required curriculum but also some activities for children with learning disabilities and advanced capabilities. Later in this chapter, you will see a list of developmental stages that will help to define what is typical in cognitive, physical, and social and emotional terms for different ages and stages.

Temperament/Personality

These inborn traits shape the child's distinct personality and approach to the world and to the product or experience we are designing. We are all made up of numerous personality traits. Combinations of these traits make us who we are. When defining a personality for a persona, it is helpful to combine positive, neutral and negative personality traits. Here is a list of different traits from MIT that will help to build different character makeups[4] (http://ideonomy.mit.edu/essays/traits.html). How would children with different personalities approach your product, service or experience in their own unique way?

Interests

Designers help nurture children's interests by exposing them to different activities within domains that they currently like. We also develop ways of making topics that they don't find interesting but are essential for their development more enjoyable. We often design for children with a range of interests from extreme to casual users. Children's interests are constantly changing as they grow. Consider how a child's interest affects the design solution. Are you designing for a range of benefits? Are you aiming for a general user or for a tailored interest? Are you encouraging the child's interest in a new topic or nurturing an interest that they already have? Observe the activities the children like, the books they read, the television shows they watch, the websites they visit and the way they spend their free time. Understand why they are drawn to these pursuits.

Childrearing Practices

Childrearing practices are embedded in the culture and determine behaviors and expectations surrounding childhood, adolescence and the way children parent as adults.[5] Parenting values of Western societies give importance to independence. In contrast, many non-Western societies value interdependent attributes such as cooperation, respect for authority and sharing. Within each culture there are also many diverse approaches. In many cases, the designer's role is to be culturally aware and sensitive to their audience. First, the designer must recognize their own culture's attitude towards the topic and value the diversities of other cultures. Research the specific culture you are designing for and participate in cultural sensitivity training if necessary.

Childrearing practices across cultures share these goals:

- To promote the child's physical and psychosocial well-being
- To provide children with the competencies necessary for economic survival in adulthood
- To transmit the values of their culture

Different practices are particularly noticed in:

- Feeding practices
- Sleeping arrangements
- Verbal interactions
- Eye contact
- Interactions between children and adults[6]

Design Prompt: Create Personas

A persona is a representation of a user. Personas answer the question "Who are we designing for?" and they help to align strategy and goals to specific user groups. Create three very different personas of children in your audience. Include diverse stages of development, personality, interests and culture. Based on unique makeups, consider how different children might approach your project. Include factors that will appeal to many facets of their makeups to add depth to your project for diverse users.

How Designers Can Improve the Lives of Kids and Teens around the World

Interview with Mari Nakano, the former design & interaction lead for UNICEF's Office of Innovation
Mari Nakano shares her experience at UNICEF, where she orchestrated the production and execution of designed materials and processes across a multi-disciplinary global team of innovators.

What should designers know about children from a global perspective?
Every child's story is different, some unbelievably beautiful and others incredibly painful. It's hard to fathom sometimes that there are children out there who deal with extreme tribulations like genocide, slavery, trafficking or something as specific as female genital mutilation, but it's happening. The more privileged and luckier ones of us need to pay attention to these stories. We also need to remind ourselves that relatively commonplace things like access to basic education, the internet or even water and toilets are challenges children and communities deal with. The world is a complicated place and it doesn't discriminate between children and adults. However, as human beings who believe children deserve to have all the things that come with childhood – happiness, play, laughter, innocence, imagination, etc. – we should leverage what we can as designers to create things that help solve the toughest challenges children face as well as create things that give children the chance to be children.

What are some of the challenges in designing for social impact?
Designers in this space are generally trained to participate in a process from beginning to end. We love to meet and know the people we're working for, we enjoy brainstorming and getting our hands dirty and we get so excited when we can start organizing and bringing clarity to challenges.

In reality, designers don't always get to participate in the entire process and that can be really frustrating and lead to being defaulted to just triaging or prettying-up a product enough to get it out into the world. We don't feel like we're being maximized and we worry that a product will suffer if parts of the design process are negated. So the two significant challenges I see

here are (1) how we foster support to execute stronger design processes from end-to-end in an organization and (2) how we get people to understand that design is not just about aesthetics, but more about smart, clear, understandable and relatable communications. Design should not be an afterthought. It never seems to work out when treated that way.

What are the most important things designers could do for kids/youth in the coming years?
I'm a big proponent of improving how we educate children and how we support teachers to be competitive and better honored.

One thing we can do is help make renowned design institutions accessible for all. At this time, design schools are a very privileged place to be. They are expensive. If we are genuinely designers for social change, let's focus on making the institutions many of us came from more diverse and affordable and staffed with more socially conscious teachers.

I think at one point or another we should all become mentors, volunteers or even teachers ourselves. And if we have the capital, provide some level of foundational support to enable people not so lucky as ourselves to study design. Kids love being creative and we have a lot of techniques to share with them to make learning and growing up fun and exciting. Spending more time with young people in general is an advantage to you. They remind you that they think differently than you. They make you more empathetic to their needs and you learn how to talk and relate to them.

The other thing to consider is making a commitment to design something for or with children, be it a toy, an activity, a technology, a safe playground in a community that doesn't have one, or a new program or an advocacy campaign for something like teen mental health. We are makers; so let's make things for youth that can make their lives better.

In what ways will new technologies affect the design for social impact field in the coming years?
The three I list below are by no means mind-blowing, but are a mix of practical and much-needed:

Technology 1: Off-line open source learning

When we know that over 2.5 billion people don't have access to internet, off-line open source learning and other information tools not dependent on the high-speed data transfer many of us are so used to, this will be essential to closing the gap between some of the most vulnerable people in the world, from small villages in the thick of the Amazons to children in refugee camps. For designers and engineers, this means we need to think about how we design great learning environments that come with deep constraints.

Technology 2: Soft robotics

When I think of technology, it's easy for me just to imagine metal, blinking lights and hard materials. There's nothing really gentle, warm or kind about that. With things like wearable technologies or soft robotics, there's perhaps an opportunity for technology to embed itself in a way that is more comfortable, culturally sensitive and human. I think there's a lot of potential to design soft robotics to address issues in the development space. The media talks a lot about how we bring aid to those in need – food, supplies, education, housing, etc., but what I think about too is how we can provide comfort, therapy, resolution, calm. Tending to human psyche is essential also, don't you think?

Technology 3: Materials sciences

It's a UN mandate, and I think humanity's responsibility, to protect the environment, so I think investing into the research, development and scaling of material sciences is imperative to the survival of our planet. From the way we package products to how we build a temporary shelter, sustainable, biodegradable or reusable materials should be part of our technological explorations.

Are you currently exploring any other new technologies?

UNICEF has been exploring the above technologies mentioned and a few others such as virtual reality as a new mode of communication; unmanned aerial vehicles and drones for surveillance of disasters or delivery of time-sensitive vaccines or blood samples; and blockchain as a form of secure and anonymized identity for vulnerable or displaced populations. There are a lot of in-depth articles and stories on unicefstories.org.

What kind of cultural and geographic shifts do designers in the social impact space need to pay attention to in the coming years?

One issue is the current refugee crisis. War, racism, politics and greed are forcing whole cultures to go on the move, which thus means there are an entirely new set of problems and challenges needing to be addressed for migrant populations and countries receiving migrants.

Another issue is environmental – with the vast shifts in climate, rising waters, pollution, etc., what species or islands do we need to target before, well, they're gone?

The world is changing faster than it ever has before it seems, and that means we need to think about how design can keep up with this pace without losing our attention to detail, beauty and clarity. Our ability to be agile, lean and iterative will be more critical than ever.

What would you advise people who are interested in a career pathway for social innovation projects specifically to make the world better for kids/youth?

The design is a way for you to help problem-solve for something you really care about. First, ask yourself what you care about. Understand it well. Experience it. Then, think about how design can help you achieve your goals and fuse your practice of design together with your knowledge of what you care about. Be good at both – understand what you care about and practice design – color, typography, layout, space, mediums, history and of course research methodologies.

I'd also like to say organizations not traditionally inclusive of design really ought to start thinking about how to recruit and integrate designers into their core teams. A creative on your team will bring a world of difference as long as you don't stifle their drive to imagine and speculate how to make things better.

What roles do designers play at UNICEF?

We have this saying on the team that the designers are "doing everything all the time." We have our hand in just about every public-facing communication piece both print and digital, but we also advise on product enhancements and user experience design. We've also spent a lot of time designing the organization and operation of our innovation team from within – from the way we lay out our office space to how we onboard new teammates, we've played a huge part in how our team interacts and plays with one another.[7]

Parenting Approaches

Modern ideas of raising children are always central to the topic when we are designing for them. We ask: How might parenting approaches affect the acceptance or rejection of our designs? For example, when creating a children's museum exhibit, there might be one child carried around and clinging to a parent throughout the experience (attachment parenting). Another parent is controlling the child's experience for them (helicopter parenting). Another is ignoring the child asking for help so that they gain independence (fostering risk and resilience). Another is texting and did not realize that their child left the room a long time ago (distracted parenting). How would each of these users interact in the space to gain greater understanding of its purpose?

Just like parents who model behaviors, as designers we each develop our view towards what we consider a design approach, similar to a parenting approach. We base it on our research, our understanding and our belief system. Sometimes, we look at how we were raised and adopt or reject those beliefs based on our feelings about it. Other times, we look at all of the research and map out a big picture of pros and cons of different methods to build a philosophy. Developing a framework to those beliefs helps us make decisions that represent our voice to share in the design.

Peer-to-Peer Learning

Sometimes, while adults are doing their best to educate children, we forget that much of their development is dependent on how they relate to a peer group. Children's cultures in many places around the world are thought of as practice cultures, practicing skills and values that they will need as adults. In hunter-gathering societies, children learn from the pack of children. Judith Harris, in a discussion of such research, noted that the widespread phrase *It takes a village to raise a child* is accurate if interpreted differently. In her words, "The reason it takes a village is not that it requires a quorum of adults to nudge erring youngsters back onto the paths of righteousness. It takes a village because in a village there are always enough kids to form a playgroup."[8]

When adults guide and instruct children, they don't get to practice the independence, social dynamics and behaviors that they learn best with peers. They try on different roles as leader and follower and practice independence, empathy and teamwork. They create activities, solve problems, and adjust to the needs and desires of those in their group.[9]

In our designs, we can prioritize peer-to-peer learning and create opportunities for children and youth to practice how to be in the world and educate themselves with other kids. Instead of involving a parent or teacher to help or supervise in a project, consider how to create interaction between peers that serves the same purpose. Co-play, co-creation and co-constructivism are all avenues to explore.

◀ Figure 2.3

Kids learn together by filling the bucket to make it tip in this pop-up play activity. Source: author photo.

Peer groups are children and youth of relatively the same age. Activities, values and shared understandings are co-constructed by peer groups through their daily activities. Every school, neighborhood and city has its own culture with consistent activities, routines, objects, values and concerns that peers share.[10] Designers have a powerful influence on shaping these interests for the positive or negative as creators of experiences (activities or routines) and products (objects) that kids adopt.

In adolescence, we see a gradual withdrawal from psychological dependency on adults. During adolescence, there is an increased need and ability for intimate friendships and romantic relationships. They need to establish a unique and autonomous identity different from that of their parents. Acceptance and participation in peer culture are an integral part of this process.

Design Prompt: Social Engagement

Look at your project through three different lenses. How do they differ?

- Intergenerational engagement: Guided or modeled by an adult
- Solitary play: Individual open-ended exploration by the child
- Peer-to-peer engagement: Cooperative and conflicting interactions among peers

Interview with Mihaly Csikszentmihalyi, author of *Flow. The Psychology of Optimal Experience and Creativity, Flow and the Psychology of Discovery and Invention,* and several other books.

Mihaly Csikszentmihalyi's theory of *Flow* is that feeling that people have when they are in a highly focused mental state of engagement. We forget everything else and are "in the groove" doing what we should be doing to bring about joy and happiness at that moment. Many children have it when they are playing. It is highly connected to creativity.

To achieve a flow state, a balance must be struck between the challenge of the task and the skill of the performer. If the task is too easy or too difficult, flow cannot occur. Both skill level and challenge level must be matched and high. The task has to be active and engaging; the brain must not be on autopilot. The activity has to have clear parameters for success, as they indicate progress and quality. The motivation for the activity must be intrinsic and support personal satisfaction.

Mihaly Csikszentmihalyi described how the state of *flow* relates to areas of child/youth development as well as some of the common themes (and concerns) in current children's culture today.

> Let me start by mentioning something Dante Alighieri wrote some 700 years ago, and that applies to everything related to flow. He said (more or less) that every creature experiences delight when it can express his or her being in action ... a hunting dog when it can smell the scent of a rabbit, a bird when it can soar in the sky, a fish when it can dive and jump under the waters of the sea. For human beings, it is more difficult to express who we are, because we are more different from each other. Some of us are good at sports but bad at dancing, or vice versa. So we experience delight in various forms, by doing different things that express the unique strengths of who we are; for instance in physical, cognitive and social and emotional development.

Physical Development

> From early infancy to late childhood the most obvious challenges we face are related to the control of the body and its processes. For an infant, being able to move his limbs, to reach for things, to turn her head, to hold and throw objects, are "first" that besides learning to control oneself and the immediate environment, provide the infant with a basis of self-control, self-esteem, and competence. As the child grows, social play and opportunities to compete physically will become increasingly important for positive development. It should be remembered that *competition* in Latin meant, "seeking together," because we learn about our strengths and limitations early in life by measuring ourselves against other children first through physical, then through cognitive performances. Later, of course, schools and youth organizations will provide opportunities for competition. But if winning becomes the primary goal, then we risk becoming addicted to outperforming others and lose the chance of getting most of the benefits of physical development.

Cognitive Development

> For human beings, the importance of physical development is rivaled by cognitive changes taking place during the lifespan. In modern societies, primary education in schools is not an option, but a necessity. Unfortunately, public education in most schools is almost entirely concerned with the *transmission* of knowledge, not so much with its *acquisition*. The unvoiced assumption is that providing children with what is considered to be indispensable knowledge at the time, is all we can expect from educators. Whether the children will internalize the experience, whether they will become life-long learners because they like the process and want more of it, is not the responsibility of the schools, but of the individual student. We know a lot about how organisms learn when they are forced or rewarded to do so; we need a lot more knowledge about how to induce self-motivated, intrinsically rewarded learning.

Social and Emotional Development

> Even less do we know about how to support the interpersonal development of children in a way that they will feel joy and growth through being with other people. Luckily most children inherit enough prosocial genetic links between feeling good and being with others, as this relationship must have been actively selected for through many generations of ancestors. The urban home living environments we have devised do not support the development of sociability in children. Most urban parents come from families with few children and lack the skills and the understanding of what their children need to benefit from the company of peers. Some private pedagogies, such as the Montessori schools, try to overcome this developmental deficiency in their pupils – for instance, by emphasizing the development of buddy-pairs where each child becomes responsible for helping and sharing with another child.[11]

Design Prompt: Designing for the Whole Child

Mihaly Csikszentmihalyi has touched on topics within the three areas of development – physical, cognitive, and social and emotional – that can use some insight from designers to improve children's lives. Use this as a model to create a framework with three areas of concern today for your audience. What are the cognitive, social and emotional, and physical development concerns surrounding your project? Develop solutions that include a whole child approach.

The Generational Perspective: Gen Z and Gen Alpha

Designers, market researchers, cultural observers and trend forecasters are looking at Gen Z and Alpha to see how they experience the world to understand better who they are, what they value and what they need. "Generation Z" were born in 1995–2010 and "Generation Alpha" since the year 2010 and until the year 2025, although the years shift slightly depending on the source, since there is little consensus among demographers about birth years. Understanding values also helps us to fit our project into a culturally relevant context. Keeping on top of this research over time will help us to see how each generation evolves differently from previous ones.

Generation Z

Who They Are

- Children of GenX and Millennials
- Independent
- More risk averse in specific activities than earlier generations
- More mature, smarter and safer than the previous generation
- A large group – 60 million American born. This group makes up a quarter of the US population and in 2020 will account for 40% of all consumers

Their Childhood

- Parents have tried to keep them safe and secure
- They have grown up in a time of hardship, global conflict and economic troubles

What They Do

- Use digital technology and media from a young age
- Social media is a significant portion of their socializing
- Careful with their money
- Want to change the world
- Cautious
- Focus on sensible careers and entrepreneurial futures
- Integrate technologies directly into their lives

- Peers are their role models
- Quickly filter information
- Social creators

What They Value

- Privacy
- Their personal brand
- Multiculturalism
- Prefer communication via images and voice control over typing and texting
- Human rights
- Hard work
- Pragmatic

Generation Alpha

This group, unofficially dubbed Generation Alpha, is the next generation of toddlers, babies and kids, and the first generation wholly born in the 21st century. There are more than 2.5 million Gen Alphas born globally every week. When they have all been born (2025), they will number almost two billion.[12] To determine what their life will be like, we look at their demographics. This group is the first to grow up for their entire lives with technology. What could this mean for you as a designer on your project?

Who They Are

- Mostly the children of Millennials (and GenXers)
- Parents (older)
- The most formally educated generation ever
- The most technology-supplied generation ever
- Socioeconomics (slightly wealthier), globally the most affluent generation ever
- Family size (smaller), one-child families have gained ground
- Born with "a chip" in their head
- Life expectancy (longer)
- The cultural mix (more diverse)

Their Childhood

We look at their parents, the millennials, to give us a sense of how they will be raised.

Millennials

- Will have frequent career changes
- Are more materially endowed

- Are more technologically supplied
- Are more likely to outsource aspects of parenting such as childcare
- Value experiences

Expenditures on Children by Families

According to the US Department of Agriculture, we know that a child born in 2015 will cost $284,570 to raise if projected inflation costs are factored in for food, shelter, and other through age 17.[13] With the expense of non-necessities and college, it is much more. Families have a lot invested emotionally, financially, and in terms of time in the development of their children to adulthood.

Design Prompt: Generational Influences on Design

Learn about how living today is different than for previous generations – socially, economically, politically or with the introduction of new technologies. How will the social, economic or political climate affect your design? What are the pros and cons of the easy access to information? What can we do with our skills and insights as designers to contribute? Look back at history to see patterns that have occurred in the past and how they might affect the future. How might Gen Z and Alpha approach a project differently depending on the cultural context they are living in?

Innovation Research and Trends for Gen Z and Alpha

Interview with Tessa Mansfield, content and creative director, Stylus, an innovation research and trends firm
Tessa Mansfield and I discussed the trend research process and her thoughts about trends related to children and youth today and in the near future.

What value does trend forecasting offer marketers of products and services?
We identify consumer trends, ranging from those triggered by social, economic or political drivers to others rooted in and catalyzed by new technologies. Demographic or psychographic trends can provide context for business when considering its future audience and what motivates them. Aesthetically driven design trends can offer strong creative direction for all types of creative professionals, from buyers or designers to visual merchandisers – each hoping to gain new visual inspiration brought together with a strong narrative.

What value does trend forecasting offer designers in various fields?
Designers share commonalities regarding their discipline. At the start of any project, the creative process involves an explorative research phase to understand the market and its broader context, to consider new relevant technologies and manufacturing techniques, and to brainstorm visually.

The best designers understand multiple disciplines and pull broad reference materials into their work. Whether you're a graphic designer, a product designer or an architect, it's critical to have a good awareness of trends and apply non-linear thinking. And regarding all industries, there are complementary sectors – the ones you look at carefully for direct correlation – versus the ones that are further afield and have the potential to provide broader inspiration and a more disruptive influence.

The range of our trend reporting is vast – from fast-moving blog post reports to cross-industry consumer Macro Trend reports that have a longer-term influence. To meet the timelines for manufacture and supply in design, we create Design Directions, Color & Material trends and Fashion Forecasts that look up to 18 months ahead. These visual reports are an inspirational product development tool to stimulate initial design concept work, composed of broad-ranging mood boards that look at everything from color, finish and material through to graphics and spatial design, all trended by theme.

How can designers best use trend services to build on existing trends or create new ones?
One way that our team keeps an eye on industry change is by attending global cross-industry events each

---

year – a mix of trade shows, design weeks, seminars and conferences. Over time, this allows us to see the evolution of products and ideas, which provides a reliable source of trend knowledge. As a trend service, we put our time, experience and resources into gathering this edited information for designers, so they can spend their time applying it to the project at hand.

What are important things to know about the differences between Gen Z, Millennials and Gen Alpha?

Looking at youth trends, the group to watch is Gen Z. Millennials (Gen Y) are now young adults, the youngest part of the workforce. The two groups share attributes, but we identify differences in attitude. Gen Alpha is our youngest demographic, and likely to have Millennial parents. Today's accelerating pace of change means there are critical differences between the generations.

We describe Gen Z as empowered, resourceful, determined and ethical. These proactive teen consumers are the most entrepreneurial that we've witnessed – many are even creating their own companies in response to a mainstream market that fails to resonate with them. We often cite their shrewdness as a potential result of being brought up in a recession by careful Gen X parents.

Like Gen Z, Gen Y (Millennials) feel connected to the rest of the world and are also socially and environmentally conscious. Tech-oriented Millennials enjoy a strong sense of community, thrive on social validation, and happily share their data. With pretty high expectations, they embrace personalized experiences, are multitaskers, and value work-life balance and flexibility over things like job security, which resonated more with older generations.

Gen Z were "born on the web," and our youngest demographic Gen Alpha take tech knowledge a step further. They are characterized as the "touchscreen generation," in response to understanding the gestural swipe interface even before they learn to talk. This digitally empowered generation will have extraordinary expectations of products and services as they grow up.

What about childhood do you see that remains the same/changes?

Childhood concerning physical development is much the same as it has always been. Today, children go through most of the same fundamental emotional and physiological changes as they did generations ago. Thus the same support and stimulus, from love and education to play, remain critical to child development.

If I were to pick one fundamental change that will continue to make all the difference concerning child development, it would be technology. The world these children are growing up in is radically different to the one their parents and older siblings faced. Today's youth live in a digital world that is becoming more seamless and interactive. They have global reach at their fingertips. We see digitally immersive experiences opening up to the next generation, from virtual reality to more great shifts occurring due to advances in artificial intelligence and robotics. The childhood that we remember will be dramatically different to theirs.

Many positives are emerging from digital transformation with regards to enhanced access, education and play – but there is also cause for concern, primarily around threats to personal privacy and security and, for the younger generations, the risk of open access to adult material. The challenge for parents, teachers and children will be creating balance with technology, being tactical with tech time, and sourcing sensitively designed devices, software and apps that positively aid learning and encourage play.

Can you name big ideas that are changing for children's and youth culture in the coming years?

Many of the same ideas will affect all generations in the near future – it's the application that will differ for children and youth culture. We are beginning to see connected products and the Internet of Things (IoT) take shape within the home. And wearables with sophisticated tracking tools are already starting to enable better security for children, and will soon begin to encourage self-knowledge concerning health, nutrition, etc.

Reflecting the increasing pace of change, digital behaviors will continuously evolve. Products and services designed to meet the needs of tech-savvy teens will be a growth market for innovation.

We already see significant developments in co-creation and DIY tech culture, with new tools and coding capabilities enabling Gen Z consumers to create products and services themselves that satisfy their quest for personal "betterment." Over the years this will become an even more exciting area as a broader pool of young people get access to basic coding, allowing them to personalize around their own needs.

Technology that responds to emotional factors will be even more critical in the future. We'll see more products with highly intuitive artificial intelligence capabilities, including facial/emotion recognition that can adapt according to mood. Younger generations, notably Gen Alpha, will embrace these early stages of consumer-facing robotics.[14]

◄ Figure 2.4

Playtesting a new app at Toca Boca.
Source: Toca Boca (used with permission).

Ages and Stages

Learning developmental traits and the range of abilities that are normally developing during these age groups will help to better understand how to design to meet those needs and skills. With this information on physical, cognitive, and social and emotional development, you can build profiles of different kids' needs and capabilities by combining topics from each of the lists. Those can be used for brainstorming concepts or to refine details in your designs. Always consider how you will include delayed and advanced development in your solution.

0–18 Months (Infants/Babies/Toddlers)

Infants and babies explore and learn about the world through the senses. They develop physically from not being able to control their limbs, to being able to reach and play with objects, to climbing and walking. They develop cognitive skills in thinking, communicating and solving problems. They learn to socially interact with people through facial expressions and language. What they need most of all is to develop a secure attachment to a primary caregiver.

◄ Figure 2.5

Babies learn though the senses through diverse experiences and movements.
Source: author photo.

Physical

- Learns through the senses by feel, taste, sound, sight and smell.
- Rooting, sucking, grasping reflexes.
- Lifts head when held at shoulder, moves arms actively, able to follow objects and to focus.
- Rolls over, holds head up when held in sitting position, lifts up knees, crawling motions, reaches for objects.
- Sits unaided, spends more time in upright position, learns to crawl, stands and walks, climbs stairs.
- Develops eye–hand coordination, learns to grasp with thumb and finger, transfers small objects from hand to hand.
- Puts everything in mouth, feeds self.

Cognitive

- Repeats movements to help brain growth and memory.
- Acquires communication and language. Responds with baby sounds when caregiver talks and smiles, knows own name and responds by looking when called, understands words, begins to respond selectively to words, says words like "mama," "dada."
- Realizes that people and objects still exist when they are not around.
- Can understand some emotions behind actions.
- Demonstrates intentional behavior, initiates actions.
- Has extensive visual interests, curious looking in the mirror.
- Solves simple problems (imitating from observations of others).
- Responds to changes in the environment. Repeats action that caused effect.

Social/Emotional

- Wants to have needs met.
- Develops a sense of security.
- Starts to smile and look directly at people, smiles spontaneously and responsively, expresses pleasure, laughs aloud, imitates faces people make.
- Attachment and bonding are very important. Prefers primary caregiver, more aware of familiar people and strangers, may start to show stranger anxiety.
- Play is individual or interacting with caregivers or next to other children.
- Understands that people think differently.
- Learns how to get what they want from other people, likes attention from others and will behave to get reactions.
- Likes movement, to be held and rocked.
- Likes affection. Will hug/kiss, responds to tickling, familiar people.
- Demonstrates object permanence. Knows parents exist and will return.

◀ Figure 2.6

Designers consider all possible user scenarios.
Source: author photo.

◀ Figure 2.7

Reading to children enhances language development and bonding.
Source: Autry Museum of the American West (used with permission).

18 Months to 3 Years (Toddlers–Early Preschoolers)

At this age, children continue to take in information from their senses and build independence through walking and communication. They begin to interact more with peers. Expressing themselves can be frustrating because of their command of language. They often think of creative approaches to tasks and activities. Toddlers develop and test hypotheses through repetition to learn about the world. Everything is new and exciting.

Children begin to use manipulatives to test hypotheses and solve problems.
Source: Kidspace Children's Museum (used with permission).

Physical

- Jumps, runs, climbs, hops, throws, kicks a ball, stands on one foot.
- Turns pages of a book, begins to scribble/draw (make circles and lines).
- Walks up and down stairs, two feet to one step.
- Manipulates objects with good coordination, builds a tower of blocks, begins with scissors.
- Increasingly able to manipulate small objects with hands, uses spoon effectively.
- Beginnings of bladder and bowel control.

Cognitive

- Will try every approach to solving a problem until they find the one that works.
- Will make simple plans based on past experiences.
- Is able to understand most of what is said.
- Learns how things work through using manipulatives.
- Finds objects that are moved out of sight.
- Uses order and sequence to solve problems.
- Sorts objects into simple categories (can group by type, size or color).
- Non-readers, vocabulary increases, able to form two words, three to four-word sentences.
- Uses an object to represent something real in life (teddy bear is a baby).
- Can only concentrate on one thing at a time.
- Enjoys listening to stories. Can recite stories with caregiver.
- Aware of the difference between inner "mental" and outer "physical" events.
- Asks questions, believes that adults know everything.
- Can follow simple two-step instructions.
- Has little understanding of pronouns, basic grasp of prepositions.

- Can count or say alphabet from memory, without a complete understanding of what it represents.
- Always in the moment, time sequencing not developed. Has a hard time understanding "the other day" or "tomorrow."

Social/Emotional

- Transforming from dependent toddler to becoming more independent.
- Able to play alone, moving into playing alongside other children. Then happy to play with children.
- Reading is a social activity.
- Uses own name to represent self, egocentric and concrete in their thinking, look at everything from their own perspective.
- Expressing feelings in words.
- Becoming aware of limits. Says "no" often.
- Possessive, not happy to share with other children.
- Beginning of empathy. Shows concern for other children who are crying or upset.
- Likes to do things without help from others, enjoys helping others.
- Becoming aware of gender identity.
- Continuing to develop communication skills and experiencing the responsiveness of others. Able to relate their experiences, in detail, when specifically and appropriately questioned.
- Learning to use memory.
- Acquiring the basics of self-control.
- Mimics what other kids and adults do and say, as well as tone of voice.
- Shows increasing separation anxiety (by 18 months), typically eases (by 24 months).
- Disobeys more than before; not doing what they are told to test what happens.
- Has tantrums to express frustration.

◄ Figure 2.9

Friendships begin to form.
Source: Audry Zarokian (used with permission).

► Figure 2.10

Pretend play at its height.
Source: author photo.

4–6 Years (Little Kids)

Little kids engage in experiences as a journey with no sense of time. Children only live in the present. They are wired for discovery, not focus. They are bringing in a lot of stimuli through endless repetition. They are curious and want to know about everything. Once they are settled in school, they learn to only ask useful questions. This allows them to dive deeper and learn more specifics. This open-ended curiosity creates opportunities for designers to be free to experiment with creative ideas.

Physical

- Gross motor skills development with better coordination. Likes to do things like climb, hop, run, skip and do stunts, throw and catch a ball (moving from larger to smaller balls), ride a tricycle then a bicycle, swim, skate, hopscotch and jump rope.
- Fine motor skills refined, holds a pencil correctly, manipulates clay, hand-writing, works with smaller and smaller objects, dresses and undresses, key-board and mouse skills, takes shoes on and off (ties shoes), develops one hand preference, cuts with scissors, staying on the lines, draws recognizable pictures, uses a fork and knife to cut.
- Masters daytime bladder and bowel control. May still have nighttime incontinence.

Cognitive

- Enjoys making and creative exploration.
- Problem-solving skills continue to develop; uses trial and error to tackle a problem.
- Bases their ideas on what they see and feel in the moment.
- Focused on one solution to a problem, but may try out multiple solutions to a problem.
- Shares their thoughts and approaches problems cooperatively.
- Uses language and materials to solve problems. Begins to think abstractly without actually needing to do or manipulate something.
- Begins to solve problems by applying what they observe in their environment.
- Heightened pretend play.
- Moves from non-reader to reader. Language usually develops ahead of their speech ahead of their reading.
- Vocabulary will have increased to between 8,000 and 14,000 words but will often repeat words without fully understanding their meaning.
- Coloring, writing, learning.
- Asks why? Often.
- Usually doesn't care about rules.
- Has trouble with the concepts of sequence and time. May seem inconsistent when telling a story because they don't follow a beginning–middle–end structure.

Social/Emotional

- Egocentric and concrete thinking. Gradually gains some understanding of another's perspective.
- Plays cooperatively with peers, participates in interactive games.
- Develops capacity to share and take turns.
- Is developing some independence and self-reliance.
- Tests boundaries.
- Developing ethnic and gender identity. Moves from mixed gender play to playing with same-sex peers.
- Starts to be more conversational and independent. Shows and verbalizes a broader range of emotion.
- Pretend play at its height and may confuse real and make-believe. Then beginning to distinguish between reality and fantasy although still confused about what is real (tooth fairy, Santa Claus, wizards, fairies).
- Learning to make connections and distinctions between feelings, thoughts and actions.
- In the moment and impulsive.
- Meltdowns, tantrums, protests because of changes in routine or not getting what they want. Moves into negotiation.
- Begins to understand what it means to feel embarrassed.
- Eager to learn new skills, pushing the limits to help out and accomplish tasks.
- Expresses unconditional love and forgiveness.

Enjoying playing games with a close group of friends.
Source: author photo.

► Figure 2.12

At this age, kids have a greater understanding of problem-solving and how things work.
Source: Kidspace Children's Museum (used with permission).

7–10 Years (Bigger Kids to Early Tweens)

Their world expands outward from the family as other relationships are formed with friends, teachers and mentors. Because their experiences are expanding, some situations can create stress and affect self-esteem. Kids this age are negotiating their way through the peer group "rules" about what is cool to wear and do, and what is acceptable to eat for lunch. Extracurricular activities and selected activities begin to be a big part of their lives. During these years, strengths and weaknesses, preferences and tastes appear and become clear.

Physical

- Has increased coordination, strength and endurance. Can play and be active for more extended periods without getting tired.
- Enjoys using new skills, both gross and fine motor.
- Is increasing in height and weight at steady rates.
- Uses one hand and foot better than the other.

Cognitive

- Starts to plan. Becomes more goal oriented and focused. May create a drawing of something to build or a plan for an experiment.
- Starts logical thinking. Applies their knowledge and experience to a particular situation to determine whether it makes sense or not.
- Keener metacognition. An understanding of their inner world and uses it in problem-solving.
- Increase in ability to apply logic and reasoning. Looks for the reasons behind things (asks questions for more information).
- Begins to use metamemory, or the ability to comprehend the nature of memory.
- Temporal concepts greatly improve. They start to understand the idea of the passage of time, as well as day, date and time.
- Can apply mental operations to concrete problems, objects and events. Struggles with understanding abstract or hypothetical concepts.
- Has acquired the necessary cognitive and linguistic concepts required to sufficiently communicate an event.
- Learns to mentally combine, separate, order and transform objects and actions to conserve mass and area.
- Can engage in classification, or the ability to group according to features, and serial ordering, or the ability to arrange according to logical progression.
- Understands cause and effect and makes more in-depth connections between mathematics and science.
- Can consider multiple perspectives and apply various thought-out strategies.
- Has a longer attention span (can sit and pay attention to something that interests them).
- Moves from "learning to read" to "reading to learn."
- Uses writing as a way to express feelings. Tells stories and summarizes information.
- Playing with words to make puns. Uses bad words for shock value.
- May enjoy collecting things.
- Understands and may enjoy game strategy, competition, rules (leveling-up goals).

Social/Emotional

- Increased self-expression and understanding of self in relationship to others.
- Comprehends stable identity – that one's self remains consistent even when circumstances change.

- Increased empathy and ability to interact with others outside the family.
- Has more same-sex friends, sometimes stereotypes members of the opposite sex.
- Narrowing peer groups to a few close friends based on compatibility and shared interests instead of just proximity.
- Eager to fit in and try out new personalities to see where they fit. Easily influenced by peers. Wants to fit in with peer group rules; acceptance can affect self-esteem.
- Has a strong group identity. Increasingly defines self through peers; enjoys being part of a team, group or club.
- Ability to engage in competition.
- Developing and testing values and beliefs that will guide present and future behaviors.
- Tells lies or steals, and might not yet have fully developed a proper understanding of right and wrong.
- Develops a sense of mastery and accomplishment based on physical strength, self-control and/or school performance.
- Dramatic emotion and impatience. Bounces right back to everything being just fine.
- Keeps secrets. Forms secret societies that adults know about.
- Tempted by the forbidden.
- Beginning to show signs of being more responsible. Takes better care of possessions.
- Starts to understand another person's view on topics.
- Likes jokes and riddles.

Overscheduling

William Doherty, a professor of family social science at University of Minnesota, first coined the term "overscheduled kids."[15] Parents see childhood as a time to make kids into their best selves and believe it is best to provide a diversity of experiences to kids. It is not about just getting into a team but getting on to the right team. This makes some kids anxious and robs them of free time, idle imaginary play and boredom.

11–14 Years (Later Tween to Early Teen)

At this stage, kids may be more mature physically than cognitively or emotionally. The child's changing body, the need for independence and a desire to be accepted by peers are the focus. They begin to be more aware of what's happening in the world and how it affects them personally. Verbal and body language skills typically develop both in communicating and in reading others' intentions through tone of voice, facial expressions and posture. They begin to understand how things are connected. They question everything, including their parents' authority and opinions. Friendships become more complex and important. Peer pressure can be an issue, especially for those who have problems with social and emotional skills.

 Figure 2.13

When given the opportunity, these kids enjoy exploring like younger children.
Source: author photo.

Physical

- Has increased precision, coordination, strength, agility, balance and flexibility. Starts showing uneven development in skills.
- May be a little clumsier as height and weight change quickly.
- Most begin puberty. Evident sexual development, voice changes and increased body odor. May start sexual activity.
- Has an increased appetite and needs more sleep. Energy used for growing.
- Growing fast. May complain of growing pains or muscle cramps.
- Developing hand–eye coordination, and use of tools improves.
- Displays fluid handwriting skills, increased writing speed and consistent sizing and spacing of letters, joins letters together when writing, and writes well without ruled lines.
- Draws pictures in detail. Draws three-dimensional geometrical figures.

Cognitive

- Problem-solving and thinking skills become more complex (uses old skills in new situations to solve problems).
- Pays more attention to decision-making and organizing ideas, time and possessions.
- Creative with storytelling.
- Increased ability to learn and apply skills.

- Beginning of abstract thinking, but reverts to concrete thought under stress.
- Still developing reasoning and is not able to make all intellectual leaps, such as inferring a motive or reasoning hypothetically.
- Interpretative ability, cause and effect sequences, predicting the consequences of an action.
- Can answer who, what, where and when questions, but may have problems with why questions.
- Realizes that thoughts are private and that people see others differently than they see themselves.
- Can understand how things are connected, the effects of how the behavior or mood of one person can impact everyone else.
- Develops a better sense of responsibility. Starts to understand concepts like power and influence.
- Questions things, doesn't take everything at face value.
- Thinks about how current actions affect the future. May worry about global issues like climate change and war.
- Memorizes information more efficiently. Checks their work and changes approach as needed.
- Starts thinking more logically. Begins developing a worldview, including a basic set of values.
- Wants to contribute, make money and have own money.
- Experiments with metaphors, slang and different ways of speaking.
- Is interested in having discussions, debates and arguments (sometimes just for the sake of it).
- Pays more attention to body language, tone of voice and other non-verbal language cues.
- Envisions "what if" scenarios and talks through other ways of problem-solving.
- Decision-making skills are still developing, learning about the consequences of actions.

Social/Emotional

- Seeking more independence. Starts withdrawing from family activities. Finding ways to be an individual.
- Playing with identity. Experiments with the way they dress, things they do, and mannerisms to find where they fit and express their individuality.
- Values friends' and others' opinions more, especially when it comes to behavior, sense of self and self-esteem.
- Body changes can leave kids feeling uncertain about themselves.
- Increased ability to interact with peers. Forms stronger and more complex friendships.
- Enhanced ability to engage in competition.
- Seeking out new experiences and engages in risk-taking behavior. Still developing control over impulses.
- Developing and testing values, morals and beliefs. Questions more things.
- Has a strong group identity. Increasingly defines self through peers, spending time with friends. May struggle to fit in.

- Acquiring a sense of accomplishment based on the achievement of greater physical strength and self-control.
- Defines self-concept in part by success. Begins to develop the ability to understand different points of view.
- Has mood swings.
- Getting better at reading and processing other people's emotions.
- Tests limits. Trying to figure out which rules are negotiable and which are not.
- May face peer pressure and find it hard to resist.
- Has a deeper understanding of how relationships with others include more than just common interests.
- Can be affectionate. Has a first crush or pretends to have crushes to fit in with peers. Starting to develop and explore a sexual identity. This might include romantic relationships, or going out with someone special.
- Extends their way of thinking beyond their personal experiences and knowledge and starts to view the world outside an absolute black–white or right–wrong perspective.
- Developing resilience, although may be sensitive to other people's opinions and reactions; thinks the whole world is watching.
- Develops a sense of pride in accomplishments and an awareness of weaknesses. Hard to accept failure.
- Keeps secrets. Having secrets is more important than the secret.
- Has better knowledge of what's appropriate to say in different situations. Can be silly and curious but also rude, argumentative and selfish.

◀ Figure 2.14

Playing and exploring social relationships.
Source: KaBOOM! (used with permission),

14–18 Years (Teenagers)

Adolescence is an emotionally and physically challenging period for all teens. Physically, they may be our height or taller, but they still have a lot of social and emotional growing to do. They like to be involved with friends and activities

With more independence, teens experiment with risk-taking behavior.
Source: author photo.

that bring joy. They have mood swings that smooth out towards the end of high school. At this time, they can appreciate the positive things they have to offer friends and family and set goals for the future.

Physical

- Needs a lot of sleep and food because they're growing fast.
- Less concerned about physical changes but increased interest in personal attractiveness.
- Excessive physical activity alternating with lethargy.
- Secondary sexual characteristics.
- Is clumsy and uncoordinated because of growing.
- Has the hand–eye coordination to learn to drive.
- By the end of high school, girls are likely to have grown as tall as they're going to be. Boys often are still growing and gaining muscle strength.
- Sports focus on performance.

Cognitive

- Interacting with technology is directly in tune with their brain.
- Diversity of interests, although see themselves as specialists in different areas.
- Thinks creatively and challenges the status quo.
- Cause–effect relationships better understood.
- Starts thinking less about just their own life and more about how the whole world works.
- Increasing ability to reason, make educated guesses and sort fact from fiction.
- Thinking more abstractly, comparing what *is* with what *could be*.
- Problem-solves ways to deal with hypothetical situations.

- Begins to set their own goals for the future; takes other opinions into account but makes their own decisions.
- Understands the consequences of actions, not just today, but in the future.
- Develops a strong sense of right and wrong. Makes decisions based on following their conscience.

Social/Emotional

- Conflict with family predominates due to ambivalence about emerging independence. Is embarrassed by family and parents (14). Appreciates siblings more than parents (15). Starts relating to family better; begins to see parents as real people (16–18).
- New at living with impulses – still needs to learn how to tolerate and manage.
- Strong peer allegiances are very important. Spends more time with peers than with family.
- Fad behavior.
- Can recognize personal strengths and weaknesses. Develops a better sense of who they are and what positive things they can contribute.
- Experimentation – sex, drugs, friends, jobs, risk-taking behavior.
- Gains confidence and independence. Shows pride in successes.
- Struggles with sense of identity; experiments with adult roles.
- Mood swings; analyzes their own feelings and tries to find the cause of them. Denial of emotions.
- Rejection of adult values and ideas. Testing new values and ideas.
- May be interested in dating or be strongly invested in a single romantic relationship.
- May resist adult relationships. Dependence upon adults threatens "independence."
- Exploring sexual identity.

Challenges for Adolescents Today

Today's adolescents are more dependent than they want to be. They are reliant on their parents for car, money and emotional support. College is a huge concern for them. Since World War 2, Americans have assumed that all young people follow the same path to adulthood: going to college. This single path measures all kids by the same metrics: SATs, grade point average (GPA) and Advanced Placements (APs). With the cost of college, there is pressure to spend on the best.

We see children being protected from an early age and have imposed limitations on risk. Is this harming their ability to manage risks? What do teens think is fun? They drive fast, throw eggs at people's houses and graffiti for fun. Psychologist Dr. Michael Carr-Gregg says that alcohol, sleep, social media and the web, and early sexualization are some of the most significant challenges for teens today.[16]

► Figure 2.16

School performance is a high priority for many teens.
Source: Michael Lyn (used with permission).

Emerging Adults (Kidults) 18+

Adulthood emerges at the end of the teen years when a child leaves home for college, becomes established in their first job, or joins the military and is on the verge of independence. However, brain development likely persists until *at least* the mid-20s, possibly until the 30s, specifically in the area known as the "prefrontal cortex." At 18, the prefrontal cortex doesn't have nearly the functional capacity it does at 25.[17] From a developmental perspective, these are some of the milestones that have been reached at this stage.

Physical

- Physical maturity – usually attaining full height.
- Reproductive growth ending.
- Firmer sense of sexual identity.

Cognitive

- Moves into adult roles and responsibilities and may learn a trade, work and/ or pursue higher education.
- Abstract thought developed.
- Aware of consequences and personal limitations.
- Able to plan for the future and focus on long-range goals and concern about what to do with their lives.
- Philosophical and idealistic.
- Greater capacity to use insight.
- Secures their autonomy and builds and tests their decision-making skills.

Social/Emotional

- Separation from caregivers.
- More comfortable seeking adult advice.
- Peers are important, but now evaluates their influence and opinions instead of just embracing them without question.
- Intimate relationships are essential and greater intimacy skills.
- Acceptance of adult responsibilities.
- Executive function skills are still developing.
- Defines their values framework.

Today, many are putting off adult responsibility and commitment for experimenting with relationships and jobs to explore their identity and goals. This intermediate phase between adolescence and adulthood has developed into what sociologists, psychologists and demographers have established as emerging adults: a permanent trend moving into the mainstream. Why is this group important for designers to understand when designing for children and teens? A few of the most important reasons are:

- With the renewed interest in including play throughout the lifespan, an audience might extend beyond children into an adult market.
- You may be designing for children and their "adults," not just the children.
- Adults are often with their children; therefore, many products and experiences for children are designed to promote intergenerational engagement.
- Some children's products and services are marketed to both children and adults.

Rejuvenilles or Kidults are yet another subset of this group. They live in the adult world with a childish frame of mind. These are people who enjoy all forms of play throughout the lifespan. Some may be single; some may be married or even have kids and grandkids of their own. I have seen this as a trait many designers possess, especially those interested in designing play spaces, products and technology for kids. As Matthew Urbanski, principal at Michael Van Valkenburg Associates Landscape Architects, shared, "I was attracted to designing play spaces because of their creative potential. My mom called me Peter Pan."[18] Eric Poesch, senior vice president, Design and Development, at Uncle Milton Toys, explained to me: "I really enjoy that I have to think like a kid to be successful in my role. Otherwise, I won't be effective in creating toys that will engage kids and spark their natural curiosity."[19] Jini Zopf, senior manager, Hot Wheels Packaging Design at Mattel: "We get to be playful, relive childhood, and get in touch with our inner child."[20]

I spoke with Christopher Noxon, the author of *Rejuvenille: Kickball, Cartoons and the Reinvention of the American Grown-up*. He explained:

> In adulthood our culture has a fixation with specialization and finding a path and staying with it. Childhood is largely free of that definition – it's about imagining all the ways we can be. Being a

rejuvenile is hanging on to adaptability, that innate part of ourselves that we lose as adults.

He expanded with how Kidults know how to play: "Adults have to practice and actively work to maintain fun. Kids are born with an innate ability to find fun – no one has to teach a kid to play. Retaining that ability is a gift to people who know how to do it."[21]

We discussed the changes in recent years in our culture in this age group. He explained:

> The rejuvenile movement is better integrated in the adult vernacular. It used to be rebellious for adults to participate in things typically connected to children, now the lines are blurred. The pressure isn't there to give things up. Childhood is typically spent discovering who we are and jumping around, trying on selves, keeping open to change and possibilities. Kids learn faster. Their imagination is superior to adults. When we were growing up there was a sense that we would grow out of childhood. Rejuveniles are about trying to keep that flexibility and adaptability into maturity.[22]

Companies have caught on to how this group embraces trends in pop culture entertainment, food and fashion geared towards teens and children.

Harry Potter and Barbie inspired clothing and Disneyland vacations are enjoyed by adults. The anthropomorphism of playful products by Alessi. Williams Sonoma offers Star Wars and Marvel cookie cutters for gourmets. Just take a visit to Comic-Con and you still see the toyification of many "adult" products. Christopher Noxon stated: "Many companies market to adults the urge of discover yourself within childlike culture /products. This is also a way that companies can sell the same products to a new market and avoid regulations restricting marketing to kids."

Christopher Noxon explained how the kidult mindset relates to identity:

> With adults you have a larger diversity of interests based on the child-like stage of development they most respond to. It says a lot about a person's identity when they talk about what type of childhood things they enjoy as adults. If they like stuffed animals, 1st person shooters, collect baseball cards, play laser tag and paint ball, or skateboard. Each is a strong expression of self.

Many people who create for the kids, teens and kidult markets are committed to things they loved as kids. People who loved comic books, toys and theme parks are able to hold on to them in the form of a career. Christopher Noxon described:

> This is a natural evolution to take something you loved as a kid and make it part of your adult life. It you want to carry those loves over into your full identity and not just participate in them in your off hours you become a producer of things that reflect what comes naturally for you.

Childhood favorites from foods to characters remain a part of identity and enjoyment for Kidults.
Source: author photo.

Kidults may also enjoy participating in shared experiences and playing with kids as equals. At children's museums, kids' sporting events or kids' concerts there are clear differences between the caregivers with a kidult mentality and typical adults. Those who are fully engaged in the experience and with their kids do it for their own enjoyment. Christopher Noxon added:

> Play-along parenting is the complicated dynamics that happens when parents become playmates. Kids don't need parents to be a friend, they need parents but it's a mistake to think that means that parents can never play with their kids. The fact is you don't undercut your role as an authority figure by playing. You enhance it. When you play with your kid, enjoying and engaging together, you can in effect strengthen your bond and intensify your position as a loving parent.[23]

Design Prompts for Kidult Design

Can you add motivational play value to your project, which currently has no play?

Can you turn utilitarian objects into playthings by removing mechanical barriers?

Can your product or service be labeled for universal appeal and not "for adults" or "for kids".

Can your designs for adults be influenced by toys, games, children's experiences or preschool educational methods?

Can the technology and interface be intuitive and inviting as if it were for a 3-year-old?

Can your childlike design promote intergenerational engagement?

Notes

1   Gopnik, Alison 2011
2   Cherry, Kendra 2017
3   Benyamin, Beben 2015
4   Ideonomy
5   Small, Meredith F. 1998
6   beststart.org
7   Nakano, Mari 2017
8   Harris, J.R. 1998
9   Gray, Peter 2016
10 Corsaro, William A. 2017
11 Csikszentmihalyi, Mihaly 2017
12 Michel Carter, Christine 2016
13 United States Department of Agriculture 2017
14 Mansfield, Tessa 2017
15 Doherty, William 1999
16 Carr-Gregg, Michael
17 Mental Health Daily
18 Urbanski, Matthew 2017
19 Poesch, Eric 2017
20 Zopf, Jini 2016
21 Noxon, Christopher 2017
22 Noxon, Christopher 2017
23 Noxon, Christopher 2017

Chapter 3

Physical Development

Kids' bodies are continually growing, changing in size, strength and abilities. When designers know what kids are capable of physically, we tailor our products, services and environments to suit their limitations and strengths. Designers create for the physical requirements of children's bodies, growth, comfort, functionality and ergonomics, and to prevent injuries.

Designers of all disciplines are involved in projects that support a child's physical development. For example, environmental designers are challenged with creating safe spaces for kids to participate in active play to help with motor skill development, strengthen their bodies and keep fit. Interface and interaction designers are challenged with ergonomic topics related to the use of digital technology. Product designers tackle mobility issues and gross motor skill development when designing strollers, scooters, bicycles and sports equipment. We choose safe materials and consider the sensory experience. Graphic designers communicate messages about health and safety and consider fine motor skills in packaging. Throughout our projects, designers ask: "How can we support healthy physical development with the safest solution for children and adolescents?"

Growth

Kids are experts at growing; it's what they do naturally. Physical growth refers to an increase in body size (length or height and weight) and in the size of organs. Infants and young children experience rapid brain development in the first few years. Children develop at varied rates. Designers consider children's different patterns of growth at different ages and use sources on ergonomics to make design decisions. We also consider strength and endurance to make sure they have obtained the ability to perform specific tasks intended in our designs.

There are four stages of physical growth in human development:

- Infancy (birth to 2 years old)
  Babies and toddlers grow rapidly and put on weight and then lengthen. After this time, growth slows.

- Early childhood (3 to 8 years old)
  During these years, growth in height and weight is sporadic, and muscle tone increases while body fat decreases. Younger children tend to grow more in their extremities.
- Middle childhood (9 to 11 years old)
  Children tend to grow off and on until early adolescence. Puberty may begin.
- Adolescence (12 to 18 years old)
  Adolescents quickly grow again following puberty. Their growth primarily affects the spine.[1]

Measurements: Length, Height and Weight

Doctors measure length in children too young to stand and height once the child can stand. Infants typically grow about 10 inches during the first year, and the child's birth weight is tripled by year one.

Throughout the prepubescent years, there will be weeks, or even months, of slightly slower growth alternating with mini growth spurts in most children. The average weight gain for children aged 5 to 10 is between 5 and 7 pounds per year. Body mass index (BMI) is a calculation that uses height and weight to estimate how much body fat someone has and can be used to determine if a child is overweight.[2]

▶ Figure 3.1

Over the first year babies grow rapidly.
Source: author photo.

Strength and Endurance

Muscle strength is how strong the child is, and muscular endurance is how long the child's muscles can work. Strength increases between the ages of 6 and 10 years old. Muscular endurance helps maintain proper posture and fitness. Improving strength and endurance contributes to a higher metabolism, which in turn reduces the risk of obesity. When a child has good strength, they are more likely to have better joint health, which reduces the risk of serious injury.[3] Muscle strength is essential for specific tasks and is considered where strength is needed for pressing buttons, squeezing or carrying weight. Endurance is considered when a child must perform a task for a prolonged amount of time.

The Senses

The inputs that the senses receive help kids understand and explore the world around them. Senses are part physical, part cognitive. Building in a sandbox and interacting with water improve motor skills, raise awareness of how the world works and contribute to language acquisition. As they grow, children learn how to take in and process all this sensory information at the same time and focus their attention on particular sensations while ignoring others. Their brain interprets sensory information and tells them how to interact with the world.

Babies' senses develop over time, and all of the senses are physically functioning like those of adults by early childhood. What they are lacking is the learned meanings of sensory inputs. When we design for young children, they might not see what we see or hear what we hear because of their lack of experience in interpreting the inputs. As children learn meanings behinds sounds, colors, temperatures and textures, they can react appropriately to sensory inputs.

The senses are also closely connected to social skills and interactions with others. The senses capture information that can be brought into long-term memory. Years later, many of us remember the smell of our childhood home or the tunes of our favorite songs. If designed well, engaging several of our senses in an experience can potentially offer depth of experience and understanding. When unnecessary or too much sensory information is exhibited, it can produce distractions. Children can become overwhelmed with noise, light, touch or other sensory experiences. Designers knowing just the right amount of stimulation that a range of children can handle can help make the design more inclusive.

Design Prompts: The Senses

- Create a baseline of existing sensory experiences related to your project and expand to others that are unexpected.
- Consider each of the senses individually and develop solutions for each. Then explore how they can collectively work to support the project objective.

- Develop the message you want to communicate, the touch points in the experience and how the senses can be used to reinforce that message.
- Use the senses in your brainstorming session. Awaken the senses that are dormant.
- Keep a senses diary – track the experience and create a senses graph. What does it tell you?

Sight

There is a strong connection between visual perception and cognition. When designing something visual, we think of factors related to seeing, color, visual composition, light and form. Babies see high-contrast patterns and are drawn to faces. Young children look for patterns in the world that help them build schemas, literacy skills and memory. We all need eye-to-eye contact with others to get an emotional response.

Central vision (what we are directly looking at) is used to look at details. Peripheral vision (the outer parts of the visual field) gives us a general idea of what we are seeing. Movement captures our attention. Color influences our perception and decisions.

Research with kids and the sense of sight:

- Use humor and create visual puns and see if they can "get the joke."
- Draw a range of illustration styles from realistic to abstract. See what they gravitate towards and why.
- Have them organize color paint chips by value and saturation. Have them describe the colors.
- Design characters that are human, animals and objects. Have them express the personalities of the characters.
- Create an optical illusion around your topic. What do they understand about the image?

Touch

Our sense of touch is used all over our body, but some areas have more receptors, such as our fingers, face, lips and feet.

There are four types of receptors:

- Sensing vibration
- Sensing tiny amounts of slippage
- Sensing stretching of the skin
- Merkel endings sense the finest kinds of textures (fingertips, lips)[4]

Kids use touch to explore and learn. They are fascinated by the feel of the movement of water. They enjoy playing with materials of different textures – clay, water, sand, rice, rough, smooth, hot and cold. There is a special system for

◀ Figure 3.2

Touch tanks in aquariums allow visitors to interact with the animals and habitats to better understand them.
Source: Brian Boyl (used with permission).

feeling social and emotional touch. Holding hands and giving hugs help them feel safe and loved. There is also a system that makes pain hurt. When designing something tactile, think about material, temperature, surface, form and textures and how the body interacts with it. Can you describe the feeling?

Researching with kids and the sense of touch:

- Have them feel things with their hands or, to be silly, with their feet
- Have them feel hidden objects inside a box and have them describe them or guess what they are
- Make sample swatches of different textures to touch
- Compare touch in the analog and the digital world
- Create drawings, models, finger painting, or use clay and other sensory materials
- Have them touch something with their whole body and then describe what it feels like

Touch, Gaze and Attachment

The senses of touch and sight are vital to bonding after birth. Skin-to-skin contact and looking into a baby's eyes are key to bonding and helping build healthy relationships. Body massage decreases infants' stress and enhances their feelings of well-being and emotional security. Loving touches promote growth in young babies. Premature babies who are massaged three times daily are ready to leave the hospital days earlier than babies who do not receive massages. Holding premature babies helps them to develop more rapidly.[5]

Hearing

Ears are continually taking in information about the environment and telling us how to act. Although ears automatically hear, it takes practice to learn how to listen. Listening is essential for kids to learn, get information from others, and keep safe. Especially in young children who cannot read, sound can be the input to signal a task and promote action. Loud noises grab attention. People become habituated to sounds. Hearing and language development are closely connected.

Researching with kids and the sense of hearing:

- Listening games, patterning exercises and exploring an environment can help us understand what they hear.
- Go on a hearing walk – discover things on the way.
- Listen to music together and have them interpret what they hear.
- Create a drum circle and play with rhythms.

Smell

Some smells are yummy. Others can evoke comforting, scary and exciting feelings. Smell has a strong connection to memory. Do you remember that smell of a new box of crayons? Kids don't distinguish good from bad smells until they are around two. There is an entire industry wrapped around scents building associations with environment and products. These smells are used to evoke memories, emotions and associations with a brand or a product. As the technologies become more readily available, scents will be used help kids learn and remember.

Researching with kids and their sense of smell:

- Stop and smell the roses, or any flower, and have them describe the scent.
- Have them describe the smell of a gas station, a skunk. How are they different?
- Have them paint with spices and taste them too. How does the smell influence the image?
- Hide scented objects or liquids in plastic smelling bottles and have them describe the scents and connect them to a memory.

Taste

Encouraging kids to taste different foods and food combinations will help them to develop healthy eating habits. Babies are naturally drawn to sweetness, and their tastes change over time. Taste and texture are connected. Does a pithy apple taste as good as a crunchy one?

Researching with kids and their sense of taste:

- Have them close their eyes, taste and describe (sour, sweet, salty, bitter, fruity)
- Have them explore their taste buds by applying different flavors with a cotton ball to different areas of their mouth
- Have a tea party – talk about taste, smell, texture and sound
- Add several different tastes to a plate. Have the children take a bite of each one. What does it remind them of? What do they or don't they like about it?
- Always make sure there are no allergies in the group.

◀ Figure 3.3

Taste preferences develop from what kids are fed in their first few years.
Source: author photo.

The Other Senses

The vestibular sense, or movement and balance sense, gives us information about where our head and body are in space.

The proprioception sense tells us where our body parts are relative to each other. It also gives us information about how much force to use in specific activities, allowing us to crack open an egg without crushing it in our hands.

In my interview with author Richard Louv, he explained: "Scientists who study the human senses no longer talk about five senses, but about nine or ten, conservatively – and some scientists describe as many as 30 human senses."[6] How, as designers, can we use the confluence of all of these to promote playing, learning and growing?

◀ Figure 3.4

When walking on a log, children gain a sense of equilibrium in relation to space and gravity. Strongly developed balance allows them to feel in control of their bodies.
Source: author photo.

Sensory Processing Disorder

This is when people with sensory processing issues can be oversensitive or undersensitive to sights, sounds, textures, flavors, smell and other sensory input. Consider these kids as extreme users on both ends of the sensory spectrum. How can your project best support their needs?

Design Prompt: Co-creation Using the Senses

Use the senses as a catalyst in the co-creation process with children. Explain each of the senses to them. Have children think about times when their senses helped shape an experience. How did the senses help? Was it what they saw? What did they hear? The smell? How did they feel? Something that tasted good or bad? Ask them to describe the same object or environment using each of the senses.

Hearing Screening Device for Infants

Interview with Neeti Strittmatter, co-founder of a company that developed a hearing screening device to test infants for hearing impairments to be used in India.

How was developing a project for the medical field different from other projects you have worked on?
Projects in healthcare demand a high level of sensitivity to patient privacy, hygiene, comfort and emotions. Testing is usually difficult. When developing a device, getting approvals for clinical trials and having access to medical experts for research will make or break a project.

Can you explain the story behind the project?
Projects like this for children can have a significant impact. Many parents in India don't know about new-born hearing screening. By the time the child is tested, they are too old, and it is too late for the intervention to make a difference – and you end up with poor outcomes.

What did you find the challenges in designing a product for babies as compared with a general user?

- Babies are tiny.
- They are hard to access.
- Trying the device out on users was extremely difficult because we were looking for newborns and infants.
- The team tested the product on ourselves all the time. In fact, we used to take turns in the group to wear electrodes on our head so that we could test our device easily. This worked to try out the functional prototypes.

- We used a life-size doll of an Asian baby for testing the visual and ergonomic prototypes.

Can you explain the input that you got and how it was used in your final design?
We learned that current devices on the market do not perform in a low-resource environment. The costs of the devices are too high. The proper environment and infrastructure for conducting the testing (like a sound-proof room) do not exist. There are not enough skilled technicians to perform the test. So we designed around those challenges.

It became apparent quickly that developing the device and software wasn't going to be enough. The more significant question was, whether the babies would get the intervention they needed if they screened positive? Do we have a system in place for that? We started the network for specialists (ENT and speech) to help with providing the next step after screening.

Meanwhile, the parents need to be made aware of doing the tests. To give you context – most people in India use wall calendars (keeps track of all the festivals). This is how the calendar (wall hanging) for new mothers came about. I aligned child development milestones with the calendar and added easy checks for periodic reminders for the family as part of the calendar. For example, if the kid doesn't turn their head to the pressure cooker in the first 3 months, then ask your doctor to do a hearing screening test. With the central gov't push towards institutional births, the calendar can be given out as part of a new-mother care package.

The hearing screening project has many facets of the system – the education, the product, the interface, the business model. How did you participate as a designer in each role?

Creating a new business model is just like designing anything else. There is a lot of research, versions and drafts, with quick testing and learning with parents as well as doctors, to validate the decisions. I looked at data published in papers or journals, other companies' numbers, etc. For creating our financials, I got estimates on product development, did calculations on cash flow, assessed time needed, resources required, created realistic market penetration numbers, created scenario plans for revenue expectations.

For the design of the product – it was a lot of sketching on templates of the human head to work under the constraints. I made some foam and clay models for quickly prototyping to work through the form and ergonomics. We got some models 3D printed to make visual prototypes that would explain the concept and test it.

For the design of the interface, I kept things as simple as possible. Understanding what and when they needed to learn information, from technicians and doctors, was vital. Again, as we showed it to people, we would get new insights and would go back and make changes.

Do you have any insights that you can offer from this project about designing for behavior change?

The design of the calendar as a visual reference of reminders to make sure people were identifying developmental milestones for the child is very important. Life gets in the way of thinking about bigger things. How do we bring things front and center – get people to care about it? They don't know how urgent this is. They all care about their children – they just don't know what needs to be done.[7]

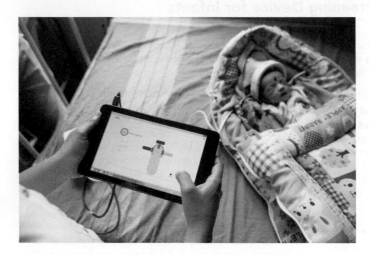

◄ Figure 3.5

Discovering a hearing deficiency early will help with mental development, social development, education, employment, and contribution to society in the long run. Source: Rolex Awards/Ambroise Tézenas (with permission).

Skills

Motor Development

Have you ever seen a 1-year-old walk and fall down over and over again? What about a preschooler trying to tie his/her shoes? Do you remember being an awkward and clumsy teenager? These experiences all involve gross and fine motor developments of the body that allow children to learn how to interact with the physical world. Motor learning consists of complex processes in the brain that occur in response to practice or experience. Practice results in changes in the central nervous system and muscle memory such that all kids acquire skills in daily living such as feeding, dressing, mobility, drawing and writing. Nurturing these skills even further allows individuals to excel in sports, the arts, music, dance, keyboarding and penmanship. Additional factors that influence motor development include growth, genetics, muscle tone, gender, teachers or coaches, race, family position and additional social influences.[8]

As designers, we consider both gross and fine motor skills when designing objects that children interact with. Gross motor skills are considered when designing spaces kids use or in designing sports equipment. Fine motor skills are considered when the child must perform a task such as squeezing a bottle, pinching a clip or pressing a button.

Considering the skills related to the motor skill physical development of the child performing the tasks is essential in developing an age-appropriate engagement. The interest level of the product must be timed to the skill level of the child. For example, I have observed 4–6-year-olds play with an electric train set that is designed for a 4-year-old. It says four and up on the box, but the design has several flaws related to motor skills. The connection design of the tracks requires detailed fine motor hand skills to put the tracks together. It is also challenging for a child of that age to put the cars on the tracks. This causes a lot of frustration for the children, and they lose interest in the train set and move on to other toys that allow more playtime. When these children finally have the hand skills to play with the train set, will they still be interested in toy trains?

Gross and Fine Motor Skills

Gross motor skills involve the larger muscles in the arms, legs and torso. When an infant is born, they don't have much control of their movements, but as they develop, they explore the world through their body, play and movement. They form body memories and gradually develop gross motor skills for activities including walking, running, throwing, lifting, balancing and kicking. These skills also relate to body awareness, reaction speed, balance and strength. An Olympic gymnast is a master of gross motor skill control.

Regular movement experiences help children to develop movement control, coordination and strength. Children also develop a sense of where their body is in space.

Fine motor skills are generally thought of as the movement and use of hands and upper extremities. They develop later as the whole body starts to move and become more stable. These skills involve the coordination of small muscle movements, for example in the fingers, usually in coordination with the eyes. These abilities start with primitive gestures such as grabbing at objects and move on to more precise activities that involve precise drawing and writing. We use fine motor skills to reach, grasp, manipulate objects with our hands and pick up items, and for activities such as playing an instrument.

Motor skill development has a logical development progression. My son's pediatrician explained motor skill development to me like this. "Imagine a baby gaining control of their body from the top down from the head to toes. First, they gain control of their head and neck, then torso and arms, then legs. Then the details fill in and strengthening occurs." This is why babies learn to hold their heads up before they learn how to crawl. In preschool, emphasis is put on developing fine motor skills for school readiness.

Muscle memory is the ability to repeat a specific muscular movement with improved efficiency and accuracy that is acquired through practice and repetition. It is a connection between physical and cognitive skills and usually associated with motor skill development. Just as the brain is ready to learn academic skills at particular times, there are windows of opportunity to develop muscle memory so that movements naturally occur and tasks can be performed without thinking about them.

The Body and Motor Skills

- To perform a task or movement, our brain sends signals to our motor units (individual nerves and collections of muscle fibers) at precise intervals to orchestrate the contraction of muscles throughout our body.
- As they mature, nerve cells make more connections, and the muscles of the body get stronger.
- Large muscles develop before small muscles.
- Muscles in the body's core, legs and arms develop before those in the fingers and hands.
- Children learn how to perform gross (or large) motor skills such as walking before they learn to perform fine (or small) motor skills such as drawing.
- The center of the body develops before the outer regions. Muscles located at the core of the body become stronger and develop sooner than those in the feet and hands.[9]

Areas of Physical Literacy Development

Locomotor (mobility) skills: Involve transporting the body in any direction from one point to another. These skills start with spontaneous arm and leg movements during the fetal and newborn periods, followed by rolling, crawling then pulling to standing and balancing upright, then, finally, running, jumping and more sophisticated forms of mobility such as galloping, skipping and swimming.[10]

Body management skills: Balancing the body in stillness and in motion. Examples are static and dynamic balancing, rolling, landing, bending, stretching, twisting, turning, swinging and climbing.[11]

Object control and manipulative skills: Require a child to control an object using a part of the body or using an implement. There are two types of object control skills:

Propulsive – sending an object away (throwing, kicking, striking, batting)

Receptive – receiving an object (catching, dribbling a ball, receiving a Frisbee)[12]

Physical Skill Development through Play

To develop motor skills, children need the time, space and opportunity to move their bodies in big ways and small. Physical movement, dancing and active play provide opportunities for gross motor skill development. Dexterity toys and constructive and creative play provide opportunities for fine motor skill development. Digital play provides opportunities for developing hand–eye coordination when children visually follow an object on the screen. Hand–eye coordination is essential as children begin school, when they learn how to use their hands and eyes while writing and drawing.

When designing for the capabilities of children, learning where the target demographic is at with their fine motor skill development skills is always a challenge. It is hard to evaluate the efficacy of a design unless there is a proto-type available for a range of kids to physically test.

▶ Figure 3.6

While developing a drawing tool to help develop motor skills and creativity, ergonomics were tested and prototyped with children of different ages and abilities.
Source: Katarzyna Burzynska (with permission).

▶ Figure 3.7

Playing with a hula hoop practices gross motor skills.
Source: author photo.

Tippi Tree, a Dexterity Toy

Interview with product designer Devin Montes, creator of Make Anything/3D Printing Channel.
The pieces of Tippi Tree can stack together in dozens of orientations, making it an exciting toy where the child must discover what can and cannot be done.

Can you describe your process of coming up with the perfect pieces to stack? How did you know when you got it right?
I realized that hooks that use leverage to hold in place might serve for engaging gameplay and I quickly switched to 3D printing to test physical prototypes. From that point, it was mostly a matter of changing tolerances so pieces could stack together but still be precarious.

What did you learn from playtesting with kids?
Playtesting was very useful in refining the game. I quickly realized that my original rules led to very short games and children would get disappointed when one mistake ended the game. Based on those observations I modified the rules and also added extensions for the tree base so that the trees could grow larger, which increased the "fun factor." Observation also showed me the value of Tippi Tree as an open-ended construction toy. I developed Tippi Tree as a turn-based game, intending it to be played by the rules, but I was surprised to see how much kids enjoyed simply building things from the stacking bricks. That's a significant realization.[13]

◄ Figure 3.8

Tippi Tree.
Source: Devin Montes (with permission).

Stages of Progression

Gallahue proposed that children move through a developmental progression in the acquisition of motor skills.

Reflexive Movement Phase (Birth to about 1 Year of Age)

Primitive reflexes are necessary for survival and include reflexes that provide nourishment, such as rooting and sucking. Infants engage in reflexive movements, involuntary responses to environmental stimuli such as touch, sound and light. Postural reflexes maintain body orientation and include stepping and body righting.

Rudimentary Movement Phase (First 2 Years)

Basic motor skills acquired in infancy – reaching, grasping and releasing objects, sitting, standing and walking – are acquired and form the foundation for the fundamental phase.

Fundamental Movement Phase (2 to 7 Years)

Children gain increased control over their movements. They develop and refine motor skills such as running, jumping, throwing and catching. Their control of each skill progresses in isolation from other skills before reaching a mature stage where they are able to combine them with other skills as a coordinated movement.

Specialized Movement Phase (Begins at about 7 Years of Age and Continues through the Teenage Years and into Adulthood)

Complex movement activities for daily living, recreation and sports pursuits are further developed. For example, hitting a ball with a tennis racquet is a specialized movement.[14]

Designing for Motor Skills Development

- Start with proper form or technique.
- Use repetitions to influence the mind and body with the technique.

- Build up from a solid base of fundamental skills to the more complex skills over time, utilizing developmental windows.

Designers Can Learn from Occupational Therapists

To learn specifics about the sensory processing, visual perception, fine motor and gross motor skills needed for a project, talk to a pediatric occupational therapist. Research their approaches to teaching these skills.

Techniques developed for children with special needs that include step-by-step breakdowns, and methods of teaching these skills to those with challenges, can also help designers with awareness for a general user.

Exercise Toys for Children with Juvenile Arthritis

Interview with Shirley Rodriguez – Monsta toys
As a student, Shirley Rodriguez developed and won several awards with this project, including an IDEA gold award. We talked about her inspiration and her process.

How did you develop ergonomics to fit a range of children's hands?
For the tools, the different shapes were modeled out of foam at various sizes and then tested with different aged children. During testing, it was essential to observe the natural way children played with the toy without any guidance. This allowed for specific sizes to be ruled out as it afforded wrong hand placement.

What role do the materials choices play in the design?
The materials the toys are made out of are essential for functionality, customization and safety. Conductive silicone converts the toy into a giant stylus when squeezed. The gels inside the toys offer resistance when squeezed based on the child's strength. Finally, the use of silicone makes the toys resistant to drops and safe for the tablet.

What value does the digital interaction bring to the project?
The interaction between the toys and the digital game adds customized therapy and allows for a range of

motion exercises that the toys alone cannot provide. It is also able to track and adjust to each child's needs. For example, it can measure and calibrate the game to each child's hand size, keep track of progress, and adjust the therapeutic routine based on set parameters by their doctor. On the user experience side, the digital game is fun and always changing, unlike typical physical therapy routines.

How did you connect exercises used in occupational therapy to the hardware and software?

There are two types of exercise games: range of motion and strengthening. Range of motion games are played with the hand directly on the tablet. Multiple touch points can track fingertips and palm location. Strengthening games offer an element of surprise, as the tool partially covers the tablet and then something happens. All of the exercises are based on typical therapy routines that the children would be doing in-clinic, but digitized.

What was the most exciting thing you learned in your research for this project?

I was surprised to discover the lack of pediatric products for children, especially for chronic physical conditions and for rehabilitation. Currently, adult products are being retrofitted for children and are often very intimidating or non-ergonomic which results in children not using them. Learning this first hand from parents, doctors, occupational and physical therapists helped fuel my motivation to create Monstas.[15]

◄ Figure 3.9

Strengthening games are played with the tools on the tablet.
Source: Shirley Rodriguez (with permission).

Writing and Drawing

While I was lecturing on the future of play, a 5th grade teacher asked: What is the role of the pencil today and in the future? When I speak to parents and teachers, they always question whether it is really necessary for kids to focus time on writing and drawing skills in a technology-based world. At least for now, writing and drawing are two skills that need to be practiced from a physical development perspective to improve fine motor skills and hand–eye coordination. They develop cognitive abilities and offer the opportunity to learn abstract concepts. The concept that letters build words, develops language and reading skills. Drawing develops visual communication skills. From a social and emotional perspective, we can look at children's drawings as representations of reality, reflections of their feelings, not presentations of reality. Writing and drawing exercises force children to coordinate their brain impulses with their physical reactions.

Children Learn How to Write in Stages

Prephonemic Stage

- Random scribbling – The starting point is any place on the page.
- Controlled scribbling – Progression is from left to right.
- Circular scribbling – Circles or ovals flow on the page.
- Drawing – Pictures tell a story or convey a message.
- Mock letters – These can be personal or conventional symbols, such as a heart, star or letters with extra lines.
- Letter strings – These move from left to right and progress down the page of actual letters. They have no separations and no correlation with words or sounds.
- Separated words – Groups of letters have space in between to resemble words.

Early Phonemic Stage

- Picture labeling – A picture's beginning sound is matched to a letter.
- Awareness of environmental print – Environmental print, such as names on cubbies, is copied.
- Transitional stage spelling or invented spelling – First letter of a word is used to represent the word.

Letter-Name Stage

- Beginning and ending letters are used to represent a word.

Transitional Stage

- Medial sound is a consonant.
- Medial sound is in correct position, but the vowel is wrong.
- A child hears beginning, medial and ending letters.
- Phrase writing develops.
- Whole-sentence writing develops.

Conventional Writing Stage

- Transitional stage spelling (or invented spelling) is replaced by full, correct spelling of words.
- Cursive writing.[16]

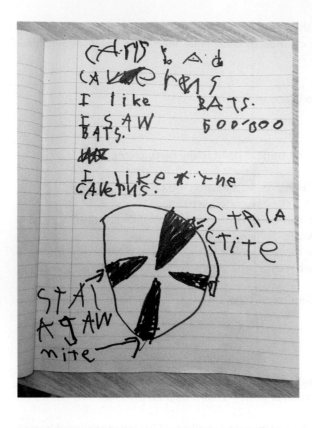

Writing takes many years to master.
Source: author photo.

Drawing Stages

It is widely believed that there are six stages of drawing. These include:

Scribbling stage (2–4 years): Children go through a process of understanding that their physical actions can dictate the marks they make. Initially, this is random scribble but develops into a more controlled activity.

Pre-schematic stage (4–7 years): Children begin to use shapes and symbols to explore relationships and their environment.

Schematic stage (7–9 years): Children develop a "schema" or consistent way of portraying an object or person or environment. These images generally exhibit their knowledge of something.

Realistic stage (9–12 years): Children begin to focus on detail and realistic features in their drawings. They are conscious of their peers and the level of detail in their drawings.

Pseudo-naturalistic stage (12–14 years): Children begin to focus on the final product, on whether the image looks good and is pleasing to peers and parents. They begin to have an understanding of three-dimensional space in their drawings.

Crisis of adolescence/artistic decision (14–17 years): A conscious decision must be made to continue drawing and engaging in visual thinking. Images created become highly individualized.[17]

Drawings by children at different
stages.
Source: author photo.

Mobility

As children are developing their locomotor skills, they need help to seamlessly
move from point A to point B. There are many parental assistance products
(strollers, baby carriers) designed around aiding a child's mobility to carry them
or push them around. The mobility of the body is the beginning of freedom.
Simple toys that toddlers can ride on for play evolve into sports gear that they
can ride for fun and fitness, such as scooters and bicycles. They may show an
interest in a particular sport, such as skateboarding, and the fun outweighs
the mode of transportation. Electric skateboards and bikes have become a
viable commuter option in recent years. For some children with special needs,
mobility is an everyday challenge. Wheelchairs and other mobility products
have improved their freedom and independence.

Safety and mobility go hand in hand. Once a child is mobile, childproofing
the environment with locks and gates has reduced injuries and deaths. Car seat
laws and requirements have improved safety. Bike helmets have prevented and
reduced the impact of head trauma.

Personal Care

Children have a healthy drive to be independent and do things on their own
and take care of themselves. Good personal hygiene habits enable kids to stay
healthy, free from illnesses and diseases caused due to bacteria. Caring about
the way they look is important to feel good and gain a healthy body image, self-
esteem and confidence. As children's bodies develop, they require more privacy
in their tween and teen years, and new skills need to be learned.

Being Clean – Looking Good

Hand washing: Clean hands prevent kids from spreading germs. Hand washing is taught in the early toddler years and reinforced in preschool and grade school.

◀ Figure 3.12

SoaPen makes hand washing creative and playful. The child draws in soap to encourage hand washing habits.
Source: SoaPen (with permission).

Bathing: Bath time is a multisensory experience, bonding, time for touch and experience in learning about fluid dynamics. When a child can stand in a shower stall and tolerate the feeling of water hitting his body from above, he can start taking showers. There is a slow transition from 5–8 years old when children have supervised showers.[18]

Hair: Kids need to keep their hair clean and brushed. Lice are a common problem for children. Haircuts can be severe for the sensory sensitive child. This can cause anxiety and fear. Noisy hairdryers, buzzers, clipping scissors and light tactile experiences can be unmanageable for some children.

Teeth: Lower front teeth usually begin to appear by the age of 5 to 9 months and all 20 of the baby teeth by 2½ years. Permanent teeth replace baby teeth between the ages of 5 years and 13 years. The American Dental Association recommends that children brush their teeth twice a day for two minutes at a time. The aim of brushing babies' teeth is to teach them to tolerate the oral sensation of something in the mouth and tongue movement. For toddlers/preschoolers, tooth brushing and flossing are done by an adult, and they are beginning to clean parts themselves. For school-age children, the aim is independent brushing and flossing of teeth.

Using the Potty

A child's biology, skills and readiness will determine when they can take over their toileting. Many parents encounter challenges and needs for assistance.

Pre-potty (10 months+): Use words and meanings for bathroom-related functions.

Potty training (18–32+ months): Recognizing patterns, establishing a routine, practicing, positive reinforcement, repetition and accidents are often a part of potty training.

Nighttime training: Many children are potty trained in the day but take longer for nighttime training. It is achieved when a child's physiology supports this. Many girls are nighttime trained at around 6 years old, with most boys around age 7.

Independent Dressing

Clothing and footwear designers are continually challenged with functionality as children grow. Dressing and undressing involve the development of gross and fine motor skills and skills related to the cognitive mapping of their bodies. Children's gross motor skills must be developed to have proper stability, postural control-waistbands balance, and bilateral coordination in order to dress and fasten clothing. They need to have hand–eye coordination and be able to control visual input and control their hands at the same time. Children's fine motor skills involve fastener systems and have the following sequences of development.

- Velcro
  1) Opening Velcro attachments
  2) Pressing down on Velcro attachments (adult line up)
  3) Securing Velcro
- Snaps/Zippers
  1) Pulling apart snaps
  2) Lining up and clipping together snaps
  3) Unzipping large zippers
  4) Zipping up large zippers (adult hooked and held)
  5) Lining up
  6) Hooking and doing up zippers
- Buttons
  1) Undoing large buttons
  2) Doing up large buttons
  3) Undoing smaller buttons
  4) Doing up smaller buttons
- Tying waistband
  1) Untying waistband at the front
  2) Tying waistband
- Laces
  1) Untying shoelaces
  2) Putting laces in eyelets in shoes
  3) Untangling knots in shoelaces
  4) Tying shoelaces

- Buckles
  1) Undoing
  2) Doing up buckles

Health

Health is such a broad topic that when we think about it from an international perspective, we see broad inequities in this area around the globe. In the US, we speak about more exercise for overweight children to prevent diabetes, while other children in the world are dying from starvation and lack of clean water. Any issue you choose around improving children's health anywhere in the world is a noble effort. Teaching children about healthy lifestyles and promoting a positive body image are essential throughout childhood.

Basics for a child's good physical health:

- Nutritious food and balanced diet
- Adequate shelter and sleep
- Fitness, exercise to develop strong muscles and bones and to maintain a healthy weight for height status
- Immunizations
- Healthy living environment

Exercise/Sports

Exercise and strength building are essential for motor and brain development. Growing children need approximately 1 hour of physical activities daily to grow, maintain a healthy weight and become strong. The Surplus-Energy Theory hypothesizes that active play allows kids to release pent-up energy that has collected over time, so they are ready to return to academic learning. Many schools have eliminated recess and physical education. Non-profits are stepping in to add PE classes, holding different physical exercises, play activities or skill building, like riding a bike.[19]

There is an overall interest and emphasis today on exercise that emphasizes wellness (healthy eating and exercise) over weight loss. Instead of big blocks of time focused on fitness, how do we facilitate a daily routine in children's lives to prioritize health?

Developmental Benefits of Sports

Children who participate in sports are less likely to drop out of school and become involved in drugs and alcohol activity, while they also excel in academic performance and sociability.

Benefits of Sports

- Develops specific skills, stronger muscles and bones
- Helps with weight control

- Improves endurance and keeps the heart healthy
- Boosts self-esteem and confidence
- Strengthens perseverance
- Develops teamwork, leadership, responsibility and healthy competition
- Teaches discipline
- Provides guidance
- Concentration skills[20]

▶ Figure 3.13

In addition to physical skills, sports build cognitive and social and emotional skills.
Source: Michael Lyn (with permission).

Designing a Bike Program

Interview with Penny Bostain
Penny Bostain coordinated a bike program for The Bradshaw Institute for Community Child Health & Advocacy (B.I.) of the Children's Hospital of Greenville Health System (GHS).

What did you learn from coordinating the bike program?
Many children from 9 to 12 years old cannot ride a bike yet, from the most athletic kid to the heaviest kid. When I've randomly questioned the kid who is the most athletic, their answers are the same, "no time, I play soccer or baseball, or I ride 4 wheelers." When I ask the heavy kids, I always get the same responses, "no one ever showed me how or I don't have a safe place to ride."

Why is it important for children and youth to learn how to ride a bike?

For physical development: Bicycling helps improve muscular strength and endurance as well as cardiovascular endurance; bicycling is a lifetime physical activity that helps develop coordination, balance and kinesthetic awareness in particular parameters.

For cognitive development: There are many ways that active education is helping children learn and retain important concepts in all subject areas. Physical activity soothes the brain – activates the cortex where learning occurs.

Social and emotional development: Learning to ride a bike develops independence, persistence and determination – the reward of sticking with something that is difficult and eventually mastering it, serves children well whenever they encounter difficult tasks throughout life; builds confidence, feelings of freedom, increases responsibility and is a person's first vehicle. Group rides with friends and or family members promote social play. We teach youth bike etiquette on trails as preventing crashes can be very much dependent on cooperation between people you meet or who want to pass or whom you want to pass.

What bike skills do you think are most important for kids to develop? Why?

- Wearing a bike helmet, knowing how it should properly fit, and being able to adjust your helmet without the help of an adult.
- Riding a bike that fits, not one that you will grow into.
- Making thoughtful decisions on clothing – bright colors, shoes that tie.
- Balancing on two wheels – we require a child to ride slowly first, and then they can ride fast.

- We require that you show us you can stop using the brakes on the bike, if they're two brakes you have to use both.
- Signaling by keeping the right hand on the handlebars at all times, using the left hand to make the appropriate signals for turning right, left, and stopping.
- Rules of the road, riding w/ traffic, following traffic signs, avoid hazards.[21]

◀ Figure 3.14

The primary goal of the bike program is to prevent injury and disease and promote healthy child development, and they collect physical data pertaining to child health and wellness. In order for a teacher to use the program, the curriculum is matched to existing South Carolina State Department of Education standards of learning in physical education.
Source: Bradshaw Institute for Community Child Health & Advocacy (with permission).

Sleep

Sleep is the primary activity of the brain during early development. Circadian rhythms take time to develop, resulting in the irregular sleep schedules of newborns. By the age of 2, most children have spent more time asleep than awake, and overall, a child will spend 40% of his or her childhood asleep. Sleep is especially important for children as it directly impacts mental and physical development.

Benefits of sleep:

- Promotes growth
- Helps the heart
- Affects weight
- Helps beat germs
- Reduces injury risk
- Increases kids' attention span
- Boosts learning

Non-rapid eye movement (NREM) or "quiet" sleep: During the deep states of NREM sleep, blood supply to the muscles is increased, energy is restored, tissue growth and repair occur, and important hormones are released for growth and development.

Rapid eye movement (REM) or "active" sleep: During REM sleep, our brains are active, and dreaming occurs. Our bodies become immobile, breathing and heart rates are irregular.

Each family has their own setup for sleep. Some babies sleep in their own spaces from day one, while others sleep with parents in their bedroom for a period of time before they move into their own sleep environment. In some cultures, families sleep together continually.

There are safety issues around sleep for small babies who do not have complete control of their bodies. A substantial effort has been put into building awareness around sudden infant death syndrome (SIDS) and suffocation of small children. There are recommended ways to dress a baby safely for sleep, setting up a crib and products surrounding the baby and co-sleeping, that are continually changing, so look into the latest scientific research. Between 20 and 30% of children have experienced sleep problems, say Oliveiero Bruni and Luana Novelli.[22] As many as 40% of kids have sleepwalked at least once, usually between the ages of 2 and 6, according to the National Sleep Foundation. An additional 6% of children may have night terrors.

How much sleep do kids need? Parents of infants are up on and off all night reacting to their babies' sleep patterns. Learning to fall asleep is a skill. Many children need to learn how to soothe themselves to fall asleep.

1–4 weeks old: 15–16 hours per day
1–12 months old: 14–15 hours per day
1–3 years old: 12–14 hours per day
3–6 years old: 10–12 hours per day
7–12 years old: 10–11 hours per day
12–18 years old: 8–9 hours per day[23]

Dreams are complex and dynamic, and their function may be multi-purpose, serving biological, psychological and cognitive needs all at once. All dreams are collections of emotions, events, images and impressions that we experience during sleep. Kids, like adults, dream what they know.

Nightmares: Frightening dreams that usually result in waking the dreamer from sleep.
Sleep terrors (also known as night terrors): These are intense periods of fright during sleep. They are more common in children than in adults.
Recurring dreams: These are dreams that, over time, return again and again for the dreamer.
Lucid dreams: Though sleeping, the dreamer is aware he is dreaming and may be able to control some aspects of the dream.

Dreams evolve in story and narrative and vary in intensity throughout childhood. In a study on childhood dreaming, some researchers believe that infants don't dream, as they have not experienced enough of life to dream. Others think they do, as they have proportionately more REM sleep, the phase where most dreaming occurs. In his research David Foulkes found that toddlers' dreams

(2 and up) tend to involve little more than a setting: a child taking in the scene, without action or characters to speak of. Over the next few years, dreams tend to become a little more complex, in pace with social understanding, moving beyond just simple settings. When something does happen, it tends to happen *to* the child, rather than the kid acting on his or her surroundings. Foulkes has argued in his research that kids don't begin acting on their surroundings in dreams until around age 7 or 8, when their sense of agency is developed enough to support them as protagonists in their own narratives. In the teenage years, dreams are often similarly vivid and formative, grappling with quintessentially teenage themes of identity and sexuality, but there are fewer of them. By late adolescence, dreams start to settle into what we usually expect to find among adults.[24]

Can we learn about children by asking them to explain their dreams? Maybe, but it could be fascinating, humorous or scary, among other things, depending on the cast of characters and the plot. If you decide to include research about dreams in your process, be respectful and sensitive to the child's emotions. Very young children have difficulty recalling their dreams. At age 2, some children have the communication skills to communicate their dreams.

Eating and Nutrition

Which foods do you best remember loving as a child? Broccoli? Spinach? Probably not. Many of us have nostalgic feelings towards fun foods that today would be considered unhealthy. This contradiction of good for you and fun foods poses some of the most significant challenges for children and eating. What children need to eat to meet fundamental nutritional goals leading to healthy growth and development is often different from what they desire to eat. Children need to be exposed to foods over and over again, since they initially reject many healthy foods. Offering the foods 8 to 10 times may result in children accepting them.

What a child eats in the first few years has a lasting impact on food preferences, leading to better health over a lifetime. During the most rapid growth phases, infancy and adolescence, children have high nutritional needs. Children need different amounts of specific nutrients at different ages and need to snack more often than adults. Food and emotions are connected, and it is essential for kids to have positive eating experiences.

Parents model eating behaviors, and when parents eat healthily, kids learn to do the same. Parents make food choices based on other factors than nutrition, such as taste, convenience and cost. Additional factors that influence food choices for children are packaging, familiarity, peer influences, media, plating, easy access, priming taste expectation and naming.[25]

Kids' Understanding of Nutrition

Young children connect food with being hungry, not with being healthy. It is either appealing or not; it tastes good, or it doesn't. From an early age, kids learn about the food pyramid and the importance of eating a balanced diet.

They know fruits and vegetables are healthy but don't know why. Making the concepts of foods relatable to their lives helps them understand them better. Most kids want to be strong (from protein) and have a lot of energy to run fast (from carbohydrates).

Vitamins and minerals, nutrients in food, and digestion and chronic disease risk are abstract ideas, and most kids can't understand abstract concepts until they are developmentally ready (around 7 or 8). School-age kids need to be presented with concrete ideas, not abstract, to understand nutrition.

Examples of Abstract and Concrete Nutrition Concepts

Abstract concepts	Concrete concepts
Vitamins and minerals	Whole foods
Nutrients that can't be seen or touched	Diversity of foods
Classifying foods by nutrients	Everyday foods versus sometimes foods
How food affects health	Classifying foods (meat, milk, plant foods)[26]

Designers can help by engaging children in a positive and healthy eating experience.

- Consider the social aspect/role modeling of food
- Food time needs to be fun time – conversations and experiences should be positive
- Teach about vitamins and effects at an appropriate cognitive level
- Creating positive products, environments, experiences and messages around food
- Designed products could help them to be independent
- Create solutions for hunger, malnutrition and access to clean water
- Teach them about healthy foods, fun foods and food safety

An example of creating and engaging program to promote healthy eating is the family dinner project (see Figure 3.16 page 188).

Learning to Eat

The American Academy of Pediatrics (AAP) recommends feeding babies only breast milk for the first 6 months of life. Infants can only consume liquids, because they are only physically capable of sucking. After that, they recommend a combination of solid foods and breast milk be given until a baby is at least 1 year old.[27] At first, caregivers feed pureed foods and smooth cereals, and as they get older, children can practice feeding themselves. Finger foods give them a sense of autonomy. In the early phases of eating, choking is a concern for many parents. Chewing does not begin until about 6 months of age, and it is not well developed until about 3 years of age. Young children tend to prefer food that is easier to chew. Softness and hardness, chewing and swallowing are all factors. Babies and toddlers usually follow a pattern of initially rejecting food with an unfamiliar taste and texture.

Baby's first taste of food.
Source: author photo.

The taste for sugars is inborn. Many of children's first accepted foods tend to be sweet fruits and juices. A preference for salt starts at 4 months of age. In general, kids like some acidity but not excessive sourness or bitter tastes. Some children begin eating behavior issues due to taste and texture (picky eaters). Eggs, peanuts, tree nuts, milk, soy milk, wheat and shellfish are common allergies in children. They can range in severity from mild to life-threatening.

For younger children, a sense of "play value" and diversity of color on the plate can make food more appealing. Playing with food sometimes signifies the child is full or not interested in eating. Young kids generally don't like food mixed together or touching each other, such as stews and casseroles, although, in contrast, children eat combinations of foods or flavors that would not appeal to adults, such as applesauce combined with spinach or ketchup with carrots. A young child does not develop sufficient motor control to manipulate eating utensils until age 5 or 6.

Older children/teens like to explore and experiment with a variety of textures, spices and cuisines. Kids with braces like to eat smaller, thinner and softer food. Food preferences and special diets (vegan or vegetarian) may be a way to express identity. Some may have control, eating disorders or body image problems.

DIY Food

If they grow it or cook it, they will eat it is a current philosophy used in teaching nutrition and practical skills to children. Gardening and cooking can be viewed as play for young children and as a hobby as they get older.

Gardening: While gardening at school or home, children see how the farming and growing process works, and they practice patience. Many popular chefs, from Alice Waters to Jamie Oliver, have popularized the school garden movement. Gardens allow the kids to get dirty, experience the plant life cycle and learn about seeds, fruits and vegetables and how these are planted and cared for while the plants grow, and ecosystems, including bugs and composting.

Cooking: Cooking and eating together gives adults and children time to talk about proper nutrition, food safety, the proper way to use tools (such as knives), and why you chose the ingredients you're using, which lays the groundwork for healthy eating later on.

Gardening and Cooking Timeline

Toddlers: Can play in the dirt, dig holes and help water plants. In the kitchen, they can play with safe cooking tools (wooden spoons and spatulas) and pots and pans. They can learn simple skills such as washing food and pouring mixtures into bowls.

Preschoolers: Young children can pick fruits and vegetables and plant seeds. In the kitchen, they can move from pretend cooking to learning how to cook for real. They can get hands-on experience in mixing, cracking eggs, and observing how the meals they eat are put together. If they have a play kitchen or pretend food, keep it close by so they can create their own recipes as parents cook.

Young school-age kids: Can garden and cook for themselves with a few reminders, a little help and some few basic recipes. They can prepare their own breakfasts and lunches. They use their math skills when combining ingredients for recipes. Make everything accessible. Teach children how to use tools and safety.

Tweens and teens: Some might like to have their own garden or join a community garden or club. They may cook to satisfy their hunger; others might cook for the whole family. Some appreciate the chance to explore in the kitchen and improve their cooking skills, or might be interested in exploring cooking as a hobby, such as baking or exploring different cuisines.

Design Prompts: Food

- Visit a local school during lunchtime. Observe the kids before they eat, while they eat and after they eat. Notice the nutritional aspects, their behavior with the foods and the social side of lunch.
- Join a family for dinner or observe families at a child-friendly restaurant.
- Create a game to teach kids about nutrition.
- Have kids design their own lunchbox with their favorite foods in it.

Reproductive Health

Sexual Development

Boys begin puberty between 9.5 and 14 years old. Girls begin puberty between 8 and 13 years. Menstrual periods on average start between 10 and 16.5 years old. Girls are experiencing puberty earlier than previous generations. Along with developing physically, children and teens run the risk of early pregnancy. Kids need to understand their body changes and feel comfortable asking questions or raising concerns about hair and body growth, menstruating, and how to take care of themselves hygienically. In many places around the world, girls stay home from school because of the embarrassment of having their periods, limiting their education.

◀ Figure 3.16

The Family Dinner project Community Dinners provides support for families who want to change their dinner habits. The toolkit provides checklists, agendas, games, conversation starters and food preparation advice.
Source: Family Dinner Project (with permission).

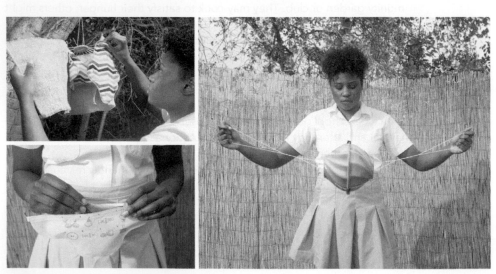

▲ Figure 3.17

Flo was designed for low resource environments to encourage girls to attend school when they have their period. When away from home, a girl wears a lightweight carrying pouch under her garments where pads can be comfortably hidden. Back at home, girls privately clean used pads with the affordable and small-sized washer/dryer unit.
Source: Mariko higaki, Tatijana Vasily, Sohyun Kim (with permission).

Interview with Mariana Prieto and the Divine Diva Project/Ideo.org

How does the approach to social impact projects differ from other projects that use design thinking?
Projects defined explicitly as social impact have a big responsibility to how the end result creates long-term change. Usually, there is one goal in mind and if that goal isn't achieved that project would not be considered a success in the traditional sense (even if you have thousands of users at the time but the social goal was not met).

What are the most important things designers could do for kids/youth in the coming years?
Design for the future of these kids and youth. Small actions today, what they learn while they are young and what they are exposed to now will impact the decisions they will make as adults.

Can you explain the story behind the Divine Diva Project?
The goal of the project was to lower the teen pregnancy rates in Zambia through a new approach to family planning services for teenagers.

Who was on the project team, what were their backgrounds and how did they bring their expertise to the project?
The first team on the ground from the IDEO.org side was mostly made up of designers. Two of them were product designers, one UX designer and someone that isn't a designer but is extraordinary at many things – a wild card "swiss army knife." A few months later the team grew with the addition of a writer/storyteller, behavioral scientist, and photographer/videographer. The Marie Stopes Zambia team pushed it forward and made it a reality.

How was each of these research methods used in creating the project?
The research strategy was created by the initial team members in the studio in San Francisco, and we split it up into two separate field trips. The first one for design research and the second one with rough and rapid prototypes to test.

Concept testing/Observational research
During the research phase, we set up pop nail salons to test different ways of talking about contraception to see what made the most sense to them in a safe environment. We casually talked about birth control with each other and broached the subject when they asked us questions so we could learn what was interesting to them and how to best talk about it. We realized that while doing nails, you don't need to look at someone in the eyes, which made it easy to ask hard questions weaved in superfluous comments. "Do you want glitter? How long have you been on the pill? Is blue glitter ok?" This observation was so powerful that the nail salon environment made all the way to the final concept of the Diva Centre.

Design research
We also did interviews but realized that if we asked a teenage boy/girl about their experience, they would probably not tell us the entire story. So we conducted a design research workshop to teach them how to do it on their own. We gave them worksheets and sent them home with homework so they could interview other friends. In the end, the teens in our workshop felt comfortable answering questions because they were speaking about someone else's experience removing the fear of judgment, while that individual's identity was left anonymous to everyone in the group.

We had visits to the clinic with and without them to understand both sides of the issues. The teen experience and the clinicians' experience separately. The interaction she had with the receptionist working behind the counter at check-in, and with the nurse in the room was critical to the success of the initiative.

What were the biggest surprises?
One of the research findings that surprised us was how open and public the solution needed to be. Because the subject was uncomfortable for teens, our initial thought was to make it discreet and subtle, but through prototyping, we found that *because* it was so uncomfortable, it needed to be public. A girl told us that if she got caught with something that looked suspicious in her bedroom her mother would be upset, but if it were colorful, bold and vibrant it would blend in with how she already dressed and the things she had. That's part of the reason for all the bold and vibrant colors of the Divas – so they blend in.

▲ Figure 3.18

Design of the communications materials for the Divine Diva Project.
Source: ideo.org (with permission).

Do you have any insights that you can offer from this project about designing for behavior change?

The most foolproof way to do behavior change is to find the current behavior and design around it. The diva center would not have worked if we insisted that girls go out of their way to a new location. We needed to find where they go now, the places where they are already comfortable and designing around that. Getting them to talk about contraception is not behavior change – it was a need because they already wanted to talk about it. They just didn't have the right outlet for it. All we did was find the need and a current behavior we could build on and intersect the two.[28]

Notes

1 American Association for the Advancement of Science
2 Graber, Evan G.
3 Kid Sense, *Strength and Endurance*
4 Stromberg, Joseph 2015
5 Scholastic
6 Louv, Richard 2017
7 Strittmatter, Neeti 2016
8 Study.com
9 Gallardo, Ramil 2014
10 Weise, Marin L.; Adolph, K.E. 2000
11 Pangrazi, Robert P.; Beighle, Aaron 2016

12 Thompson, Alison 2004
13 Montes, Devin 2017
14 Frost, J.L, Wortham, S.C., Reifel, S. 2010
15 Rodriguez, Shirley 2017
16 *Stages of Writing Development*
17 Child Art (Weise; Adolph)
18 Geller, David
19 Johnson; Christie; Wardle
20 Health Fitness Revolution 2015
21 Bostain, Penny 2017
22 Bruni, Oliveiero; Novelli, Luana.
23 Sleep Foundation
24 Romm, Cari 2016
25 Savage, Jennifer S., Fisher, Jennifer O., Birch, Leann L. 2007
26 Davidson, Jill Camber 2009
27 American Academy of Pediatrics, *Benefits of Breastfeeding*
28 Prieto, Mariana 2016

Chapter 4

Cognitive Development

As designers, we always say we need to get into the heads of our users to gain a greater understanding of their needs. This provides knowledge and inspiration for our creative process. When designing for children, knowing how the brain develops and how thinking processes work is one of the best ways to do that. As children grow and learn through experiences, they are able to integrate the new with the old, building complex cognitive skills. From the earliest stages of development, gene activity interacts with events and experiences with others and in the environment to produce growth. The child's treatment by others, parent's preferences, and a child's choice of actions and activities affect their development. Like any toolkit, the brain can be developed and formed with networks and tools that kids can use whenever they need them.

Cognitive development combines psychology and neuroscience and focuses on information, auditory and visual processing, conceptual resources, attention, memory, logic and reasoning, language, learning, creativity and other aspects of brain development. They are the skills the brain uses to think, learn, read, remember, pay attention and solve problems. All areas of cognitive development and learning are important, and some may have a specific relationship to your project.

Psychological theories and scientific research on brain development help us understand and integrate age-appropriate development into our process. By understanding different strengths and weaknesses in cognitive development in children, we can design the best solutions to support them. It is often not enough just to state that a project focuses on cognitive development. Even though, as designers, we are not scientists, when conducting research, we need to be very specific about the areas of cognitive development we are designing for. What design features will enhance that development, and how? Is the project intended to help children process information, remember or build their problem-solving skills? How can we use what we know about cognitive development to enhance a play experience, learning, creativity and growth?

Brain Architecture

Gaining knowledge of the structure of the brain and how it works gives us the ability to design and support the brain's function and the child's understanding. A child's early interactions directly affect the way that the brain is "wired."[1]

Minimally intrusive brain scanning of babies in real time has helped scientists better understand development.
Source: Isabel Cristina Quiroz, Birkbeck College's Babylab, UK (used with permission).

The basic architecture of the brain is constructed through an ongoing process that begins before birth and continues into adulthood. Simpler neural connections and skills form first, followed by more complex circuits and skills. The first 5 years are particularly necessary for the development of the child's brain, and the first 3 years are the most critical in shaping the child's brain architecture, laying the foundation for future learning, behavior and health throughout life.

In their first few years, children have a high degree of neuroplasticity (the ability to adapt). By age 6, much of the brain has adapted to learning and behavior patterns, although new connections are continually made and pruning occurs throughout childhood and adolescence and into adulthood. Kids' minds are flexible enough to work continuously to build new circuits and refine commonly used neural pathways. In teens, the wiring system of the brain becomes more intricate, with circuits intertwining with other circuits to allow all of those skills to work together.[2] Since the repetition of experiences builds neural networks, consistency solidifies positive physical, cognitive, and social and emotional behaviors in children.

Forming of Neural Pathways

Each neuron takes in information, branches out and creates multiple connections to other neurons, creating pathways of circuits. They pass information through synapses, the gaps between neurons, by means of neurotransmitters (brain chemicals). When a connection is made and the circuit is activated, it powers a response. Some brain circuits, like the ones for breathing and circulation, are biologically dependent. Other channels are "activity dependent," such as play, which promotes brain development through the repetition of actions and strengthening connections. As a child grows, connections are reduced through pruning. The links that are not stimulated by mental or physical experiences are discarded, allowing the brain circuits to become more efficient.[3]

Windows of Opportunity

Brain development is non-linear, although there are prime times for acquiring different kinds of knowledge and skills. Understanding when the "windows of opportunity" occur for skills to be developed quickly and spontaneously, we can target the appropriate timeframe to achieve the best responses to development and the most significant impact. No matter how much we try to teach a child something, they will not understand until their brain is capable of receiving and understanding the information. The brain is primed for learning, and a child's attention is naturally drawn to the area of development that their brain is working on. Once the window has closed, however, the same information/skills are harder to learn. For example, children can learn many languages by merely being exposed to them during the language window of opportunity before age 5, and they can learn those languages simultaneously. It is critical for children from birth to be around people speaking to them and with them to take advantage of the time when their brain is primed for it.[4]

Windows of opportunity exist in almost every area of development, from depth perception, to potty training, to learning to play an instrument. In your project, understand when the windows of opportunity are for your topic and develop a method or framework to build opportunities at a time when the child will most easily absorb that information. An example of a framework can be seen in the Montessori Method. Maria Montessori recognized and categorized 11 fundamental sensitive periods of development from birth to age 7. These include sensitivity to movement, emotional control, math patterns, vocabulary, order, small objects, letter shapes and sounds, music, writing and reading.[5]

Learning

Learning is knowledge acquired through observation, experience, study, instruction or practice. Kids are always learning by reacting and adapting to their environment. Babies and young children find everything new and exciting. Psychologist Alison Gopnick explains, "What we see in the crib is the greatest mind that has ever existed, the most powerful learning machine in the universe."[6] When a toddler uses a crayon for the first time, it is magical. As kids get older and continue to learn, the magic fades away a bit, and we can help them learn by engaging them in new ways. We can create engaging stories, environments, tools and interactions to help the content stick. Exactly how does learning occur? There are many thoughts about the brain and how it learns, and in this section, I will go over just a few.

Play is a vehicle for learning.
Source: Anji Childhood Education Research Center (used with permission).

Principles of Child Development and Learning

- Learning is a result of an interaction of maturation and experience. Early experiences, motivation and approaches shape learning.
- At different ages, children think in different ways. Children become increasingly capable of handling more complex and abstract ideas, symbolic or representational capacities, and more sophisticated problem-solving.
- Social learning, social interactions and modeling enable cognitive growth.
- Children have a variety of ways they learn and advance through rigor and when they are challenged.
- Children actively construct meaning. Play is an essential vehicle for learning.[7]

Growth Mindset

After studying the behavior of thousands of children, Dr. Carol Dweck coined the terms "fixed mindset" and "growth mindset" to describe the underlying beliefs people have about learning and intelligence. When students believe they can get smarter, they understand that effort makes them stronger. Telling children they are smart encourages a fixed mindset, whereas praising hard work and effort cultivates a growth mindset.[8] This research has added to the belief that intrinsically motivated activities are important for learning and growing.

Intelligence

Intelligence is defined as general cognitive problem-solving skills. Among psychologists, there isn't a consensus as to what constitutes intelligence, although it is widely accepted that there are different types of intelligence. Most children come to understand the world in many ways and have preferred modes of learning and expressing that learning. Harvard psychologist Howard Gardner developed his Theory of Multiple Intelligences,[9] which set forth distinct types of intelligence and claimed that there need be no correlation between them:[10]

◀ Figure 4.3

Linguistic individuals like to talk and tell stories.
Source: author photo.

Physical-Bodily-Kinesthetic: Using one's body to solve problems and express ideas and feelings.

Social-Interpersonal: Understanding and interacting with others.

Solitary-Intrapersonal: Turning inward with a well-developed self-knowledge and using it successfully to navigate through the world.

Verbal-Linguistic: Using words, either orally or written, in an effective manner.

Logical-Mathematical: Understanding and using numbers effectively. Thinks abstractly, conceptually and reasons well.

Aural-Musical: Sensitivity to rhythm and sound.

Visual-Spatial: Perceiving physical space in an accurate way, as well as communicates though drawing and models.

◀ Figure 4.4

Visual/spatial individuals like to build and make.
Source: Anji Childhood Education Research Center (used with permission).

Learning styles are approaches to learning. Although popular currently, there is little evidence to support the idea that matching activities to one's learning style improves learning. Instead, use what we know about them and look at the content that is being taught and the best way to teach it. Harold Pashler and his colleagues explain: "The obvious point is that the optimal instructional method is likely to vary across disciplines."[11] In other words, it makes disciplinary sense to include kinesthetic activities in sculpture and anatomy courses, reading/writing activities in literature and history courses, visual activities in geography and engineering courses, and auditory activities in music, foreign language and speech courses.[12]

Design Prompt: Intelligences

How would users with different intelligences encounter, experience and gain from your project?
Considering different intelligences, how can the content be taught in ways that enhance learning?

Cognitive Development Theories

Many phenomena described in classic theories still attract the interest of current researchers. For designers, gaining an understanding of these theories offers a base of knowledge that continues to be built on. In the past few decades, there have been many advances in understanding how the brain works through analyzing soft tissue by MRI, developing neural networks as an analogy, and advances in the treatment of medical conditions. Cognitive development has come to the forefront and will continue to be an exciting area of research in the coming years. Through learning about theories of cognitive development and looking at the most up-to-date scientific research, we can design for natural thinking processes in children and youth.

Constructivist Theories

Constructivism is a theory about how humans learn and states that children need first-hand experiences. The premise is that we construct our own understanding of the world through experience and reflection rather by passively absorbing. To learn, we must ask questions, explore and assess what we know. When we experience something new, we compare it with previous ideas we already know and accommodate accordingly, either changing what we believe or ignoring it because we find it irrelevant. Jean Piaget and Lev Vygotsky have probably had the greatest influence on our ideas about how young children learn. Piaget focused on the way an individual child acts upon objects in the environment in order to build mental models of the way the world works. Vygotsky looked more closely at the way children acquire knowledge through interaction with more experienced people, and at the role of play in the process.[13]

Designers can use the constructivist approach in our work to promote learning by

- Developing tools, games or interactions that prompt kids to recognize their current knowledge base.
- Constructing experiments or having kids create their own to build on that knowledge.
- Providing opportunities for them to reflect on their experiences through discussion, writing or drawing to understand the concept better.

◀ Figure 4.5

Interactive "learning by doing" is one way children construct learning about the world around them.
Source: author photo.

Piaget's Model of Cognitive Development

Jean Piaget's four stages of development form a foundation for our thinking about stages of cognitive development. The model is based on the idea that learning and development follow sequences. Although newer research has challenged Piaget's model, it is useful to understand as a basic framework to build on.

Sensorimotor Stage: Birth through about 2 years. During this stage, children learn about the world through their senses and the manipulation of objects.

Preoperational Stage: Ages 2 through 7. During this stage, children develop memory and imagination. They are also able to understand things symbolically and to understand the ideas of past and future.

Concrete Operational Stage: Ages 7 through 11. During this stage, children become more aware of external events as well as feelings other than their own. They become less egocentric and begin to understand that not everyone shares their thoughts, beliefs or feelings.

Formal Operational Stage: Ages 11 and older. During this stage, children are able to use logic to solve problems, view the world around them and plan for the future.

Schemas

In Piaget's theory of development, children construct a series of schemata, based on the interactions they experience, to help them understand the world. This framework suggests that the brain recognizes patterns and the schema helps organize and interpret information (patterns). Schemas put little bits and pieces together and allow us to take shortcuts in interpreting the vast amount of information. The assumption is that we store these mental representations, apply them and adjust them as needed. Early schemas provide the basis for later learning and development of abstract thought.[14]

For example, a child might have a schema about going to the grocery store. The pattern of behavior includes entering the store, getting a shopping cart, sitting in the shopping cart, going up and down the aisles, choosing food from the shelves, placing the food on the conveyer belt and paying the cashier. This type of schema is called a "script." Whenever they are in a grocery store, they retrieve this schema from memory and apply it to the situation. Schemas are also built around animals, objects and people. A dog is a dog because it has four legs and barks. A ball is a ball because it is a sphere. Keep in mind that sometimes schemas exclude information, so we only focus on things that confirm pre-existing beliefs and ideas. A woman may be a mommy until it is explained that not all women have children.

▶ Figure 4.6

These children are accommodating their existing schema to adjust to the new situation.
Source: author photo.

Assimilation is the process of using an existing schema to deal with a new object or situation. Accommodation occurs when the existing schema (knowledge) does not work and needs to be changed to deal with a new object or situation, such as when a child visits a grocery store and the carts are different. Or, visiting a garden store for the first time, they may adjust the supermarket schema to fit the new environment and process. Maybe there are no shopping carts, and they walk down the aisles. Instead of gathering food, you purchase plants. They now have a new schema for a garden store. Young kids act out schemas in pretend play by playing with a toy kitchen or train set.[15] When designing for kids, we need to consider their familiarity with life experience. Have they seen what you are presenting to them before? What is their schema? Mental model? Conceptual model? When kids see a shadow, do they know what it is?

Mental Models and Conceptual Models

A mental model represents a person's thought process for how something works (i.e., a person's understanding of the surrounding world). They are based on incomplete facts, past experiences and intuitive perceptions. They help shape actions and behavior, influence what people pay attention to in complicated situations, and define how people approach and solve problems.[16] A conceptual model is the model that is given to the person through the design and interface of the real product.[17]

- Kids are forming mental models and have had different life experiences and, therefore, have different mental models. Designers should investigate what the target age group's current understanding and mental models are with task analysis, observations and interviews.
- The design of the conceptual model should fit the user's mental model, so the user finds the product /service easy to use. User testing could confirm this.
- If there are multiple user groups, then the mental models of all of the user groups should be considered in the design of the conceptual model.
- Sometimes a user's mental model may need to change to match the conceptual model, and additional training will be needed.[18]

Illustrator Terri Fry Kasuba explains the importance of speaking kids' language. "I continually question, Do they know what I am referring to? For example drawing a telephone. I like drawing telephones like they looked like in the 'old' days pre-cell phone and I REALLY like drawing the rotary dial phones. I wonder if kids today even know what that is?"[19]

Design Prompt: Schemas

Understanding children's existing schemas through a typical developmental scenario is a good starting point for a designer. What schemas can a child this age be expected to have based on experience? How might the scenario be adjusted for a better experience and expand on a mental model?

Although this toddler probably doesn't have a phone like this at home, he knows what to do with it.
Source: Autry Museum of the American West (used with permission).

Vygotsky's Social Development Theory

According to Lev Vygotsky, essential learning by the child occurs through social interaction with a skillful tutor MKO (more knowledge other), a parent or teacher. Current views include peers and digital tutors as MKOs. Vygotsky saw children as social learners, an expansion of Piaget's theory, which sees the child as a lone learner. The tutor may model and/or prompt behaviors by providing verbal instructions for the child. Through regular interactions, children learn from others how to solve problems and then internalize them to guide their behavior.[20]

The zone of proximal development (ZPD) is an area of learning that occurs when a person is assisted by an MKO. When educators (and designers) understand what children are able to achieve alone and with assistance from an adult, they can develop plans to teach skills in the most efficient manner possible, giving students a gradual release of responsibility to perform tasks independently. Cooperative learning exercises are used whereby less competent children develop with help from MKOs within the ZPD.[21]

Information Processing Theory

This area of cognitive psychology sees the individual as a processor of information, in much the same way as a computer takes in information and follows a program. A computer codes (changes) information, stores information, uses information and produces an output (retrieves info). As designers, our role is to communicate information and develop content, platforms and structures in order to communicate that information. Understanding the way kids process information can be used to make decisions on the best methods for communication and processing.

According to this model, attention, short-term memory and long-term memory are developing between the ages of 2 and 5. Auditory processing, which is critical for proper reading skills, is developing between the ages of 5 and 7. Logic and reasoning also become more established after 5 years of age as a child becomes better able to make connections between ideas.[22]

Serial processing is when one process is completed before the next starts. Parallel processing is when some or all processes involved in a cognitive task(s) occur at the same time. The human brain has the capacity for extensive parallel processing. It is influenced by conflicting emotional and motivational factors.[23]

How Do People/Kids Process Information?

- Designers must understand what makes people attend to information and causes people to concentrate or become distracted.
- People process information best in story form. Tell a story with a clear narrative and a beginning, middle and end.
- People process information in bite-sized chunks. Breaking down text into chunks, bullets, pictures and short sentences will help them process and remember it better.
- People decide quickly and then validate that decision.
- We only consciously process a fraction of the information that enters through our senses.
- We only see what we want to see.[24]

Some Activities Require More Brain Power than Others

The three different types of mental resources (below) require different mental loads.

- Motor: (least effort): pressing buttons, moving the mouse, etc.
- Visual: seeing things and taking them in

- Cognitive (most effort): thinking, remembering, making decisions[25]

Strengths and Weaknesses

Everyone has different cognitive strengths and weaknesses. For example, some people have solid memory skills, while other people excel in logic and reasoning. Cognitive strengths and weaknesses have an impact on whether we are successful – or whether we struggle – when it comes to thinking and learning. Cognitive profiles, however, are not set in stone. They can be changed.

The process begins with identifying weak skills through a cognitive assessment and then strengthening them. Many parents can easily identify their child's strengths and weaknesses. They search out tools and experiences to build on their strengths and further strengthen their weaknesses.

Design Prompt: Strengths and Weaknesses

How would a child with strengths in your topic area approach your project? How would a child with weaknesses in the same area approach it differently? What can you include to match the skill level and make it enjoyable for more people?

Perception

Perception consists of the complex interactions between external sensory stimuli and prior knowledge, goals, motivations and expectations. All manipulate our perceptions. As a result, interpretations vary drastically from person to person. When we take action, it is swayed by some task or goal in mind, and we want to take steps that bring us closer to achieving that goal or accomplishing that task.[26]

We take in something through the senses and then need to process what we take in order to derive meaning from it. When designing for children, we consider questions such as: What does a toddler actually perceive is happening in the story when watching TV? Our brains need to find meaningful patterns in our environment in order to make decisions about what to do and how to respond. Understanding how people take in stimuli and what kids comprehend at different developmental stages is essential when we are trying to communicate information, teach, entertain, educate or promote a response. It can influence anything from a logo design to how to interact with a product or app or which format is better to execute a story: 2D or 3D animation or live action.

The Building Blocks Necessary for Perception

- **Sensory Processing:** Accurate registration, interpretation, and response to sensory stimulation in the environment and the child's own body.
- **Attention:** The ability to focus on valuable visual information and filter out unimportant background information.
- **Discrimination:** The ability to determine differences or similarities in objects based on size, color, shape, etc.
- **Memory:** The ability to recall visual traits of a form or object.
- **Spatial Relationships:** Understanding the relationships of objects within the environment.
- **Sequential-Memory:** The ability to recall a sequence of objects in the correct order.
- **Figure-Ground:** The ability to locate something on a busy background.

- **Form Constancy:** The ability to know that a form or shape is the same, even if it has been made smaller/larger or has been turned around.
- **Closure:** The ability to recognize a form or object when part of the picture is missing.[27]

Visual-motor integration allows our eyes and hands to work together in a smooth, organized and efficient way. Visual-motor skills enable a child to coordinate their eyes and hands to learn to draw and write.

Visual and spatial processing is the ability to process incoming visual stimuli, to understand spatial relationship between objects, and to visualize images and scenarios.

Attention

Attention is the ability to sustain concentration on a particular object, action or thought, and the ability to manage competing demands in our environment. It can be just momentary, such as turning around to notice a poster on the wall, or it may be for a sustained period of time, such as focusing to build a complete model airplane. Most educators and psychologists agree that the ability to sustain attention on a task is necessary for achievement. There are several different types of attention that designers consider.

Sustained attention is the ability to focus on one specific task for an extended period without being distracted. Attention span is measured by sustained attention. Difficult or essential tasks require a great deal of attention. As designers, we consider sustained attention when we want to keep our user involved in an experience and develop approaches to keep them involved and focused. Understanding how long most children can sustain their attention at different developmental stages helps us to help them achieve their goals.

Selective attention is the ability to select from many factors or stimuli and to focus on only what is essential while filtering out other distractions. Designers think about selective attention when we want to draw people into our experience or want them to focus on one activity or goal in the midst of many within the experience. What makes us attend to one thing rather than another? Why do we sometimes switch our attention to something that was previously unattended, and how many things can we attend to at the same time? Understanding basic principles on where the eyes are quickly drawn to and how the brain filters information can offer inspiration for methods of selective attention. This knowledge is vital for designing, because many times our role is to get people's attention to communicate a message or seek input to create a response.

Alternating attention is the ability to switch focus back and forth between tasks or activities that require different cognitive demands. Sometimes, as designers, we are challenged with having our users alternate back and forth between tasks, or they may have many subgoals within a single task. More focus in one area means less attention elsewhere.

Divided attention (multi-tasking) is the ability to process two or more responses or react to two or more different demands simultaneously. At times, multi-tasking is necessary to complete a goal; for example, when we need to complete two or more tasks simultaneously, such as pressing a button and speaking at the same time. Although we say today's kids are good at multi-tasking, we are still in the process of proving it. Children and multi-tasking will be continually studied, and we will learn more about brains that have been wired to multi-task.[28]

Attention Spans of Children

If you have spent time with small children, you know that they jump from activity to activity very quickly. Older children are capable of more extended periods of attention than younger children. As designers, we often ask: "How long should the child be expected to pay attention at a specific age?" It depends on the content, their interest in a task, or how novel it is. It is harder to pay attention to uninteresting tasks, or when distracted. They are more likely to stay interested when they're comfortable with the task or project and feel successful.

This means that we, as designers, can create more exciting engagements for tedious tasks, so people will attend to them longer. Typical estimates for sustained attention to a freely chosen task range from about 3 minutes for a 2-year-old child, to a maximum of around 20 minutes in older children and adults. When focus is needed for more than 20 minutes, people "reattend" to the same activity.

For How Much Time Can Kids Focus?

3 years old: (3–8 minutes alone) on a single exciting activity.

3½ years old: (15 minutes alone) if there are a variety of exciting choices.

4–5 years old: (7–8 minutes alone) on a single activity, (10–15 minutes) when engaged, (2–3 minutes) on a task chosen by an adult, such as getting dressed or picking up toys.

5 years old: (10 or 15 minutes) Most children can ignore minor distractions. Without interruption for (10–25 minutes). A small group of children can work or play together.[29]

7+ years: (30 minutes including reattending) Mixing up inputs such as visual, auditory and kinesthetic will hold their attention longer.[30]

How to Get a Kid's Attention to Learn

- The brain recognizes patterns
- People are drawn to stories
- Things that are new get attention
- Danger gets attention
- Movement and motion get attention
- Faces attract and guide attention
- Food gets attention
- Provide opportunities for the unexpected
- Don't present too much information at once
- Don't provide too many choices at once

◄ Figure 4.8

Bubbles have the ability to get kids' focused attention for a very long time.
Source: Ana Dziengel (used with permission).

Behavioral Science and Designing for Kids and Teens

Interview with Susan Weinschenk
Susan uses research in brain science and psychology to predict, understand and explain what motivates people and how they behave, and applies it to design. Author of *100 Things Every Designer Needs to Know about People* and other books.

How do you see designing for kids as being different from designing for adults?
In every project, we look at the target audience, the context and the environment. When working on a project for kids, we need to go deeper.

It is easier to think about designing for adults because we are adults. When we become adults, we forget what is like to think like a kid or to be a kid. We perceive as adults. We cannot "see" as a kid. You have to dive into the target audience research with an open mind and assume you don't know what they think because you have never had this experience. You need to test ideas more thoroughly. There are more chances to get the mental models wrong.

This is highlighted when we are designing things that are brand new to today's world that weren't around when we were younger. The mental models of the world are changing and adults today don't have them. Even if you are 19 or 20 years old, you never had the experience of kids today like interacting with today's new technology. So you can only imagine it.

Can you highlight cognitive skills that differ between children of different ages?
Attention, perception, thinking, memory and motivation have more in common than one would think. These are rooted in biology. They are part of being human.

They are present at birth or close to birth. For example, food and danger grab attention.

Observing patterns, making decisions, problem-solving, reading, impulse control, self vs. other, categorization, rules and socialization are tied to what people are capable of understanding and doing at a particular age or developmental stage. It is essential for designers to know how humans change developmentally. You can't teach kids something until that part of the brain is turned on.

For example, babies don't have a sense of self vs. other. When kids are 7 to 10 years old, they like rules, especially games with rules. Everything is about following the rules. Younger kids question the rules and their purpose. They can't really grasp the concept of rules because the developmental stage for rules hasn't kicked in yet. Kids under 7 don't really understand categories very well. Once kids get into grade school socialization becomes a priority. In the tween and into the teen years it is all about their social network as a dominant force.

Do you have a list of topics designers should know about when creating design for a children's product or service?
- Know what is biological vs. learned.
- Hone in on the age group. Students have a tendency to group all kids together. I have heard many times students say I am designing for kids three to twelve. That is too large a target.
- Keep in mind that you are also designing for the parents and teachers, who might say, "Isn't it too simple, or too complicated." Having tested the

project in demos or trials and having proof in videos to share with parents and teachers will make it more convincing. You will have the evidence to support your decisions.

- Specifically in UX design don't blend the parents' (or teachers') experience with the child. A blended experience usually doesn't work for either. Design the adult experience and the children's experience separately. For example have the parents start and the product and do what they need to do. Then have them hand it off to the kids and have them do their part. Find the places where you need to go back and forth and then design the connectors. The parent has to start the app and set up the sound, level of difficulty and so on, and then can give the device to the child to use/play/watch. The parent may need to interact again to unlock features, download more videos, and so on.

How can designers use what we know about the brain and behavior to improve education and learning for children?

I have served on the local school board, and unfortunately, the research on learning doesn't get integrated efficiently into the school system through professional development programs. Some of the research that should be considered is:

- There is a limit to how much time people can pay attention, and then the glucose in the brain is used up. The 20-minute rule is suitable for adults. This is about how long they can focus and pay attention. It would be less for kids because their brains are not as big. Building in breaks and changing things up will strengthen learning.
- Repetition is essential for both children and adults, but for kids, it is vital. Kids love repetition while adults might get bored of it. Repetition strengthens neuron connection for long-term memory. Repetition also helps with muscle memory, both large and small muscles. For example by playing music over and over in a sequence, eventually, your fingers will play the song without thinking about it.
- Schemas are important. Think about all the random bits and how all the pieces go together to make one thing. For example, the face has eyes, eyebrows and eyelashes. Once you have enough pieces, the whole schema comes out. Schemas are different at different ages, and we need to consider those as well.

I learned from a mentor that there is fundamental science to how you teach different topics. The way you teach something depends on what you are teaching. Facts, principles and concepts all should be taught in different ways.

Facts: The best way to teach a fact is to repeat it over and over in different ways. For example, if you want to remember that the capital of New York is Albany repetition is the way to learn that.

Principles: To learn a principle you need to connect it to an experience personally. The brain connections that are made on that Ah-ha moment experienced first. For example, if you want to teach that designing for kids relies on taking developmental stages into consideration show a video or observe kids of different stages first. This allows the student to experience the principle. Then you can state the principle and discuss what it is the students experienced. And then lastly you can talk about where and when to apply it.

Concept: First you state the concept. For example: What is a mammal? Then you outline the rules of the concept. A mammal has a backbone, is warm-blooded, females produce milk for their young … etc. Then show examples. A cat is a mammal because … a horse is a mammal because … and a human is a mammal because …. Then you show what is not the concept and explain why. A snake is not a mammal because ….

Do you have suggestions for ways to use storytelling through characters, words, sounds and pictures to enhance education, learning and memory?

I am huge fan of stories. There are different brain chemicals that come into action at different points in a story.

- If you want to grab audience attention, produce conflict early on.
- Make us interested in the story by connecting it to something kids have personally experienced.
- You want to develop characters that can do things kids can do so they can put themselves in the place of the character.
- If the character is somehow like us – looks like us, is a child, they will care about the character.
- The characters and stories can be universal or can relate to a specific market. For example, when developing a game for kids with disabilities the characters and story could be universal to that audience.[31]

Higher Mental Processes

High-level cognitive skills include memory, creating, conceptualizing, categorizing, planning, logic and reasoning, and problem-solving. Research at the University of Chicago and the University of North Carolina at Chapel Hill shows that children begin to show signs of higher-level thinking skills as young as the age of 4½. "Overall, knowledge is necessary for using thinking skills, as shown by the importance of early vocabulary, but also inhibitory control and executive function skills are important contributors to children's analytical reasoning development," explained Lindsey Richland.[32]

Executive Functions

This includes abilities that enable goal-oriented behavior, such as the ability to plan and execute a goal. Some are cognitive abilities; others are an interplay between cognitive and social and emotional skills. This process of understanding the concept that all actions cause a response or have a consequence takes a long time to mature.

Flexibility: The ability to quickly switch to the appropriate mental mode.

Theory of mind: Insight into other people's inner world, their plans, their likes and dislikes.

Anticipation: Prediction based on pattern recognition.

Problem-solving: The process of defining the problem, then generating solutions and picking the correct one.

Decision-making: The ability to make decisions based on problem-solving, incomplete information and emotions (ours and others').

Working memory: The capacity to hold and manipulate information in real t ime.

Emotional self-regulation: The ability to identify and manage one's own emotions for positive performance.

Sequencing: The ability to break down complex actions into manageable units and prioritize them in the right order.

Inhibition: The ability to withstand distraction and internal urges.[33]

The average executive function skills in the children we are designing for are vital for designers to understand because they play such an essential role in the ability of children to understand and learn from the information we are trying to communicate or activities we are engaging them in.

Design Considerations for Executive Function Skills

What level of self-motivation does the child possess on getting started with a task and goal setting?

How aware are they of the process involved in your activity?

Does the child have the ability to self-organize a sequence of steps in order or realize that there are subgoals in a task?

Do they match a strategy to a problem?
Can they reflect on past experiences to plan for the future?
Do they see the "big picture" of a task or situation?
How high is their tolerance for failure or frustration?
Can they quickly shift perspectives?
Do they live in the current moment?
Can they think about the future?

Memory

Memory is a fundamental capacity that plays a vital role in social, emotional and cognitive functioning. Our memories form the basis for our sense of self, guide our thoughts and decisions, influence our emotional reactions and allow us to learn. Knowing how memory works can help designers create solutions that correspond to the natural abilities of the users to save their effort and boost usability.

There are three networks for memories in the brain:

1. Encoding (sensory memory): The encoding system takes in the data for a short moment when we perceive it with our physical senses, such as hearing, vision or touch, and holds it in short-term storage. How well we attend to the new information affects how well we remember it.

2. General knowledge, long-term memory: This includes two main kinds of memories:

    Semantic memory: General fact memory such as who we are, or that one must say "please" before asking for a snack. Long-term memory holds facts such as school-related learning, so it can be recalled when asked.

    Episodic memory: This type of memory stores personal events and feelings associated with those events. An example of this kind of memory is your last trip to the supermarket. Sometime between the ages of 2 and 4, children develop memory regarding the details of specific events.

3. Recall: The recall system taps our executive function skills to "find" information and pull it out efficiently. This includes organizing what we know about a topic, checking and forming associations, and holding and comparing information.

Recalling a fact, event or object that is not currently physically present (a mental image or concept) requires the direct uncovering of information from memory.

Recognition is the association of an event or physical object with one previously experienced or encountered and involves a process of comparison of information with memory; for example, recognizing a known face. It is known to be superior to recall.[34]

The relationship between working memory and long-term memory is significant for designers to understand. For example, knowing how long an individual can hold information in their brain and use it to promote an action can help designers convey the appropriate timing in interaction design. Is it best to have the user recall from memory, or would recognition from a prompt be better? School-age children are better at remembering than younger children. Experiencing more of the world, they pull from previous experience when encoding and recalling information. To remember, they use mnemonic devices or memory strategies. Creating humorous lyrics, devising acronyms, chunking facts (breaking long lists of items into groups of threes and fours) and rehearsing facts (repeating them many times) help children memorize increasingly complicated amounts and types of information.[35]

Childhood Amnesia

Most adults can't remember much before the age of 3 and remember very little of life before the age of 7. But memory matters a lot during early childhood to build on for further development. We retain information from infancy, so why is it that we cannot remember it? Psychologist Nora Newcombe of Temple University in Philadelphia suggests:

> It may be because that's when the hippocampus starts tying fragments of information together and episodic memory may be unnecessarily complex at a time when a child is just learning how the world works. The primary goal of the first two years is to acquire semantic knowledge, and from that point-of-view, episodic memory might actually be a distraction.[36]

Children and teenagers have earlier memories than adults do. This suggests that the problem may be less with forming memories than with maintaining them. Language also plays a role. To some extent, a child's ability to verbalize about an event at the time that it happened predicts how well they remember it months or years later. When parents reminisce with very young children about past events, they implicitly use and teach them narrative skills – what kinds of events are essential to remember and how to structure talking about them in a way that others can understand. Reminiscing has different social functions in different cultures, which contribute to cultural variations in the quantity, quality and timing of early autobiographical memories. Adults in cultures that value autonomy (North America, Western Europe) tend to report earlier and more childhood memories than adults in cultures that value relatedness (Asia,

◀ Figure 4.9

At 3 years old she will be able to tell you all about her day at the park shortly afterwards, but in coming years she will remember very little about it, as the memory fades away.
Source: author photo.

Africa). Even when children can't remember specific events from when they were very young, their accumulation has a lasting effect on development.[37]

Autobiographical Memory

Autobiographical memory is our memory of our own life experiences. It enables us to learn from the past and to make plans for the future. Children collect stories about their life and use them to develop a personal identity and a sense of continuity and direction in life.

Working Memory/Short-term Memory

One of the brain's executive functions, working memory holds on to new information so we can turn around and use it immediately as even more information is arriving and needs to be incorporated. It also helps the brain organize new information for long-term storage. Working memory plays a vital role in concentration and following instructions and affects many aspects of learning.

There are two types of working memory. Auditory memory records what you're hearing, while visual-spatial memory captures what you're seeing. Working memory is responsible for many of the skills children use to learn to read. Auditory working memory helps kids hold on to the sounds letters make long enough to sound out new words. Visual working memory helps kids remember what those words look like so they can recognize them throughout the rest of a sentence.[38]

Design Considerations for Memory Skills

- Most children won't remember much before the age of 3 years old, although the closer they are to that age, the more they will probably remember.
- Consider the speed of thinking and time to react in the design.
- Use narrative and storytelling to help with memory.
- Should the user recall from memory, or would recognition be better?
- Recent events are more easily remembered in order (especially with auditory stimuli).
- Recall decreases as the length of the list or sequence of topics to remember increases.
- There is a tendency to remember the correct items but in the wrong order.
- Where errors are made, there is a tendency to respond with an item that resembles the original item in some way.
- Repetition errors are relatively rare.

Tips for Designers and Helping Kids Remember

- Concentration. To remember something, a person needs to attend to it. Otherwise, it will never move from short-term memory to long-term memory.
- Designing personal associations with memories (such as a particular piece of music playing at an vital life event) can bring people back to the details of that event at a later time. Building an association with something they already know increases the chances of memorizing better.

- Repetition enables people to transfer information from short-term memory to long-term memory by activating the memory multiple times. Kids' tolerance for repetition is greater than adults'.
- Use visual representations. Visual memory, system diagrams and icons are better than words. Designing a visual system for classifying facts may make them easier to remember later on.
- Learn cooperatively. Have the child recall, speak, draw, engage or teach others.
- Include memory games.
- Use mnemonic devices. Recognizing is more natural than remembering. Classify easy-to-remember categories or number the sequence.
- Connect emotion with the information.
- Use simple concrete language.
- Use multiple sensory inputs for different types of learners: Link visual and auditory content, smell, taste and touch.
- Stimulate different types of memory.
- Break information into small chunks.
- Save effort for the user with recognizable patterns and symbols.
- Provide consistency.

Categorizing

Grouping things together to create patterns is a mental skill that saves time, is practical and keeps us safe. Classification is a primary process that children can use to develop logical and mathematical reasoning abilities. When children classify, they are using information about what is the same and what is different. At first, children classify items based on how they look, sound and feel. Later, children learn that some things belong together because of their purpose; what they do or how they are used. While most young children can differentiate the attributes of objects, they need opportunities to learn how to use and complexify this capability.[39]

◄ Figure 4.10

This open-ended light exhibit can be used to practice classification, patterns, math, problem-solving and creativity. Source: author photo.

Miller's law, based on his 1956 research *The Magical Number Seven, Plus or Minus Two*, states that the number of objects an average person can hold in working memory is about seven plus or minus two, depending on the type of information. Later research by Richard Shiffrin and Robert Nosofsky, *Seven Plus or Minus Two: A Commentary on Capacity Limitations*, gives an average seven for digits, six for letters and about five for words.[40] If designers divide the tasks or information into appropriate categories and sized chunks, it will be easier for children to use and remember.

Problem-solving

Almost every situation starts with a problem and needs a well thought-out solution to solve it. We look at the current state of something to arrive at the future desired state.

Problem-solving includes remembering, hypothesizing and decision-making. Children learn about the world through problem-solving. They overcome obstacles by generating hypotheses, testing those predictions and arriving at satisfactory solutions. We all have different approaches that we use to get to that solution.

The process they use to get to the solution is most of the time more important than the solution, since the problem may change, but the strategy may remain. Kids use problem-solving in learning and interpersonal relationships. In order to grow and become independent, kids are continually encouraged to solve their own problems. When designing for problem-solving, it is helpful to look at how children and youth solve problems at different ages and the steps they take, and to come up with solutions that best support them.

The process of problem-solving includes

- Problem recognition: Defining the problem
- Problem labeling: Identifying the central issue that needs resolution
- Problem cause analysis: Determining the cause of the problem
- Optimal solutions: Generating alternative solutions
- Decision-making: Identifying, prioritizing and selecting alternatives for a solution
- Action planning: Implementing and follow-up on the solution
- If the problem is solved – fantastic. If not, define alternate solutions.[41]

Mathematics: Numerical Magnitudes

In the 1950s, W. W. Sawyer described mathematics as the "classification and study of all possible patterns." Other mathematicians who share Sawyer's view have shortened the definition even further: Mathematics is the science of patterns.

Principles and Standards for School Mathematics outlines five process standards that are essential for developing a deep understanding of mathematics: problem-solving, reasoning and proof, communication, connections and representation.[42]

Babies

Begin to predict cause and effect, sequence of events
Classify things in simple ways
Start to understand size and relationships
Begin to understand words that describe quantities

Toddlers

Understand math language such as how many, and how objects relate
Begin saying numbers, see repeated patterns in daily life
Explore measurement – filling containers
Match shapes

Preschoolers

Recognize shapes
Organize by color, shape, size and purpose
Compare and contrast using classifications such as height, size and gender
Count to 20 and count items in groups
Understand that numerals stand for number names
Use spatial awareness to put puzzles together
Start predicting cause and effect

Kindergarteners

Add by counting fingers on one hand and starting with six on second hand
Identify the larger of two numbers and recognize numbers up to 20
Copy and draw symmetrical shapes
Can use basic maps to find hidden treasures
Understand concepts of time – morning or days of the week
Follow multi-step direction words like first and next
Understand words like unlikely or possible

1st and 2nd Grade

Predict what comes next in a pattern and create own pattern
Know the difference between two- and three-dimensional shapes – cubes, cones, cylinders
Count to 100 by ones, twos, fives and tens
Write and recognize the numerals 0–100 and the words from one to twenty
Do basic addition and subtraction up to 20
Read and create a bar graph
Recognize and know the value of coins

3rd Grade

Write math problems with pencil and paper
Work with money
Do addition and subtraction with borrowing
Understand place value well enough to solve problems with decimal points
Know how to do multiplication and division with help from fact families
Create a number sentence or equation from a word problem

4th and 5th Grade

Start applying math concepts to the real world (cooking)
Practice using more than one way to solve problems
Put different types of numbers in order on a number line
Compare numbers using greater than and less than signs
Start two- and three-digit multiplication
Complete long division with or without remainders
Estimate and round

Middle School

Begin basic algebra with one unknown number
Use coordinates to locate points on a grid, also known as "graphing"
Work with fractions, percentages and proportions
Work with lines, angles, types of triangles and other basic geometry
Use formulas to solve complicated problems and find the area, perimeter and
    volume of shapes

High School

Understand that numbers can be represented in many ways (fractions,
    decimals, bases and variables)
Use numbers in real-life situations
Begin to see how mathematical ideas build on one another
Begin to see that some math problems don't have real-world solutions
Use precise language to convey thought and solutions
Use graphs, maps and other representations to convey information[43]

The Scientific Method

Designers are familiar with the iterative process of testing hypotheses and discovering a solution. Scientists use a similar process to do their experiments. The scientific method involves backing up and repeating steps over and over to test a hypothesis; to discover cause and effect relationships by asking questions, carefully gathering and examining the evidence, and seeing if all the available information can be combined into a logical answer. Young children naturally explore the world as scientists; grade school children begin to formally learn this process. The experiments and processes complexify as they build on learning.

The scientific method for kids:

- **Make observations:** Ask a question or questions about something that you observe: How, What, When, Who, Which, Why or Where? Possibly something you can measure, preferably with a number.
- **Do background research:** Explore the current body of work in the field and see what other people have done, what we know, what mistakes have been made and build on that.
- **Construct a hypothesis:** It is an educated guess to attempt to answer your question with an explanation that can be tested. A reasonable hypothesis allows you to make a prediction that can be tested.
- **Test the hypothesis by doing an experiment:** Is a prediction accurate? Is your hypothesis supported or not?
- **Analyze the data and draw a conclusion:** Repeat experiments several times to make sure that the first results are accurate, or retest by changing variables.
- **Communicate your results:** Share the results with others in a final report or presentation to learn and get feedback.[44]

Reasoning

This is the ability to consciously make sense of things by applying logic systematically, thinking through problems and applying strategies for solving them. Reasoning is associated with thinking about cause and effect, truth and falsehood, and good or bad, among other things. Children's ability to think, reason and use information allows them to acquire knowledge, understand the world around them and draw conclusions. Persuasion of others based on a logical argument is essential for competence and success in school, in friendships and in the home environment. Many theories of development point to age 6 as the time when children begin to actually "reason".[45]

There are many kinds of reasoning, deductive reasoning based on certainty and inductive reasoning based on probability being the most common.

Inductive reasoning is when you look at a collection of facts and circumstances and then make an educated guess or induce what is likely to happen. It moves from the specific to the general. For example, when you are reading a story to a child, they are putting together the pieces in their head to reason through what might make a logical solution.

Deductive reasoning is when you look at an end result and then figure out – or deduce – how it came about. Deductive reasoning is the process of reasoning from the general to the specific. For example, since all squares are rectangles, and all rectangles have four sides, all squares have four sides.[46]

Play automatically induces reasoning. It leads children to think about pretend worlds, where anything is possible, and to reason about those possibilities rather than to limit our thoughts just to things that are true in the immediate here and now. According to evolutionary psychologist Peter Gray,

It is through natural play that children practice reasoning. Play naturally leads us to think of things as they *might be* rather than just as they currently *are*. In the playful state of mind, it is easy for anyone to imagine and think about a world in which people can fly, in which time machines can transport us to the past, or in which all cats bark. Young children are masters of play, so it is no surprise that they can solve counterfactual syllogisms in the context of play.[47]

Kids reason differently than adults. A 5-year-old may assume we can do something impossible in a completely charming way. A 9-year-old may reason that wizards are real. A 16-year-old might reason that extremely risky behavior is OK if all of their friends are doing the same thing. Considering that children can't think the way we do, it is up to us to think the way they do. How children reason comes into play when designers are thinking about how to structure the design, contextualization, problem solution, and the software and interface design. For example, they may consider how to organize design activities to achieve the user experience or performance requirements. How do we make it make sense to the user? What is the easiest way for them to synthesize information? What kind of logic will they typically follow? Atypically?

Planning and Reflection

With the products and tools we make, designers can help children plan consciously, coordinate actions, evaluate their progress, and modify their plans and strategies based on reflection and evaluation. Many little kids live in the moment and don't think far into the future. As they grow to about 6–7, they learn to plan (a thinking skill), which helps them develop strategies to accomplish goals. For example, when a child gets a crazy idea for a project, they first decide on the materials they will need, gather and arrange the supplies, and then take a step-by-step process to completing the project. They can plan what they want to pack for lunch or save money for something special. Alongside their increasing ability to reflect on themselves, children also develop the ability to take in the perspective of others.[48] Reflecting with others in small or larger groups can help them articulate and hear what they are thinking. If we are creating a product, service or experience for reflection, consider ways to bring the thoughts to the forefront, so they are obvious to the user.

Metacognition

Metacognition is thinking about one's thinking. It refers to the processes used to plan, monitor and assess one's understanding and performance.

Metacognitive practices increase children's abilities to transfer or adapt their learning to new contexts and tasks. They become aware of their strengths and weaknesses and figure out how to expand that knowledge or extend the ability.

Preassessments: Examining their current thinking and what they already know about a topic.

Identifying Confusions: Acknowledgment of what they don't understand and is confusing.

Retrospective Post-assessments: Recognize conceptual change in thinking, such as how it changes (or does not change) over time.

Reflective Thinking: Providing a forum in which children monitor their own thinking through writing or discussion.[49]

Creativity

Kids seem to have no limits to their creativity, because they seem to have a novel fresh approach in everything they do, from acting out fantasy adventures and using loose parts to making their own playthings. Whenever I bring a group of designers to play with kids, they agree that kids are inherently creative and a good reminder of how many of us, even in creative fields, lose this ability as we get older and conform to more expected solutions. Alison Gopnik, in *The Philosophical Baby*, asserts that babies are born experimental scientists who take in scrolls of information by trying things on their own and tweaking as they go. They create in their own unique ways before they learn the intended way.[50]

Creative intelligence is often thought of as a trait that is hard-wired in us from birth – we are either left- or right brained. Instead, I prefer to think of it as a thought process. Some of us have a predisposition to become more creative than others; I am a believer that creativity can be taught but requires a partnership between genetic inheritance and responsive adults willing to nurture and stimulate. In their daily life, children are learning to compromise, negotiate, come up with solutions and think of ideas. They are practicing the critical cognitive skill known as divergent thinking. They are learning to envision multiple solutions, real ones and imagined, and choose the best outcomes. Risk-taking and the frustration of getting stuck and pushing through to uncover a new solution helps kids become innovative.

Resourcefulness, and the ability to meet challenges in a variety of ways, is a by-product of creative intelligence. Creative people are more flexible and better problem-solvers, which makes them more able to adapt to technological advances and deal with change as well as take advantage of new opportunities. If we start nurturing children's natural creative thinking abilities when they are young and encourage them to develop them further as they grow, the world will have more innovative thinkers, which can be applied to all areas from social impact and politics to the sciences.

Divergent and convergent thinking describe two stages of the creative thinking process.

Divergent thinking involves the creation of a diversity of new ideas; it is a method or thought process used to experiment and explore many new possible solutions. It involves the breaking up of old ideas, making new connections and enlarging the limits of knowledge. Divergent playful activities instill the idea that there can be numerous creative solutions to a problem. Creative play allows children to develop divergent thinking: the ability to see things not for what they are but for what they could be. Developing divergent thinking skills and imagination through pretend play, constructive play or creative play can strongly influence creativity.

Convergent thinking is the opposite process, whereby we start with a variety of previously generated ideas and try to filter, integrate, condense or select a subset of these as the finished product. Convergent problems require children to organize pieces of disparate information to arrive at one correct answer. An example of this is when a child decides to use the broom handle to reach under the couch to get a ball.

Both stages are important for the generation of new ideas. The divergent stage makes sure that we have explored all the possible options, and the convergent stage ensures that we have chosen from those options the most appropriate solutions given the context.[51]

The Arts

As a child, did you like to imagine characters and draw and write stories about them? Did you enjoy acting in a school play? Were you a member of a school band? Many of us who are now professionally engaged in an artistic field were encouraged throughout childhood to find and follow our creative interests. As adults, this participation affects the manner in which we live on a daily basis. For children, the arts help them to express themselves and make sense of the world. Dr. Elliot Eisner, a professor of education at Stanford University: "The arts teach children that problems can have more than one solution and that questions can have more than one answer. The arts celebrate diversity and multiple perspectives. There are many ways to see the world."[52]

The arts produce a broad spectrum of benefits:

- Problem-solving, perseverance, focus
- Fine and gross motor skills
- Creativity: Enhancing creativity in all areas
- Confidence: Improved emotional balance and emotional expression
- Promoting self-esteem, dignity and motivation
- Non-verbal communication, aesthetic awareness
- Cultural exposure, understanding history and culture
- Visual literacy: Learn to appreciate beauty, perception and the level of effort of the artist
- Collaboration, a more prepared citizen, and appreciation of diversity
- Fostering a love of learning, raising student attendance, reducing student dropout
- Critical thinking: Receiving constructive feedback
- Appreciation of creative expression or inventiveness of an idea

Creativity, Imagination and Innovation

Imagination is about seeing the impossible or unreal. Creativity is using imagination to unleash the potential of existing ideas in order to create new and valuable ones. It requires our focus to be on things that *might* be possible, but we can't be sure until we explore them further. Innovation is taking existing, reliable systems and ideas and improving them.[53]

Play and Creativity

All forms of play can be expressive and creative. Think about the creative potential a cardboard box holds. It could be a spaceship, a robot, a house or a car, depending on the day, mood or interests. Creative play is when one plays with imagination to transcend what is known in the current state, to create something new.

The experiences children have during their first years of life can significantly enhance or squash the development of their creativity.

- Creative experiences can help children express and cope with their feelings. A child's creative activity can help teachers to learn more about what the child may be thinking or feeling.
- Creativity fosters mental growth in children by providing opportunities for trying out new ideas, and new ways of thinking and problem-solving.
- Creative activities are a process of self-expression that helps acknowledge and celebrate children's uniqueness.

A NASA study was conducted to better understand the ability to come up with new, different and innovative ideas to problems. They found that 98% of kids aged 4 to 5 were "creative geniuses" and demonstrated spontaneous creativity. This percentage sharply dropped to 2% for adults. Those aged 6 and upwards showed evidence of the creativity-stifling effect of education whereby grown-up expectations are placed on students' creations, limiting the creativity by imposing an expectation.[54] Many researchers believe that we (some designers and content creators) have fundamentally changed the experience of childhood in a way that impairs creative development. Toy and entertainment companies are thought to feed kids an endless stream of prefab characters, images, props and plot-lines that allow children to put their imaginations to rest. Children no longer need to imagine a stick is a sword in a game or story they've imagined: they can play Star Wars with a specific light-saber in costumes designed for the specific role they are playing.[55] Designers can help encourage the development of creativity rather than stifle it.

We can encourage kids to:

Understand creativity: What is it? How is it different from imagination, invention or talent? What happens in our brains when we are creative?

Improve creativity: Encourage open-ended projects and process over final product. Encourage curiosity and seeking answers rather than them giving the right answer.

Apply creativity: Have kids demonstrate their creativity in projects that support divergent thinking and out-of-the-box solutions.[56]

How to promote creativity in our designs

- Rely on intrinsic motivation, not external rewards.
- Emphasize process rather than results.
- Provide materials and space to explore.
- Create room for open-ended solutions.
- Eliminate step-by-step instructions and focus on process.
- Give children space to explore and fail.
- Design forms and functions with ambiguity.
- Avoid using something in only one way. Model kits and single-use toys rob kids of chances to think on their own.
- Activate their senses. Take them to a place where they have never been.
- Allow them to brainstorm ideas without critique.
- Value varying ideas and opinions. Allow them to disagree.
- Remove external constraints – showing them how do something can reduce flexibility in thinking.
- Encourage exploration. Show respect for the creative effort.
- Model creativity in the home, at school and throughout the day.
- Have them document and display their ideas – keep a maker journal.[57]

▶ Figure 4.11

Emphasize the creative process instead of the final results.
Source: Ana Dziengel (used with permission).

Timeline of Artistic Development

Early Childhood

Preschoolers explore self-expression and the relationship between themselves and the world around them. They will begin working in small group activities, fostering their cooperative skills and communication abilities, and participate in art making, theater, music or dance instruction.

Music and Movement: Immerse babies and toddlers in a musical environment with simple movement activities such as musical games, dancing while holding the baby, or singing or playing an instrument for the child. Preschoolers learn with their whole body. Encourage improvisational movements and learning basic dance, including repetition of simple concepts. By 3 years old, children can further listening skills by identifying a beat, melody and instruments. Encourage manipulating instruments, learning repetitive songs or developing their own.

Visual Art: Have preschoolers creatively manipulate diverse materials for drawing, painting and sculpture. Use their whole body and move into tools that kids can easily operate with limited motor ability. They are very social and talkative, and arts experiences can enhance collaboration. Talking about their work can support reflection and metacognition.

Theater and storytelling: Drama, theater and story-acting allow children to create fictional spaces where they can role-play and practice communication, social, and emotion regulation skills.[58] Structured call-and-response activities can be balanced with less structured fantasy play. Many kids this age love putting on open-ended improvisational performances such as plays or puppet shows.

Elementary School

Many young children live in the moment and create just to create. They view the arts as entertainment – it is something to do that is fun. Unfortunately, as kids get older, they may gain or lose self-confidence based on their artistic development in relation to their peers.

Music and Movement: Dance instruction introduces more sophisticated techniques, involving diagonals, curves, twists and asymmetry. Kids may participate in a specific style of dance class such as ballet, tap, improv or hip-hop. More girls than boys pursue dance as they get older. By five, most children have built a foundation that has prepared them for music instruction to further their understanding of music. Piano, drums and violin are common instruments early on. By age 10, the child will have a variety of skills associated with their instrument of choice. They'll also have the physical strength to learn bigger instruments, such as brass or large string instruments, that require strength and stamina. The goal of lessons evolves from gaining experience with music to improving performance ability.

Visual Art: Fine motor abilities have greatly improved, and elementary kids may use a broader range of visual art tools, including finer pencils and paintbrushes, which make use of their fully developing fine motor coordination. Their creativity, divergent thinking and storytelling abilities are enhanced.

Theater: At this age, kids advance in language acquisition primarily through memorization and learn meanings behind language intonations. This is a good time to focus on memorizing lines, developing observation skills for little details and body movement sequences.

Middle School

At this age, kids start to specialize; they self-identify as an artist or someone who is not creative. Many kids stop engaging in the arts if they are not interested or are not exposed to it though educational programs.

Music and Movement: They gain experience with playing instruments and improve performance ability and creativity. They may take classes in music appreciation. If they enjoy music, they spend their free time listening to music. Singing voices are developing in girls and changing in boys, and they are challenged to cater to those changes. Dance is usually participated in as an extracurricular activity, since it tends to be awkward in a formal context for this age of boys and girls to interact physically.

Visual Art: Kids develop increasingly sophisticated creative strategies, skills, procedural knowledge, and craftsmanship in art making. They may apply design literacy while exploring an expanded range of media, including traditional and new media. In art making and appreciation, they can consider aesthetic judgment about the meaning and purpose of art and diverse historical and cultural heritage.

Theater: Theater production is a vehicle for learning organizational skills and logical thinking, from designing and constructing sets, to rehearsing scenes and remembering stage blocking. Group discussions offer an opportunity to verbalize different ways of portraying various roles and explore abstract thought. They can explore beyond the plot and become aware of the subtext and motivations of the characters. Through theater appreciation, they become aware that human nature does not change and begin to see the similarities that all people share.

High School

The individuality, originality and emotional expressiveness of the creator are of great interest to adolescents as a vehicle of expression. Art provides opportunities to perceive, analyze, interpret and evaluate human experience. Design thinking and creative problem-solving also appeal to their advanced cognitive skills. Those who are considering artistic fields as a career focus on preparing and packaging work to apply to college or specialized programs.

Music and Movement: Music appreciation includes learning how to listen and understanding different genres of music and studying music theory. In instrumental music, they may perform individually or in school marching bands and orchestras. Vocal music students participate in individual and choir training. Dancers condition and use their bodies for self-expression through selected genres.

Visual Art: Experiencing art production and studying art history enhance both making and critical thinking skills. Concepts of realism and abstraction can be further explored at this age, though they generally value representation over abstraction. Art is a method for experimenting with self-exploration, individuality, adult issues, media, materials and techniques. They can more clearly articulate their objectives and process.

Theater: At this age, theater becomes a means of establishing their identities and expressing their knowledge, opinions and insights through writing and speaking. Voice enhances all verbal communication and increases volume and enunciation. They acquire an appreciation of plays written for adults, musical theater and Shakespeare based on their expanded appreciation of their cultural significance. They can bring their storytelling and writing skills into playwriting.

Creativity, a Family Project

Interview with Ana Dziengel, developer of the creativity blog Babbledabbledo.com

Ana Dziengel has a BArch in architecture and an MS in industrial design. She has worked as an architect and product designer for many years. Her blog focuses on informal learning and connects families together through creativity.

What is your take on learning tactile skills and digital skills?

Kids gravitate towards learning methods based on their personalities.

I have two kids that are opposites. My son loves games, is creative and draws. He does school assignments in 3D modeling programs and has play dates via Skype. His personality is drawn to the digital world. My daughter is more interested in tactile activities and always has a hands-on project in the works. As a family, we have no device days to limit screen-time, but we try to nurture the kids and their interests.

How do the projects you develop and promote help kids in their social and emotional development?

Doing creative things together nurtures family relationships. It is helpful for family bonding, with kids getting focused attention from their parents. This gives them the tools to explore and boosts their confidence. Also while they are making, they are thinking and developing their motor skills too.

Your blog's positioning says that you inspire curiosity and the creative spirit. How does that manifest in editorial content?

My design experience influences the nature of the projects. I choose projects that are open-ended, process oriented and child led. When parents take over a project, it kills the creative spirit. It sends a message that the adult doesn't value the child's ideas.

The visual aesthetic of your blog is very strong. How did you go about developing this in relation to your online branding?

As a designer, I realize that the better you present the content, the more enthusiastic the response will be. The projects are user-friendly. Twenty steps to complete a project are too intimidating, so I make them simple. I use materials that are accessible. I want to set up folks for success. I also realize that if you make a project too good it will be intimidating. When making a project I have kids do one, and I do one to get an idea of different ways to complete it. I shy away from complicated and enjoy the beauty of mess.

How does your content differ across different age groups of kids?

I try to service a wide swath of age groups and love projects that can be enjoyed by the whole family. It is fun when preschoolers and older kids work together. I've also noticed that age groups are not so cut and dry; some 4-year-olds can do the same thing as an 8-year-old. I do however make distinctions when safety is an issue, such as projects that use magnets or electronic components.

Can you explain from your point-of-view the importance of informal learning within the family?
When parents are doing projects with kids, they are modeling creative behavior. Many kids don't have the opportunity to try projects like these at school. By doing this with them, they are guaranteed to have this experience.

How did you go about building your blog's audience?
You need to build a core community who shares your point-of-view. Solve their problem. Inspire them. Tap into what they are feeling. What tools do they need to succeed? I've learned that my blog's audience wants to be creative with their kids and projects are the means to achieve that goal. It's also important to nurture relationships with other bloggers online, to have a network of like-minded people. They will help you, and you help them.[59]

▶ Figure 4.12

When a project is kid led, they become aware of what they can do.
Source: Ana Dziengel (used with permission).

Language Development and Reading

Children do not need to be taught language communication. Toddlers pick up words and learn to talk by stringing them together through the back and forth of language play and social interaction. Without the social interaction component, it is difficult to acquire language through other methods. The course of language development is very similar across children and across languages, suggesting a universal biological basis to this human capacity. In general, understanding language precedes speaking precedes reading. Children's understanding of language is important for designers to consider when choosing methods of communication (visual or auditory), choosing vocabulary to communicate an idea or action, and thinking about the complexity of the relationship between words and image when designing for a reader and a non-reader.

Phonological Development

Phonemic awareness refers to the ability to identify, compare and manipulate the smallest units of spoken words, phonemes.[60] During the first year, children are more sensitive to phonemes in their native language and are less sensitive to acoustic differences not relevant to their language. This tuning of speech is the result of a learning process in which infants form mental speech sound categories around clusters of frequently occurring acoustic signals.[61] Caregivers are highly encouraged to speak to and read to babies often, so they hear the sounds of the words. Children often make sounds and play with sounds by themselves in a playful way to practice. Phonemic awareness and vocabulary skills are the best predictors of reading comprehension.

Children's first sentences are combinations of content words and are often missing grammatical function words (e.g., articles and prepositions) and word endings (e.g., plural and tense markers). As children gradually master the grammar of their language, they become able to produce increasingly long and grammatically complete utterances. They often ask questions in a playful way, not necessarily to get an answer but to practice a back-and-forth call-and-response pattern. Children's language play can help the child consolidate his or her growing understanding of linguistic sounds, words, grammar and meanings.

0–3 to 4 months: Coos, cries, smiles.

4–6 months: Babbles are more speech-like, include many sounds such as /p/, /b/, /m/.

7 months to 1 year Imitates different speech sounds, longer groups of sounds, begins saying words such as bye-bye, mama, dada – increased children's brain response to their native language.

1–2 years: Uses more words each month, by 21 months likely to know about 100 words, puts two words together into phrases, asks questions like "Where Doggie?"

24 months of age: Begins to put two, then three and more words together into short sentences.

2–3 years: Has words for almost everything, uses two to three words together, is more easily understood.

3–4 years: Says sentences with four or more words, talks about activities and/or people. The development of complex multi-clause sentences.

4–6 years: Uses clear voice, detailed sentences, sticks to a topic, uses appropriate grammar, says most sounds correctly. Estimated at 14,000 words. Grammatically complete and fully intelligible sentences.[62]

7–11 years: Uses language to predict and draw conclusions. Uses long and complex sentences. Understands other points of view and shows that they agree or disagree. Understands comparative words. Keeps a conversation going by giving reasons and explaining choices. Starts conversations with adults and children they don't know. Understands and uses passive sentences.

11–14 years: Uses longer sentences; usually 7–12 words or more. Knows how to use sarcasm and when others are being sarcastic to them, and more subtle and witty humor. Able to change topic well in conversations. Shows some understanding of idioms. Talks differently to friends (slang) than to adults.

14–17 years old: Can follow complicated instructions. When they don't understand, will ask to be told again. Easily swaps between "classroom" talk and "breaktime" talk. Tells long and very complicated stories.[63]

Preliterate children see a jumble of letters.
Source: author photo.

Baby Sign Language

Baby sign language – a modified version of American Sign Language – can be an effective communication tool and bonding opportunity. Baby sign language might give a child the ability to communicate several months earlier than those who only use vocal communication.

This might help ease frustration between ages 8 months and 2 years when children begin to know what they want, need and feel but don't necessarily have the verbal skills to express themselves.[64]

Words Kids Learn First

Most children learn common nouns (people, places and things) first, expanding their vocabulary to verbs (action words), prepositions (locations), adjectives/adverbs (descriptions) and pronouns (possessive) This gives them a breadth of words to start to form phrases. Children typically begin to speak in phrases when their vocabularies are close to 50 words (about 18 months).[65]

Reading

While kids are naturally wired to speak, reading and writing are acquired skills for which the human brain is not yet fully evolved.[66] Oral language is a foundational skill for reading and writing. When taught, children typically learn to read at about age 5 or 6 and need several years to improve their skills. Sophisticated reading comprehension takes an additional 8 to 16 more years of schooling.[67]

Reading is the product of decoding and comprehension. To learn to decode and read, children must be aware that spoken words are composed of individual sound parts (phonemes). When children acquire the alphabetic principle (spellings systematically represent spoken sounds) and detect the seams in speech, unglue the sounds from one another, and learn which sounds (phonemes) go with which letters, they can learn to read. Good comprehenders link the ideas presented in print to their own experiences and can summarize, predict and clarify what they have read, and many are adept at asking themselves questions to enhance understanding. They apply these skills in a rapid and fluent manner, possess strong vocabularies and syntactical and grammatical skills, and relate reading to their own experiences. Language, vocabulary and reading skills are enhanced through adult–child interaction.

Timeline for Reading Skills

0–36 months

Begin to reach, look at, touch and turn pages of books
Make cooing and sounds to respond to familiar stories
Adult reading to child supports language and reading development
Name familiar pictures such as dog, cup, baby
Answer questions about what they see
Pretend to read by turning the pages and making up stories (24+)
Recite words and recognize cover of favorite book

3–4 years

Know the correct way to hold and handle book
Recognize that the words tell a story
Understand that words are read from right to left and top to bottom
Start hearing rhyming words
Retell stories
Recognize about ½ of the letters in the alphabet
Recognize and "read" familiar labels, signs and logos
Start matching letter sounds to letters
Can recognize own name and other often seen words

5 years

Read some sight words
Use story language and vocabulary in play and conversations: "the dump truck is here," said the man
Begin matching words they hear with words on the page
Recognize and match letters to letter sounds
Identify the beginning and ending and sometimes middle sounds/letters in words like cat or sit
Sound out simple words
Tell the who, what, when, where, why and how of a story
Put a story in order either by retelling or by pictures
Predict what happens next in a story
Begin writing or dictating their own stories

Sound out new words using phonics and word families
Start reading or ask to be read books for information as well as entertainment
Answer basic questions about what they read

6–8 years

Recognize up to 200 sight words
Use context clues to decode unfamiliar words
Go back and reread when a mistake is made
Start answering questions to think about what they've read
Start writing stories using inventive spelling
Imitate the style of favorite authors' writing

9–13 years

Move from learning to read to reading to learn
Read with purpose
Explore different genres
Recognize words without hesitation
Put the events in a story in order
Read out loud accurately and with inflection
Identify and articulate the main idea
Summarize what has been read
Understand similes, metaphors and other descriptive devices
Find meaning in what has been read

14–17 years

Relate events in the story to their own lives
Compare and contrast different reading materials
Discuss character motivation
Make inferences/draw conclusions about the story
Support a thesis using examples from a story
Identify examples of imagery and symbolism
Analyze, synthesize and evaluate ideas from texts[68]

Design Prompts: Reading

How might social interaction enhance communication and language development in your project?

How can the relationship between image and text tell a story to children at different stages of language development? Reading levels?

How can auditory and other forms of multimedia be used to increase learning and reading?

◀ Figure 4.14

Signs in the playground provide prompts for parents to promote literacy with their children.
Source: author photo.

Notes

1   UNICEF
2   National Scientific Council on the Developing Child 2008
3   Center on the Developing Child, Harvard University, *Brain Architecture*
4   Sample, Ian 2014
5   Age of Montessori
6   Gopnik, Alison, Meltzoff, Andrew N., Kuhl, Patricia, K. 2000
7   Ormrod, Jeanne 2011
8   MindsetWorks
9   Gardner, Howard
10 Lane, Carla
11 Pashler, Harold, McDaniel, Mark, Rohrer, Doug, Bjork, Robert (2008)
12 Chick, Nancy, *Metacognition*
13 Hearron, Patricia F., Hildebrand, Verna P. 2009
14 McLeod, Saul 2015
15 Piaget's theory of cognitive development
16 Carey, Susan 1986
17 Weinschenk, Susan M. 2011 *100 Things Every Designer Needs to Know*
18 Weinschenk, Susan M. 2011 The Secret to Designing an Intuitive UX
19 Kasuba, Terri Fry 2017
20 McLeod, Saul 2014
21 McLeod, Saul 2012
22 Learning RX
23 McLeod, Saul 2008
24 Weinschenk, Susan M. 2011 *100 Things Every Designer Needs to Know*
25 Shaffery, Joseph 2016
26 Turrell, Andrew 2011
27 Kid Sense, *Visual Perception*
28 The Peak Performance Center
29 Speech Therapy Centres of Canada 2013

30 Mersch, John

31 Weinschenk, Susan 2017

32 Harms, William 2013

33 Michelon, Pascale 2006

34 The Human Memory

35 CliffsNotes

36 Shouse, Benjamin 2011

37 Shinskey, Jeanne 2016

38 Morin, Amanda, *5 Ways Kids Use Working Memory to Learn*

39 Micklo, Stephen J. 1995

40 Shiffrin, Richard M., Nosofsky, Robert M. 1994

41 Frans, Peter 2014

42 The National Council of Teachers of Mathematics 2000

43 Morin, Amanda, *Math Milestones*

44 Science Buddies

45 Eccles, Jacquelynne S. 1999

46 Surbhi, S. 2017

47 Gray, Peter 2008

48 Eccles, Jacquelynne S. 1999

49 Chick, Nancy, *Metacognition*

50 Gopnik, Alison 2009

51 Complexity Labs, *Divergent and Convergent Thinking*

52 Oddleifson Robertson, Katrin

53 Christensen, Tanner 2015

54 Land, George 2011

55 IMDb

56 Burns, Will 2017

57 Carter, Christine 2008

58 Nicolopoulou, Ageliki 2009

59 Dziengel, Ana 2017

60 University of Oregon Center on Teaching and Learning

61 Seidi, Amanda; Cristia, Alejandrina 2012

62 Moats, L.; Tolman, C. 2009

63 Talking Point

64 Hoecker, Jay L. 2016

65 Teach Me to Talk 2008

66 Liberman, Isabelle Y.; Shankweiler, Donald; Liberman, Alvin M. 1989

67 Moats, L.; Tolman, C. 2009

68 *Milestones on the Road to Reading Success*

Chapter 5

Social and Emotional Development

Children learning to manage their emotional state and interactions with others is considered to be a significant component of learning and healthy development. The social parts are the intrapersonal processes, including the ability to establish and maintain positive and rewarding relationships, to accurately read and comprehend emotional states in others, and to develop empathy for others. The emotional parts are the interpersonal processes, the ability to identify and understand one's feelings, to regulate one's behavior, and management of strong emotions and their expression in a constructive manner. Childhood is not always happy, as the real world that they are living in is not.

Throughout childhood, kids' brains are learning how to process social and emotional information, and they are forming the habits that they will use throughout a lifetime. Designers can nurture and support children in learning and managing life's ups and downs with creative, engaging solutions in social and emotional learning products, services, media content, interactions, books, apps and engagement opportunities.

The Importance of Social and Emotional Learning

Emotion and cognition work together, jointly informing the child's impressions of situations and influencing behavior.[1] Social and emotional learning is now being recognized within education as an area of focus because of reduced conduct problems and risk-taking behavior, and improved test scores, grades and attendance.

Children are more successful in school and daily life when they practice:

- Self-Awareness: Understanding one's personal goals, emotions and values
- Self-Management: Can regulate emotions and behaviors and have decreased emotional distress
- Relationship Skills: Have more positive and rewarding relationships with peers and adults
- Social Awareness: Understand the perspectives of others and relate effectively to them
- Responsible Decision-Making: Make constructive choices about personal behavior and social interactions[2]

Child's rainbow happy face chalk drawing.
Source: author photo.

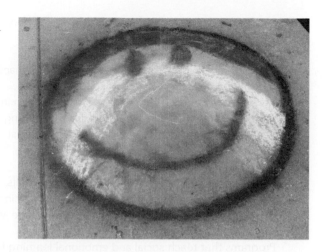

It's easy to convince people that children need to learn the alphabet and numbers ... How do we help people to realize that what matters even more than the superimposition of adult symbols is how a person's inner life finally puts together the alphabet and numbers of his outer life? What really matters is whether he uses the alphabet for the declaration of war or the description of a sunrise – his numbers for the final count at Buchenwald or the specifics of a brand-new bridge.

Fred Rogers, Speech to the American Academy of
Child Psychiatry, October 1971

Executive Function

Executive function includes the cognitive skills we all use to analyze tasks, break them into steps, and keep them in mind until they are complete. Often tied to self-regulation, the brain needs this skillset to filter distractions, prioritize tasks, set and achieve goals, and control impulses. They enable positive behavior and allow us to make healthy choices. Ani Holloway, a school psychologist at Glenoaks Elementary School in Glendale California, says:

Today, in a media-driven multi-tasking society it is becoming more and more difficult to focus and keep information organized. Teaching kids to solve problems in life, make decisions, manage behavior and have a better attachment to others leads to better academic performance, improved attitude and behaviors, fewer negative behaviors, and reduced emotional stress.[3]

Executive function and self-regulation skills depend on three types of brain function:

- Working memory governs our ability to retain and manipulate distinct pieces of information over short periods of time.
- Mental flexibility helps us to sustain or shift attention in response to different demands or to apply different rules in different settings.
- Self-control enables us to set priorities and resist impulsive actions or responses.

Children are born with the ability to develop these skills. Establishing routines, modeling social behavior, and creating and maintaining supportive, reliable relationships provide children with practice over time with decreasing adult supervision as they develop into adulthood.[4]

Programs that teach social and emotional learning focus on:

- Learning about emotions
- How to stop and think
- Resilience
- Stress reduction
- Conflict resolution
- Respect
- Responsibility
- Self-esteem
- Mindfulness

◄ Figure 5.2

Building social and emotional skills can enhance self-efficacy, one's belief in one's ability to succeed. It can play a major role in how one approaches relationships, goals, tasks and challenges. Source: author photo.

Mindfulness

Being "in the present moment" allows most people to be calmer and make better decisions. We all have a limited pool of mental resources and when they are depleted self-control becomes impaired. Mindfulness programs teach children to breathe, relax and how the brain works.
 Benefits of mindfulness in education:

- Better attention and focus, better test taking
- Emotional regulation
- Compassion
- Adaptive skills
- Calming
- Resilience[5]

When children play, they practice these skills of self-control and learn how to regulate their own emotions and feelings. They exercise their verbal and non-verbal communication, body language, and release of feelings, and build self-confidence. Activities that foster creative play and social connection teach them how to cope with stress, and provide opportunities for directing their actions. Interaction with family and friends through games and sports highlights healthy competition, teamwork, and the effects of winning and losing. Self-expressive creative projects help kids to recognize and learn to accept their feelings.

Body Language and Facial Expressions

Children can be taught to express themselves and better understand what people are communicating with their bodies. Roleen Heimann, co-director of New School West preschool in Los Angeles, California, explained: "We talk with the children about body language. They learn to read it and use it. We might say, 'What is his body telling you? What do they want? How do they feel?' and have them reflect on that."[6] Intuition picks up body cues before language and often overrides what people are saying. They can learn when their body feels happy or uncomfortable around someone else. By learning to understand the body language of others, children can express that they care about how they feel or give them more space.[7]

 All people experience seven basic emotions, no matter what culture they come from: joy, sadness, contempt, fear, disgust, surprise and anger. Although we feel emotions, they are also expressed physically through gestures, facial expressions and crying. It takes experience to recognize our emotions and read them in others.[8]

Design Prompts

In pairs or groups act out emotions with your body and communicate without saying a word.

 Develop methods to use body language and facial expressions instead of language as a method of communication in your project?

◀ Figure 5.3

Non-verbal communication in many forms, from facial expressions to hugs, conveys feelings and meaning.
Source: Terri Fry Kasuba (used with permission).

Autism Spectrum Disorder (ASD)

ASD is characterized, to varying degrees, by difficulties in social interaction and verbal and non-verbal communication, and repetitive behaviors. ASD can be associated with intellectual disability, difficulties in motor coordination and attention, and physical health issues. Some people with ASD excel in visual skills, music, math, and art. Autism appears to have its roots in very early brain development. However, the obvious signs of autism tend to emerge between 2 and 3 years of age.[9] In recent decades, the number of children diagnosed with autism has greatly increased in the US, and there is much debate over whether this an increase, greater awareness or semantics.[10] Designers who are creating inclusive designs often consider how to support children and youth with autism in their solutions.

Qki, a Preschool Edutainment Property for Social and Emotional Learning

Jin Hyung Kim uses characters and storytelling to produce apps, toys and related merchandise to support social and emotional learning.

How do you teach social and emotional learning in the character designs and personalities?
Social education builds the cornerstone of kids' personality and their value system. My characters have bumps on their bodies, which represent their defects. Those bumps change to show their emotional states that arise from reacting from other characters. It is impossible to get rid of the bumps, but they could be controlled. That is their goal.

How is social and emotional learning represented in the storyline?
The characters experience similar difficulties with kids' real lives and face all kinds of emotional turmoils, mistakes and successes. The story acknowledges a diversity of personality in diverse solutions to the problems.

You have worked on many video games and story properties for teens and young adults. How do you think about design differently for that age than designing for preschool children?
Content for teens focuses on their accomplishments such as saving worlds, going after monsters, becoming the greatest warrior, transforming into a bigger self. To support this designs are edgy, violent, realistic and detailed. For preschoolers, I focused on mimicking kids with the proportions. Also, kids are more flexible, so the stories are more dreamy and unrealistic.

The Qki characters offer children social and emotional learning through digital games, toys, bedding and clothing, and other lifestyle products.
Source: Jin Hung Kim (used with permission).

Design Prompts: Social and Emotional Development

- Use pictures or video of emotions with facial expressions to communicate them.
- Model realistic and responsible behaviors in characters and stories.
- Ask questions such as "Is he happy now?", "Is he angry or sad?", "How is he feeling?", "Why is he sad?", "Is he sad now because he lost his ice cream?", "Does that make him happy?"
- Create products and experience that promote both independence and connections among people.
- Design around the universal feelings. Use group activities, and action to express emotion.[11]

Self-esteem

Perceptions, expectations of themselves, and how a child is thought of and treated by parents, teachers and friends shape self-esteem. The closer a child's perceived self (how they see themselves) comes to their ideal self (how they would like to be), the higher the self-esteem. Designers can help build children's self-esteem by developing content that supports the areas that build self-esteem.

Self-esteem comes from a sense of …

- Security
- Belonging
- Purpose
- Personal competence and pride
- Trust
- Responsibility
- Contribution
- Making choices and decisions

- Self-discipline and self-control
- Encouragement, support and reward
- Accepting mistakes and failures
- Family self-esteem[12]

Motivation

Children's natural curiosity drives them to explore their surroundings and try new things. From their repeated experiences of seeing their actions affect their world and the people in it, children begin to see themselves as capable and having control. This helps children to feel good about themselves and builds their self-confidence. Children who are confident are motivated to engage in more experiences and learning. They expect to have successful and positive experiences because they have done so in the past.[13]

Many young children are given plenty of encouragement, often have relatively high self-esteem and are optimistic that they can learn a new skill, succeed and finish a task if they keep trying. Successful educational and entertainment platforms enable kids to explore with intrinsic motivation as compared with extrinsic goals. Ultimately, the best play and learning you can find is kids wanting to play for their own reasons, boosting self-esteem.

Body Image

Kids with a positive image of themselves feel more comfortable and confident in their ability to succeed, don't focus on food or weight, and have the energy to enjoy physical activity. Kids with a negative body image feel more self-conscious, anxious and isolated. They are at greater risk for obesity and for eating disorders.[14] Jacqueline Harding, advisor to Professional Association for Childcare and Early Years (PACEY), said: "By the age of 3 or 4 some children have already pretty much begun to make up their minds – and even hold strong views – about how bodies should look."[15]

Young children engage with some of the more extreme body portrayals through entertainment media and advertising and in the form of toys such as dolls and action figures. Caroline Knorr, senior parenting editor at Common Sense Media, explained, based on a 2015 study:

> It's problematic because dissatisfaction with body image leads to lower self-esteem, lower expectations about what you can become. It also can lead to destructive behavior. Preschool shows with female characters have more exaggerated secondary sex characteristics than the male characters and there are some stereotypical roles.[16]

Designers working in these areas can have a significant impact with their choices in the characters, products, and messages and content they create. To portray a healthy body image:

- Break the myth of the perfect body with positive role models
- Focus on health not weight
- Emphasize exercise and nutrition
- Model a diversity of body types and acceptance
- Inspire self-attuned eating, trusting feelings of hunger and fullness

Self-learning – Identity

Being able to understand oneself is important to develop social relationships. In a classic experiment in 1972, Beulah Amsterdam from the University of North Carolina published a study that has influenced decades of research on self-recognition, the "rouge test." Researchers took a group of babies and toddlers aged 6 to 24 months and placed a spot of lipstick on their noses. They then placed the children in front of a mirror to see how they responded. The results suggested that babies around 2 years old have a solid *physical* or *visual* self-concept but still have little *mental* self-concept. Once young children reach this level of self-awareness, emotions like embarrassment, envy and empathy emerge. Between the ages of 2 and 4 years old is when they start to display a rapid increase in social behavior.[17]

As children learn more about their personalities, interests, what they are good and not so good at, and what makes them happy and sad in relation to others, they further define their self-concept. By age three, their *categorical self* has developed, which labels themselves in concrete and observable terms: "I am 3, a boy and have blue eyes." Many 3–5-year-olds do not develop an integrated self-portrait and are not aware that a person can have opposing characteristics such as being "good" and "bad" at the same time.

As long-term memory develops, children gain the *remembered self*, about personal events and stories that become part of their life story, and an *inner self*: private thoughts, feelings and desires that nobody else knows about unless they share this information. Their interests, personalities and group associations, such as family, friendships and cultural identity, form another highly important part of a child's self-identity.[18] Both internal and external variables can affect young children's self-concept. For example, a child's temperament can affect how they view themselves, and their ability to complete tasks or criticism from others can affect how they view themselves.[19]

Understanding individual identity and group identity is relevant to designers to help us understand our users, attitudes and belief systems. To find out what products, experiences and brands match their identity, we ask questions such as:

Who are they? (Target market)
How do they see themselves? (Self-concept)
Who do they want to be? (Aspirational self)
What group or subcultures do they align with? (Peer affiliations)

◄ Figure 5.5

At this age the baby is interested in the child in the mirror but does not recognize the child as himself.
Source: author photo.

Cultural Identity

Worldviews of the self differ. Western cultures place importance on independence and expression of one's attitudes, while Asian cultures favor the interdependent view of self and interpersonal relationships.[20] Cultural experiences and values shape the way children see themselves and prioritize what is important to them. Children learn behaviors and beliefs from the family and community. Cultural perspectives of parenting influence children's skills and values. Having a strong sense of their cultural history and the traditions associated with it helps children build a positive cultural identity for themselves. Respect for diversity is related to people's sense of belonging. When diversity is valued and respected, children are more likely to develop a sense of belonging to their community and social connections to others. This also supports children's sense of belonging and, by extension, their mental health and well-being.[21]

◄ Figure 5.6

Families teach children cultural traditions through what they wear, their body portrayal or activities they are engaged in.
Source: author photo.

Gender

Gender is one aspect of our overall identity. It is the internal sense of whether we are male or female and what is expressed. According to neuroscientist Lise Eliot in *Pink Brain, Blue Brain*, the small difference in the brain structure and hormones influences differences in behavior. Children themselves exacerbate the differences by playing to their modest strengths. They constantly exercise those "ball-throwing" or "doll-cuddling" circuits, rarely straying from their comfort zones.[22]

When kids are young, their natural inclinations are apparent; as they develop, they become more aware of cultural norms reinforcing gendered behavior. Vanessa LoBue, an assistant professor of psychology at Rutgers University, explains:

> Before the age of 5, children don't seem to think that gender has any permanence at all. Children develop the concept of gender between 3 and 5 years old. Once children begin to think about gender as a stable trait, they also start to incorporate gender into their own identity.
>
> In grade school, their concept of gender becomes quite rigid. This is when kids start to become strict about adhering to their gender, play specifically with kids of their gender with gender-specific toys and activities. It becomes more relaxed again by age 10 and into middle childhood.[23]

As teens, they may merge their gender identity and their sexual orientation, and some play around with gender fluidity. With the greater scientific understanding of gender and transgender individuals today, many are now looking at gender not being a stable trait but a spectrum.[24]

Gender and Design for Children

There is much debate in the children's market between gendered products and gender-neutral products. The divide tends to lie around the concept of gender stereotyping. There are many companies that make products and media for children that reinforce gender stereotypes.

Even when parents promote gender neutrality, kids play around with gender by taking on different roles. Many industries, including clothing, toys, games and home furnishing, still define strict categories of boys and girls, although companies are slowly moving away from this. Many of the people I interviewed for this book from toy, home, clothing and lifestyle companies for kids that have distinctive boy and girl product lines explained that they make their decisions based on trends in consumer behavior. They all see that humanity is evolving, and recognizing diversity will impact how their company and products will change the issues of gender in the future.

Ali Otto, director of design for Baby Boy at Old Navy, explained how this specifically relates to the children's clothing market:

> I think the trend of kids clothing especially in Newborn/Layette is moving towards being more gender neutral. We purposefully design collections with the framework of Girl/Boy/Uni. Inspirational brands break gender lines through a single palette, as well as producing silhouettes that house multiple prints and patterns. This sends a powerful message to the consumer. I think it also takes the partnership with our merchants and partners to be able to provide more options in the range of color or patterns that straddle traditional gender lines. Let's not just offer girls "pink" and boys "blue." Let's offer silhouettes that the child can feel comfortable wearing.[25]

Toys and Gender

The idea behind gender-neutral toys is that encouraging kids to play beyond the boundaries of "traditional" gender roles can help them avoid gender stereotyping and falling into highly gendered roles in the future. As traditional gender roles have broken down in society, toys have only become more gendered. Melissa Hines. professor of psychology at the University of Cambridge, said: "Color-coding toys to limit their appeal to both sexes nurtures limitation rather than the possibility."[26] Roleen Heimann, co-director of New School West preschool in Los Angeles, stated: "When LEGO made bricks for girls they made them pink. We work so hard as a culture to get away from that. All children paint, play house, sew, design, build, and act out their stories. Play is interconnected, and we have to be mindful of not labeling and stereotyping."[27]

Designers can make a significant impact on this topic with their approaches to the toys they make and how they market them. Parents and educators are taking an active role in influencing their child's development with non-gendered toys. In the meantime, toy companies and distributors are thinking of new ways to bring the other gender into their "girl" or "boy" brand. Designer Charlie Hodges explains: "More and more, I see toys that encourage girls to participate in traditional 'boy' realms. I fully embrace redefining gender norms for girls. But I wonder, what can we as designers do to make it more acceptable for boys to explore realms traditionally reserved for girls?"

What do we lose when we take away attributes of gender altogether? Psychologist Peter Gray says the focus on gender-neutral toys and play overlooks gender as a crucial aspect of both children's play and their development.

> Children come into the world designed to look around, see what it is that people of their gender do, and then they want to play at those skills to become good at them – we're biologically drawn to that. It's also unlikely toys can change the state of gender equality. Rather than trying to change culture through children's toys, children's toys are a reflection of the culture children are born into. As the culture

"While redesigning the Barbie Dream house I was thinking about the responsibility of the brand and what it can say about kids and behavior and expectations. I designed a tiny house with 'rooms' that unfold to create prompts for play. This more open-ended geometry allows kids to do what they want to do. And consequently, it makes the toy gender-neutral."
Source: Charlie Hodges.

changes, children's interests also change to reflect that. So far, our culture hasn't changed regarding [gender neutrality].[28]

Designing for Gender Neutrality

- Understand the developmental similarities of all children
- Create an equal balance of activities typically thought of as being for boys and girls
- Don't include implications of gender: Use gender-neutral colors – themes – materials – finishes
- Create animals or abstract non-gendered characters
- Avoid claiming gender-specific behaviors and stereotyping
- Ask: What are the positive results that you want to create using gender neutrality?

Designing for Gender Specificity

- Research the scientific evidence on the developmental differences between genders
- Know the cultural stereotypes (colors, princesses and superheroes)
- What are the positive results that you want to create using gender specificity?
- Create positive design language around gender

Gender throughout the World

Although we see the gaps in the gender divide getting smaller in the first world, in the third world there is still a long way to go for gender equality. UNICEF and many other organizations have been working towards improving education, health and discrimination, prevention of violence and many other issues related

◀ Figure 5.8

Girls living in poverty are the best voices to tell others what they need. This affordable camera offers girls an outlet to express themselves and document their lives in a creative manner; the uploaded images are also a valuable resource for social service and humanitarian organizations that need unadulterated research and direct insight to best plan and design effective programs.
Source: Therese Swanepoel, Lori Nishikawa, Carolina Rodriguez (used with permission).

to girls. Because power structures in societies across the world mostly privilege boys and men, advancing gender equality most often requires addressing disadvantages faced by girls and women. Shifts in gender equality require not only awareness and behavior change, but also changes in the fundamental power dynamics that define gender norms and relationships.[29]

Sexual Orientation

Sexual orientation is how individuals are attracted romantically and sexually to other people – to the opposite sex, the same sex or both. Knowing one's sexual orientation is often something that kids recognize with little doubt from a very young age. Despite gender stereotypes, masculine and feminine traits do not necessarily predict whether someone is straight or gay. When teens have problems related to being gay, lesbian, bisexual or transgender, it usually isn't because of their sexual orientation or gender identity. It's usually because of a lack of support from the people they love or because they have been or are being ridiculed, rejected or harassed.[30]

Resilience, Fears, Anxiety, Attachment Objects, Stress

Resilience is the ability to cope with life's ups and downs. It is missing in many children and youth today due to being overprotected from life's stressors. Kids need the skills, the knowledge and the strategies to deal with negative outcomes when they happen. Resilience is often paired with the lack of risk protected children are not engaging in today. Paul Levine, CEO of Play Science, explains:

> Critical life skills are being stripped out of our culture. Grit and resilience are core American values. There is a consensus that kids have

Playing on a big slide involves taking risks.
Source: author photo.

been softened. These traits have been stripped out of childhood. Listening, kindness, and empathy are important and we need to design for them.[31]

Children's resilience is enhanced when they:
- Are loved by someone unconditionally
- Have an older person outside the home they can talk to about problems and feelings
- Are praised for doing things on their own and striving to achieve
- Can count on their family being there when needed
- Know someone they want to be like
- Believe things will turn out all right
- Have a sense of a power greater than themselves
- Are willing to try new things
- Feel that what they do makes a difference to how things turn out
- Like themselves
- Can focus on a task and stay with it
- Have a sense of humor
- Make goals and plans, both short and longer term.[32]

Fear-Anxiety

Experiencing feelings of fear and dealing with anxieties prepares kids to handle the unsettling experiences and challenging situations of life. Some of these fears may protect them, such as fear of fire. As kids grow, one fear may disappear or replace another.

◀ Figure 5.10

A transitional object or security blanket provides psychological comfort for children, especially in unusual or unique situations or at bedtime. Source: author photo.

Common Fears by Age

Infant and toddler: Loud noises or sudden movements, large looming objects, strangers, separation, changes in the house

Preschool: The dark, shadows, noises at night, loud noises, masks, monsters and ghosts, animals such as dogs, fear of the potty, weather, sleeping alone

Grade and middle school: Snakes and spiders, storms and natural disasters, being home alone, fear of a teacher who's angry, scary news or TV shows, injury, illness or death, peer rejection, fear of the dark, fire, bad guys, failure and taking tests, doctors and shots, bugs and animals

Teens: Fear of their safety, sickness, throwing up at school, failure in school or sports, school presentations, how they look to others, violence and global issues.[33]

Anxiety is worrying about what *might* happen, worrying about things going wrong or feeling like you're in some danger. This is a natural reaction to our fight or flight response and can be healthy. Anxiety disorders seen in children include generalized anxiety, obsessive-compulsive disorder, phobias and post-traumatic stress disorder. Play therapist Dr. Ann Bingham Newman is very concerned about anxiety:

Anxiety is an epidemic with kids today. It is the most common thing that I see. They have no downtime. With school and homework, they are overloaded. They are overscheduled outside of school with activities. Both parents are usually working and are anxious or distracted with their phones. Kids are also anxious to be on their screens – even 2-year-olds.[34]

Sensory Social Development

Children with over-registering sensory input may see the world as a scary place and have rapid shifts in emotions or anxiety. Under-registering sensory input presents other sensory processing challenges.

Children may lack emotions or reading other people's emotional clues. Both may have difficulties with how they interact with others, specifically peers.[35]

Separation Anxiety

Separation anxiety is the fear or distress that can happen when children think about separating from the caregivers they have become attached to. Separation anxiety is a normal stage of development and usually ends at around age 2, when toddlers begin to understand that a parent may be out of sight right now but will return later. Separation anxiety disorder, however, is when the anxiety exceeds what might be expected given a person's developmental level.[36]

Transitional Objects

Infants think of themselves and caregivers as one. Once a child realizes that they are a separate entity, they feel like they have lost something and are dependent on others, thus losing the idea that they are independent. The transitional object is often the first "not me" possession (a blanket or teddy bear) that belongs to the child and settles their anxieties. Richard H. Passman of the University of Milwaukee found that there is nothing abnormal about using them, and security blankets are appropriately named – they offer security to those children attached to them.[37] Often the objects they choose are readily available in their surroundings at the time they develop the need for security.

Play Therapy

Play therapy is used in diagnosis and counseling to prevent or resolve challenges in children's lives. Through play therapy, children (usually between 3 and 12 years old) work through issues and learn to communicate with others, express feelings, modify behavior, develop problem-solving skills, and learn a variety of ways of relating to others. Play provides a safe psychological distance from their problems and allows expression of thoughts and feelings appropriate to their development. The positive relationship that develops between therapist and child during play therapy sessions can help children learn more adaptive behaviors and provide a corrective emotional experience necessary for healing or social skills deficits. Play therapy may also provide insight about and resolution of inner conflicts or dysfunctional thinking in the child.[38]

Many toys, games and products are designed specifically for use in the play therapy market. Other general use toys can also be used in play therapy sessions. When creating a project around social and emotional development, designers can consider the product and environment for play therapy sessions to support this process.

Play as a Process for Healing

Interview with Dr. Ann Bingham Newman, Therapist and Child Specialist

Dr. Ann Bingham Newman is a strong advocate of building on the natural way that children learn about themselves, their relationships through play. She believes in the value that designers have to offer with consideration to content, user experience and play therapy materials.

When do you use play therapy, and how are toys used differently from games in play therapy?

I use play therapy for anxiety, bullying, anger, depression, worry, grief, divorce, reunification, self-esteem and many other problems children have. The purpose of using toys and games is to get them involved in the session to start talking. Toys and games break down the barriers, and any suitable toys and games will do that. We play Connect 4, Jenga, blocks, checkers and chess. We go outside and play ball, with hula-hoops, and basketball.

Some games are specific to the play therapy market that also are focused on bringing up particular topics related to treatment. Games that ask both serious and silly questions that you wouldn't typically discuss in a conversation, but in a game it is ok to ask.

In sand tray therapy kids use miniatures in a bed of sand to create stories and act things out. I have shelves of miniatures, trees, farm animals, barns, gravestones, people, family figures, including babies and grandparents, are very important. I also use a dollhouse with furniture.

They choose what they are drawn to and make a "picture or world in the sand" – they change it as much as they want to. I listen and watch, but I don't interject, prompt or evaluate it. When they feel done, they will tell me, and we will talk about it if they are willing. Like most play therapy is meant to be meaningful to them to work through issues.

I had a soft Frisbee that had words on it and where your thumb landed was the word you talked about. Art therapy is an emotional release too. Use different art supplies and tools.

Water is also useful to play with. Kids do pretend cooking, or sometimes we cook a recipe in the kitchen. They are acting out the things that family and friends do in everyday life.

Are there toys or games that work the best when working with children on different types of problems?

- Abuse, neglect, violence: Dolls that are anatomically correct can be good to act things out that are difficult to say or that they don't understand. The role-playing removes the experience from the self.
- Bullying, teasing, peer conflict, friendship: Role-playing, talking, play-acting. Soft swords.
- Grief and bereavement: Sand play, art therapy, bibliotherapy.
- Trauma: A tent to crawl in and books that discuss the issues through stories. Sand tray, with miniatures, trauma can surround illness and hospitalization of child or family members so police cars, fire engines, ambulances and lots of hospital figures, beds, etc. are helpful.
- Self-esteem, confidence: Soft swords help the kids feel powerful. They are also good for anger.
- Some therapists use digital tech in hero play, but I do not.

Can you explain a few things designers may need to know to make better decisions for play therapy products?

Everyone responds differently, so it depends on the kid and a particular session. It can be directed or undirected. Some kids walk in and do whatever they want. They don't want you to bother them or ask a question. They might pick cars because they like to play with cars and then act out what their specific issue is – abuse – anger. The toy doesn't need to be specific to the problem. With others, I start by saying, "Here is an activity." For example, I use a series of activities with kids going through divorce or separation designed to help them sort out their feelings and to understand this is not about them.

As the therapist, what do you look for and how do you guide the session?

I try out different approaches. We do art together. We play together and apart. It depends on the child. One week one method might work and the next it doesn't. Kids could build on their play over time. They may use the sand tray time after time to build on the same story, and each time it was a little different.

I take notes throughout the process to enhance the treatment. The goal is to come to some resolution.

Do you have any recommendations of things for designers to consider when designing the environment for play therapy?

Make it inviting and fun. One therapist I know has a place that looks like a fairyland with pretend trees and twinkling lights. A 14-year-old might question "why am I here?" but another might ignore it or like it.

It should be able to be changed up to keep it fresh, but it is also essential to keep it consistent. Some kids count on it being a certain way.

Keep the objects on display and easy to access so they see everything and can pick and choose. Objects and books need to be at their level/height – shelves, pictures on walls, etc.

I always play music that is calming and supportive of the thought processes. Color can play a part. I have some vibrant colors and some softer colors.

Some people take walks, walk a dog or go to the ice cream store. The environment does not need to be confined to space.

What can designers of toys for the general use market learn from toys designed for the play therapy market?

- Use natural materials and objects. Shells, shiny materials, sticks, leaves, rocks.
- Simplicity – keep the objects open-ended.
- Versatility – I want to have a lot of things to choose from just in case.
- Some things are only fun once – like a motorized car or airplane. Keep toys open-ended.
- Dolls of different races.

Also, the therapist chooses things that bring in their style. I like angels and worry dolls.

Do you have a wish list of toys or games that you would like to see designers develop in the coming future?

- Dolls that are soft – gentle so you can twist them. Many dolls are too hard.
- Better games: Sometimes I get excited about a game and know there is someone who went really out of their way to create it. In the end, it is disappointing. Some of these people are nerdy and mean well but don't know how to design the experience.
- Interesting miniatures that extend storytelling.[39]

Social Understanding and Relationships

Relationships with Caregivers

Primary caregivers are the initial source for social development and continue to be the most influential in a child's social awareness through role modeling. A typically developing child will develop an attachment relationship with any caregiver who provides regular physical and/or emotional care and serve-and-return interactions. Children create a hierarchy of attachments with their various caregivers and a specific attachment relationship with each caregiver. During the first months and years of life, when an infant or young child smiles, babbles or cries, and a caregiver responds appropriately with eye contact, sounds, words or affection, neural connections are built and strengthened in the child's brain that support the development of communication and social skills. By the time babies are a year old, researchers can assess whether babies are "securely attached" to (trust) their parents to meet their emotional and physical needs.[40]

Children see how their parents model emotions and interact with other people, and they mimic what they see their parents do to regulate emotions.

A child's temperament also plays a role in their emotion regulation, guided by the parenting style they receive.[41] Parent–child relationships develop throughout childhood, influenced by child characteristics, parent characteristics, parenting styles and the contexts in which families operate. Children automatically love their parents, but many risk factors can create different relationship qualities:

- Risk factors in the family and environment
- Child's behavioral problems
- Parents' warmth and control
- Emotion: Children's emotions affect parental behaviors, and parental emotions affect children's development and actions
- Attunement: Sensitive mutual understandings and interactions between children and their parents
- Parent's hostility
- Stress (challenging child behavior and parenting tasks)
- Caregivers' social factors and temperament

Psychology has outlined four major recognized parenting styles.

Authoritative

- Structure with clear, reasonable expectations and understood rules
- Consequences for disrupting this structure or breaking the rules
- Open communication style without judgment or reprimand

This is known as the healthiest environment for a child. Parents have high expectations of their children, but temper these expectations with understanding and support for their children.

Authoritarian

- Parents are demanding but not responsive
- Allow for little open dialogue between parent and child
- Children have few choices and decisions about their own life
- Parents expect children to follow a strict set of rules and expectations
- Rely on punishment to demand obedience or teach a lesson

These children are prone to having low self-esteem, being fearful or shy, associating obedience with love, having difficulty in social situations, and possibly misbehaving when outside of parental care.

Permissive

- Few rules are set for the children
- Parents avoid conflict
- Want to be the child's friend
- Bribe children with rewards

This nurturing style can lead to insecurity, poor social skills, self-centeredness, poor academic success and clashing with authority due to lack of set boundaries.

Neglectful

- Do not care for your child's needs – emotional, physical and otherwise
- Don't have an understanding of what is going on in your child's life
- The home environment provides negative or no feedback
- Parent spends long times away from child
- Uninvolved in child's life outside of home

Neglectful parenting is damaging to children because they have no trust foundation with their parents from which to explore the world.[42]

Designers create products and experiences aiming to keep the connection secure and strengthen quality time and relationships within families. Families that do things together have many opportunities to interact, communicate and build trust. Often there is a balance between the time they spend together and the independence they hope to create. Designers should always question the goals of the project and the child and parent level of involvement. To sustain interest, both parent and child must have their personal needs fulfilled.

How to design for bonding of parents and children:

- Attention = Love
- Help the parent become fully present and not distracted
- Without quantity time, there's no quality time
- Embed trust and communication

Friendships

Friendships nurture the emotional development of the child from within, the social side of the development of supporting interactions and cooperation among children.

Friendship skills include:

- Self-control
- Welcoming
- Assertiveness
- Consideration
- Play skills
- Communicating
- Helping
- Prediction
- Thinking
- Coping
- Empathy
- Flexibility[43]

Social Events, Play Dates and Hanging Out

Kids interact at social institutions such as schools, at social events such as birthday parties, in holidays and in social spaces such as playgrounds. Well-designed events programming and play spaces with activities that provide engagement can help create positive social/emotional development for children. Children learn through playing with other children; play dates, hanging out and other events give them an opportunity to do so.

- Babies have not developed social play skills but may sense the happiness of children and parents chatting.

- Toddlers observe their peers, older and younger children. They are learning the language, social rules and skills.
- Preschoolers are moving into more cooperative play, turn-taking and overall interaction.
- School age is when children can start to have social events and play dates unsupervised by one of the parents. It is a good idea that an adult is close by.
- Tweens and teens like to hang out at their friend's house or welcome their friends into their space. They also plan things to do together in public spaces; entertaining spaces within their homes and social public entertainment spaces play a big part of their lives.

Developmental Sequence of Friendships

There is a developmental sequence to the types of friendships children have, and in their complexity, depth and meaning, that evolves throughout childhood.

Infants/Babies: Children as young as 6 months get excited about seeing a peer. They smile and make noises to try to get the other baby's attention, but they tend to treat peers as objects to explore.

12 to 18 months old: Toddlers show noticeable preferences for certain peers. They imitate each other, showing they have some ability to understand another person's perspective.

18 months to 3 years: Friendships depend on the toddler's ability to understand how others feel and regulate or control themselves. Young preschoolers usually base their friendships on proximity, how often they see each other or whether they are playing with something of interest.

4-year-olds begin to become more cooperative in their relationships and play activities, and may have a special friend. Positive interactions help sustain friendships.[44]

Based on systematic interviews with children of different ages, psychologist Robert Selman offers a handy five-level framework for understanding children's friendships from age 5 and up.

Ages 5–9: They define friends as children who do nice things for them. Children at this level care a lot about friendships. They may even put up with a not-so-nice friend, just so they can have a friend.

Ages 7–12: These children can consider a friend's perspective in addition to their own, but not at the same time. They are very concerned about fairness and reciprocity, but they think about these in a very rigid, quid pro quo way.

► Figure 5.11

When children interact with each other they develop skills and learn through friendships.
Source: author photo.

Ages 8–15: Friends help each other solve problems and confide thoughts and feelings that they don't share with anyone else. They know how to compromise, and genuinely care about each other's happiness.

12 years and up: At this stage, they place a high value on emotional closeness with friends and emphasize trust, support and remaining close over time. They're not as possessive and have both close and casual friends.[45]

Bibliotherapy

Bibliotherapy is a therapeutic approach using literature to overcome problems. In addition to what they learn from family and friends, children learn about emotions early on through stories that are read or watched on TV and different forms of media. They learn points of view of characters that are different from themselves. They can see the social and emotional consequences of different actions without having to live through those consequences. They can find a hero to emulate or a peer who understands what they're going through. Reading stories about others' experiences helps children to understand emotions and offers coping mechanisms. A recent research study by Celia Brownell and her colleagues at the University of Pittsburgh suggests that reading and discussing emotions also helps children to understand them better.[46]

Interview with author/ illustrator Marla Frazee

How did you come up with the idea for this book?
I went to a clown show at my kid's school. Each kid had to come up with a clown character, their costume, and then pantomime a story that they'd come up with and set that to music. I was captivated! It was like watching a silent film. I kept thinking about it for days afterward. I explored many book ideas about clowns, but none of them seemed quite right. Then one day I was riding my bike around, and I saw a picture in my head of a baby clown holding the hand of a grouchy farmer. It was just there. I asked myself who they were and why they were together? And that was the beginning of the book.

Can you describe your brainstorming process? Methods you used?
I spent three days in a mountain cabin with my dog and hunkered down to figure it out. By the end of my time there, I had corralled the story into a tiny-sketched book dummy. I showed it to my editor Allyn Johnston, VP and publisher of Beach Lane Books, and she loved it although she thought, like I did, that it was kind of an unsettling story – in the best sense. And she wanted me to see it through. It was a compelling project for me.

Can you explain your process of developing the story and image in this book?
It is a wordless book, so the whole story is told in pictures. I did a lot of thumbnail sketches and made many book dummies, trying to figure out how to tell this story in the clearest and economical way with just images. With a wordless book, it is important that children can figure out what is happening without any adult present. But since children are experts at reading pictures, it is a matter of being as clear as possible about what each picture is conveying.

What were some challenges and turning points for decision-making?
At the end of the book the farmer and the clown part ways. It was difficult to find the right tone because I didn't want it to be too devastating. After many different sketches of possible last pages, I added a circus monkey who secretly follows the farmer home. This lightened the ending and allowed children to imagine what was going to happen once the farmer realized the monkey was there. It extended the story past the actual ending.

What were the biggest surprises?
How kids experience picture books is very different than the way adults do. Kids read pictures. Once kids learn how to read they lose some of their ability to understand images with focus and precision. By the time we are adults, most of us have a tough time just settling down and seeing pictures for what they are.

The biggest surprise to me is the depth of meaning that children can extract from this book. Even children as young as 2 or 3 have expressed, after spending time with this story, profound feelings about loss and grief and loving someone who is no longer there in some way. It is remarkable.

Anything else of importance in this project that we can learn from and bring to other projects?
What I have taken from this book is that when I am in the midst of writing and illustrating, I don't need to know what the project is about. Often awareness comes slowly during the process of making it. When you are in deep, the whole project can be mysterious. I have learned to trust that and let a project emerge almost like a dream I am having.[47]

Youth Subcultures

Adolescents are searching for identity. This is sometimes found within the context of youth subcultures. This search can be influenced by gender, peer groups, cultural background and family expectations. All young people need to feel validated and valued, and social learning and acceptance by the group help. They offer members a sense of belonging, identity, modes of expression and what they stand for outside their family, school and other social institutions.

"The two characters look a certain way on the outside but are different on the inside, which was what was interesting to me about this idea in the first place. Eventually, they each discover that the other is entirely different than they initially thought – and so does the reader."
Source: Marla Frazee (used with permission).

Many subcultures accept most of the ideas and values of the dominant culture. Some are countercultures that reject the dominant culture and seek to replace it with their own set of behaviors, values and ideas.

Designers creating products or services for youth may think of subcultures as a target demographic that they are aligning their project or brand values with. An adolescent may identify with particular values set forth by a company and choose to wear clothing or partake in activities or use media that a company promotes. Interests, beliefs, music genres, fashion, hairstyles, objects or possessions can also play a central role in identifying youth subcultures. According to Freddie Lee, professor of business and marketing at California State University, Los Angeles,

> Some of the things that define subcultures, have a foundation in consumerism. Teens need certain things to belong to a particular subculture, like a skateboard, musical instrument, bicycle, computer or surfboard. That need opens the door to peer pressure – teens may want to buy goods that other members of the group value.[48]

The way in which youth today express their allegiance to a subculture is transforming. As with virtually every area of popular culture, it's been radically altered by the advent of the internet: we now live in a world where teenagers are more interested in constructing an identity online than they are in making an outward show of their allegiances and interests.[49] Companies that are familiar with supporting a lifestyle around a subculture are experimenting with new ways of creating products and experiences that support the subculture.

Innovating for Vans, a Global Youth Culture Brand

Interview with Safir Bellali, global innovation lead, Vans

Safir developed and implemented a comprehensive innovation strategy for Vans. He represents the brand's interests to help shape the future of apparel and footwear. The insights he shared offer a well-developed framework on how to think about design for a particular subculture and lifestyle.

How were you, through your work at Vans, involved in shaping kids'/youth identity?

As part of our 50th anniversary, we revisited our culture and purpose to look towards the future. We worked on rethinking the creative vision of the brand by understanding what we stand for. The Vans brand statement is "Vans enables creative expression and inspires youth culture by encouraging the 'Off the wall' attitude that comes from expressing your true self." My interpretation of this statement is that there are many ways one can express themselves creatively. We pledge to give youth the outlets to do so, whether it is through what they wear, how they ride, what they play or what cultural references they share. This is a purpose that I upheld through everything I helped create, from protective gear for Action sports to the Customs program we recently launched.

What does being part of the skate, snowboard, bike and surf culture offer kids for social development?

"Off the wall" is a mindset for an individual. It is about finding your line, on a board, a wave, a mountain, your life. The sense of tribe is prevalent. Individuals are collaborators who grow as a tribe. We say "Vans is for anyone but not everyone."

Are there differences that boys and girls want from a brand like Vans or do they have unified goals?

We always design around the idea of inclusivity whether it's for action sports or lifestyle. For the consumer, there is a collective engagement and unified expectations. As far as product, most boy's product is available in unisex styles which girls will wear but we also have girl specific product.

Can you explain how Vans music and other events as experiences support youth culture around the globe?

Throughout my time at Vans, I had the privilege of attending numerous events around the globe: music concerts, skate, BMX and surf competitions. These events brought together people who share the same passion and culture. It's the ultimate gathering of expressive creators.

One of my favorite gathering platforms is the House of Vans and also pop-up events we've had in some unique locations. The House of Vans in London, for instance, is built across five tunnels underneath the Waterloo train station. To celebrate the brand's 50th anniversary, we had concerts around the world, Rio, Tokyo, Seoul, London, Paris and Berlin on the same day.

I was also fortunate to be involved in building an artist platform for customization and social sharing. This platform is meant to allow individuals (creatives, artists or not) to use Vans product as a canvas for their creativity through customization.

Can you explain how, through your work at Vans, you helped promote healthy playing, learning and growing?

Vans has four pillars:

- Music
- Action sports
- Art
- Street culture

> For playing: We engage in art, music and action sports. We encourage kids to get physical and find their voice on boards and bikes.
> For learning: We encourage kids to be progressive and push themselves, both physically and mentally.
> For growing: We design gear that inspires them so they can progress and push their limits.

The professional athletes Vans sponsor serve as role models for kids/youth. What are they modeling to help kids/youth with development?

The role models connect to youth to share positive values. Skate legends like Tony Alva and Christian Hosoi are idols to kids, and they inspire them to be their best selves and push the limits of what can be done.

As a creator, how do you stay on top of fresh when what is cool changes so frequently?

Authentic doesn't change with the trend. When you develop a point-of-view, and design principles stick to it. Vans has five classic styles that have been around for over 40 years. People have an emotional connection with the brand and product that transcends generations. Connect with the culture but don't follow it, create it.

In what ways will new technologies and cultural shifts affect Vans in the coming years?
Part of my job is to figure out what the new technologies mean for Vans and how to leverage them in a way that is authentic. We tend to think of technology as slick. Vans is DIY, unpolished, raw and gritty. What does that mean? As a team, we are digging into the process to find this out. We use scenarios, consumer insights and immersion trips. We are looking at the functional, emotional and social need our product is expected to fulfill.

What are some of the technical and aesthetic challenges when designing children's footwear as compared with adults?
We are developing more products for kids, so we are working on technical innovations for closure systems, new types of construction and structures to provide more support and are exploring lighter-weight materials and designs that differ significantly from our other product. For aesthetics, we have partnerships with artists but also with Disney and classic characters like Peanuts. Kids' product is meant to be fun and engaging yet still somewhat edgy. That balance is harder to strike with kids' product.

Customization is just beginning to take shape in product development. What role do you see for kids/youth in the co-design process?
With our commitment to creative self-expression, we are currently designing the product as a canvas for creativity. We have a customization platform where users generate content and upload it, and we make the product for them. No matter what your age is, being able to wear your creation is one of the most expressive and satisfying things you can do. That sense of pride and achievement is why we aim to promote.[50]

▶ Figure 5.13

"It is fascinating to see how these activities can promote a sense of belonging that maintains one's individuality. That's what's so special about action sports. It's about healthy competition, but you compete against yourself. You strive to be the best you can be at what you do."
Source: Safir Bellali (used with permission).

Peer Influences/Peer Pressure

This is when a person does something positive or negative to be accepted that they would not usually do. It is most common in middle childhood with boys. It is about balancing fitting in with being yourself. It could be used to promote more interest in activities such as sports, school, a volunteer project, having a boyfriend or girlfriend, smoking or using drugs. Children and youth who have poor self-esteem are more likely to be negatively influenced by peers.

"Two main features seem to distinguish teenagers from adults in their decision making," says Laurence Steinberg, a researcher at Temple University in Philadelphia:

During early adolescence, in particular, teenagers are drawn to the immediate rewards of a potential choice and are less attentive to the possible risks. Second, teenagers, in general, are still learning to control their impulses, to think ahead, and to resist pressure from others.[51]

According to Dr. B.J. Casey from the Weill Medical College of Cornell University, teens are very quick and accurate in making judgments and decisions on their own and in situations where they have time to think. However, when they have to make decisions in the heat of the moment or social conditions, their choices are often influenced by external factors like peers. Schools are missing an opportunity to boost learning by not tapping the teenage fixation on social life.[52]

Various types of peer pressure influence teenagers to buy things and help shape what teens ask parents to purchase. This, in turn, has a significant influence on trends. Many kids and youth imitate and follow anything which appears "cool." Others feel the pressure to have the latest electronic goods, technology and fashion or listen to cool music and wear the newest style to avoid an adverse reaction from friends. The American Psychological Association claims teen groups recognize the status of products and seek out branded products as a way to have immediate prestige with friends.[53] Most teenagers are brand conscious and may push positive or negative information about a product to peers to buy or not buy a particular item. As the designers, we have the choice to make and promote these items in a responsible way.

How can we use the influence of peers to make positive choices in many areas of their lives?

- Role model positive behaviors and messages
- Promote positive brand values, inspiration
- Create products and services that are cool in their own right
- Support kids and youth after purchase engagement

Bullying and Aggression

Bullies use superior strengths or influences to intimidate another physically or emotionally, typically to force them to do what the bully wants. Bullying almost always takes an emotional toll on the child being bullied, but the actions that constitute bullying vary. Incidents of bullying must be intentionally repeated and have a power imbalance.

There are four types of bullying, which can occur separately or simultaneously:

- Physical bullying, such as kicking or pushing
- Verbal bullying, such as name-calling or yelling
- Relational bullying, such as excluding or rumor-spreading
- Cyberbullying, which involves sending hurtful messages over digital devices like computers and cell phones.[54]

Aggression

There are two types of aggression:

> Instrumental aggression is where a child wants an object, privilege or space, and in trying to get it, they push, shout or attack a person who is in the way. Hostile aggression is where the child means to hurt another person.

Roleen Heimann, co-director of New School West preschool in Los Angeles, California, explained:

> Children may be arguing and having a conflict. In many situations, if children can't get along, they are separated. "What do you learn from that?" We spend more time with the child who was hit when we must spend as much time with the child who was hitting. How do we support him/her, too? Without labeling, we must see the behavior as meaningful and support the children working through it.[55]

A Mobile App to Promote Friendship

Interview with Natalie Hampton: Founder Sit with Us
Seventeen-year-old Natalie Hampton is a Los Angeles high school senior, anti-bullying activist, app developer and the CEO of a non-profit called Sit With Us, Inc. She won the Outstanding Youth Delegate Award from the United Nations Youth Assembly. (www.sitwithus.io)

What would you advise people who are interested in designing social innovation projects specifically to make the world better for kids/youth?
I feel like this has become my mission in life! When I am doing community outreach, I am always telling kids that they are never too young to make a difference. They need to take a look at their communities, identify problems, and then try to conceive of solutions. People probably thought I was crazy when I said I wanted to release a free mobile app around the world. After a lot of hard work, that is what I have done.

What are the most important things designers could do for kids/youth in the coming years?
Anyone who is farther along in this area can certainly volunteer time to encourage kids to bring their projects to life, and forming something like an institute allows us to share our ideas about scaling for impact.

What can designers learn by including or giving control over to kids/youth to not just consume but design their projects?
Digital natives have a significant voice when it comes to creating mobile apps, designs and marketing. There are

things that occur to us which might not to people who are not digital natives. For example, my parents initially suggested that my app contain a GPS tracking element, so that you could always find your friends on the school campus; however, I was adamant that GPS tracking is invasive and creepy, with the potential for abuse. I still believe that was the right call to make.

Do you have any insights that you can offer about designing for behavior change?
I like the idea of using technology to bring people together to socialize, and then encourage them to put down their phones. Sit With Us is a bit of a paradox that way. The hope is that people will use their phones only as an initial tool to meet up and that they will form friendships through actual conversation over a meal.

From your perspective, how does bullying affect kids'/youths' social and emotional development?
I believe that bullying and exclusion can cause very serious and long-lasting damage. So many adults have said that they still remember that fear of holding a cafeteria tray, scanning the room and not knowing where to sit. Some adults say that they are bullied in the workplace! I think that is why Sit With Us has resonated with people of all ages around the globe. It's not just for kids – people are using it in large workplaces (such as nurses at UCLA Hospital), in places of worship, at conferences and in colleges! Primarily, we are targeting both middle school and high school kids, adults are using the app as well (100,000+ users worldwide).

To most teens, developing and launching a real app would be overwhelming. What made you decide not just to keep this project an idea but to make it real?

I thought that my "little project" would be a therapeutic way of dealing with the trauma of being bullied in the past. If I could help just one person feel more comfortable at school, then it would have been worth it for me, and that is what made me forge ahead. If I sat back and did nothing, I felt like I would be just as bad as the kids who watched me suffer bullying in silence. I was a bit overwhelmed with where to start, but my parents loved my idea and promised to help me make it a reality.

How did research drive your design process?

I researched whether there were any apps out there doing the same thing and there was not – only one or two apps for adults to use for networking. I looked at anti-bullying resources on the websites of prominent organizations such as PACER's National Bullying Prevention Center, and the Making Caring Common Foundation. I also looked at a lot of academic studies about bullying and was happy to learn, as I had suspected, that student-led anti-bullying initiatives are far more effective than those led by adults.

I looked at all kinds of graphic design, knowing that I wanted something simple and cartoonish, with a redheaded figure (representing me) at all times. I also looked at how a lot of different social network apps flow and function from page to page. At that point, I drew a storyboard for each page of the app.

I have studied coding, but I didn't know enough to create a complex app with all of the features that I wanted, so I found a freelance coder to help me. He believed in my project, so he agreed to work at a much-reduced rate.

I beta tested the app in a neuroscience research lab at UC Santa Barbara. I think I was surprised by how many glitches we found and how long it took to resolve them; however, we worked it all out in about six weeks.

What was your favorite part of the process of this project?

I think the ultimate was being given the privilege to speak about it from the podium in the Great Assembly Hall in the United Nations about the importance of a peaceful and inclusive society. I will never forget that moment.[56]

◀ Figure 5.14

"I believe that every school has upstanders like me who want to take an active role in improving their community to make it more welcoming. Something as seemingly small as lunch can make huge strides in making a school more inclusive because if kids are more kind to each other at lunch, then they are more likely to be kind in the classroom, and outside of the classroom." Natalie Hampton. Source: Sit With Us (used with permission).

Children's Relations with Pets and Animals

Caring for animals can potentially boost kids' physical activity, and can likely help with learning essential life skills such as stress management, resilience, learning-to-learn, self-esteem and empathy, and manage everyday challenges.[57]

With well-developed programming, animals can be used in therapy for autism, cerebral palsy and other children with special needs. When designing children and animal interaction, it is also essential to keep in mind humane treatment and benefits for the animal. Young children do not yet have the knowledge or experience to be wholly responsible or understand the potential outcomes of their actions. When starting a project, include in the brief the benefits to the animal as well as the benefits to the child. A balanced approach can contribute to enhancing both of their lives.

Pets help:

- Teach kids about responsibility
- Encourage nurturing and empathy, cooperation and sharing
- Provide comfort, companionship and security for children on a daily basis or in traumatic circumstances
- Boost self-esteem and the connection is buffer against loneliness
- Encourage physical activity. Kids with pets have fewer allergies
- Families grow stronger and closer when focusing on activities that families do together
- Kids practice skills such as language with pets and creativity, such as imaginative play

The Benefits of Children and Animal Interactions

Interview with ethologist Dr. Francois Martin, section leader at the Behavior and Welfare of Product Technology Center at Nestlé Purina. He specializes in the study of behavior and welfare.

Are there similarities and differences among children who own different types of pets?
It is not species specific, it is about the child's relationship with the pet they care about. A single pet can meet many different needs. It also depends on who you are as a kid. For example the need to be active, to be a buddy, for comfort, reassurance or protection. The same animal can be a friend when you are sad

and listen and accept you. Dogs and cats especially adapt to context and the relationship with the child is dynamic.

Kids don't need to be an expert on how to be with a pet. It comes naturally for most kids. A dog or a cat reacts to them spontaneously.

Can you explain how a child's relationship with a pet can help with different areas of development for a general audience of pet owners?
Humans have evolved as a social species. We need others to survive, for pleasure and for reassurance. The same systems that we have that allow us to connect with

other people may have evolved for humans to have a social relationship with animals. We have a physical reaction to animals. Low-intensity physical touch of petting an animal is soothing. When petting, people are known to have lower heart rate, increased cortisol production and higher oxytocin levels.

Kids are naturally drawn to pets, and they provide emotional support. When there are animals around, they can help all areas of development including cognitive, physical, and social and emotional. Kids talk to their animals and practice language. They play with them and use their imagination. They run around with them and exercise, and they can help build self-confidence.

What types of specialized activities do you recommend for children for social and emotional development? Why?

Having a pet in the home and being around animals can be part of a well-rounded, healthy lifestyle. They help with the development of soft skills because they are practiced. If the pet is there for fun, it helps kids feel better.

For specific needs, pets can be helpful to address those needs with a curriculum or programs. For example, there are reading programs where kids read to their pet, and they forget all about their reading deficiencies, and it improves. If a child lacks self-confidence or people skills a pet can potentially help. We must know the child and which activities and pets can help. For a pet to become "therapeutic," the parent or the counselor must develop specific activities that address the child's challenges. For example, for a child with poor posture or hand–eye coordination it could be activities on horseback that specifically address these challenges; a child with anxiety may learn to cope with his stress by doing supervised relaxation activities with a dog.

How can pets help children with disabilities?

It is believed that pets act as a social lubricant. People with disabilities report that they are perceived as being more approachable and are talked to more with a pet.

When working on a specific issue for therapy, it is important to have a program that is proven along with the pet. If it is therapy in conjunction with a plan, it could help with physical, cognitive or emotional disabilities. For example, a therapy dog for autism or language skill has shown positive results with structured programs. It may help kids pay attention more, or there may be other reasons. Sometimes the family pet can be included, or specific therapy animals may be needed. It depends on what you are trying to accomplish.

Think about the goal of how to use the animal in a specific situation. When I was working with a horseback riding program, some people just wanted their kid to have fun. Riding a horse can be an enriched experience for any child and can be used for the pure enjoyment of the activity. Others want their child to work on posture and were bored with going to the physical therapist and wanted to try something new. It was an intervention for that kid. The same horse could build confidence with another kid or could help with balance and coordination. Know what you want to accomplish and come up with a plan.

What are the most important learnings for designers to consider when working on a project to boost the relationships between kids and their pets?

When developing products, programs, and experiences: Take into consideration the natural behaviors of the animal. A cat will participate until they become uninterested. Their calming presence may work when you are dealing with anxiety, energy and focusing issues. In contrast, when you ask a Labrador to do something they reward you right away. Each can be included for their strengths. Remember that animals are living beings.

Messaging and creating awareness: You can always increase safety and humane education for both the parents and the child. A 4–7-year-old really won't understand the proper behavior around a pet. They may be naturally drawn to a pet and make the pet feel threatened. Sometimes people just don't know how to behave and need to be shown.

Enhancing the relationship between the child and pet: Don't just jump into getting a pet. When choosing a pet read about the pet, speak to vets, try to interact with potential pets. Find a good fit before adopting them.

Always make sure kids are supervised: Engage younger kids to care for the animal but help them and watch them closely. If you tell a kid to feed a pet two cups of food, watch to make sure it is not one or five cups. The level of supervision will lessen, as they get older, but there always needs to be an adult involved. It is safest for the kid and the animal.[58]

Design Prompts: Children and Animals

- How does your project best support animal well-being? The child's well-being?
- Brainstorm toys, programs and interventions supporting the natural behavior of the animal

► Figure 5.15

Pet relationships create an enriched space emotionally for children.
Source: Laura Flower Kim (used with permission).

- What are the needs of the pets? The needs (physical, social and emotional, cognitive) of the children?
- How can programming support a physical product or environment?
- What level of supervision have you included?

Alone Time and Boredom

Babies and toddlers naturally play alone, since they have not acquired the skills for social play. Once a child is old enough for play dates or preschool, much of their time is spent with other kids. Grade schoolers spend most of their time with other children in school, after school, at team sports, camps and lessons. In general, modern parents, and as a result their kids, are obsessed with filling up all of their time, educating and entertaining their children. Many parents want their kids to be stimulated and fear that not being involved in enriching activities will put them at a disadvantage. Tom Mott, senior producer and interactive learning designer, affirmed: "We're starting to get to a point where nobody is bored anymore. Adults whip out their phones if they're bored. Kids see that behavior and want to do the same. Boredom can lead to wonder and experimentation and trying new things. We need to leave time to be bored."[59]

Every child needs quiet time to be bored and create a full inner life. Since kids don't gain a particular creative talent, skill or intellectual benefit from boredom, it is hard to allow time for it, especially if the child complains about being bored and nags about having nothing to do. It is harder to support a child in alone time than it is to drop them off at a class or give them something to do at home. Alone time does not include time watching TV, playing video games or on social media. It is so easy to partake in these activities from the minute kids awake. Alone time often requires prompts and materials so that they can discover how to fill their time.

Children who know how to fill their time alone rarely feel isolated or lonely. This is easier for only children than for a child with siblings, but should

◀ Figure 5.16

Alone time allows children to pursue their interests, nurture what they enjoy, and develop powers of inventiveness, observation and concentration.
Source: Eric Poesch (used with permission).

still be encouraged as a crucial aspect of the development of independence. Without it, children will always seek support in others or in external activities. As designers, we can design activities for children to do alone, but how do we allow children and families to allow for alone time? How do we design for boredom to promote some of the benefits that come from it?

What happens when children are comfortable being alone and bored?

- They entertain themselves, pursue interests, explore and discover the unexpected, and ignite their curiosity, imagination, playfulness and perseverance.
- They are able to relax and rejuvenate, daydream, let their mind wander. It allows their brain some free time from worries and expectations of the day.
- They learn how to use and manage free time effectively and find solutions to fill their time.
- They learn to be comfortable with themselves, become self-sufficient and confident, which can elevate self-esteem.[60]

Humor

Designing for kids and play is surely on the lighter side of the design profession. Humorous writing, visual representation and communication are an important part of our toolkit. For many of us, humor and levity might be one of the big draws when working on a kid's project. Not only do we have fun, but we can also be funny. With our creations, we encourage children's inherent love of jokes, riddles and puns and at the same time form an emotional connection with them to make them laugh.

Humor in our designs can make a boring subject more engaging, a serious topic lighter and easier to understand, or a playful topic more entertaining. Humor generates excitement in learning. In her book *Using Humor to Maximize Learning*, Mary Kay Morrison explains: "The surprise elements of humor alert the attentional center of the brain and humor has the potential to hook kids that are easily bored and inattentive."[61] Almost anywhere we use humor, children will accept and appreciate it.

Humor can be an effective tool in design

- Through visual puns, illustrations and graphics
- In design forms, materials or product concepts
- In communicating messages that are distinctive and recognizable
- To create positive feelings and associations
- In designing characters, funny interactions, and writing stories
- By challenging kids' ability to analyze patterns
- To appeal to kids' natural silliness

It requires us to
- Tap into what makes that particular situation funny or amusing
- Hint at the joke and keep it simple and easy to grasp
- Develop a little twist on what is expected
- Find it in everyday situations that kids can relate to
- Laugh at ourselves and universal experiences
- Enjoy and participate in the playful aspects of life

In her book *Design Funny*, Heather Bradley explains reasons to include funny in our designs:

- Funny is persuasive
- Jokes trigger a gut reaction
- Learning is easier when you are edutained
- We crave eureka moments
- It produces pleasure

Funny is social:
- Ha-ha is a high five
- Some of our best friends are brands
- Inside jokes bring us together

Funny is hard-wired:
- It is addictive
- It's in our programming
- Everybody laughs[62]

Benefits of Humor for Kids

Cognitive Benefits

- Play with language introduces new words, double meanings and builds vocabulary. Repeating riddles consolidates memory skills and helps with word recognition.

- Creativity in joking boosts divergent thinking and problem-solving.
- Humor is linked to verbal intelligence, general intelligence and abstract reasoning.

Social-Emotional Benefits

- Kids with a well-developed sense of humor gain control of their daily mood, are less likely to be depressed, have higher self-esteem, and can handle stress better.
- Children who know how to use humor in social interaction are also better at putting others at ease, making us feel more connected to them. Kids feel happiness by making friends happy.
- Humor creates a natural environment in which all communication is easier.
- Kids who can smile at their own mistakes are better able to deal with their own quirks and are more tolerant of others.

Physical Benefits

- People who laugh may even have an increased resistance to illness or physical problems. Have lower heart rates, pulses, and blood pressure; and have better digestion. Laughter may even help humans better endure pain, and it improves our immune function.
- When we laugh, pleasurable chemicals (endorphins) relax our muscles and relieve stress.[63]

Ages and Stages

Humor is subjective and largely shaped by individual taste, so it is important to know about the children you are humorous with. The specific topics children laugh at tell us which developmental tasks they are struggling with. Children who are mastering toilet training are enthralled by "bathroom" humor.[64] Neuroscientist Robert Provine studied when and why we laugh, and his book *Laughter: A Scientific Investigation* supports the theory that laughter is not only instinctual but serves as a primal form of communication. People around the world begin laughing as early as 4 months old regardless of their cultural upbringing. Social conditioning doesn't influence why and when we laugh.[65]

Babies

They know when people are laughing, smiling and happy. They are responsive to funny noises, a laugh or a smile, playful physical interactions, and stimuli. By 15 months old, babies understand that doing something unexpected is funny.

Toddlers

- Toddlers appreciate physical humor and elements of surprise (an unexpected tickle)
- A toddler's use of symbols and language leads to playing with reality. Incongruities are very funny.
- They are beginning to master the intricacies of language and will giggle when they hear a combination of words and nonsense syllables/words. They understand that the nonsense syllables are different from the words.
- They are learning that there is an order to the world. Placing a sock on a foot is not funny. Setting it on an ear is hysterical.
- Toddlers start trying to make their parents laugh. The 2-year-old with a funny hat has made a joke.

Preschoolers

- Three-year-olds enjoy sharing their sense of humor with significant adults and making up silly stories.
- Four-year-olds are fascinated by bathroom humor and are not yet completely sensitive to the effect their humor has on others.
- The mismatch between pictures and sounds (a cow that quacks). They are more likely to find an image that is out of whack funny than a pun or joke.

School Age

- They can play with basic wordplay, exaggeration and slapstick, replacing words in a sentence to see the humor in it.
- Five- and six-year-olds use humor as a way of building friendships and becoming part of the larger group.
- The excitement of playing with meanings, logic and abstractions is at its peak at this point. Around age 6 or 7, kids start to understand the language well enough to know that words can have two (or more) meanings.
- Love nonsensical situations, especially when they involve grown-ups.
- Kids often find playful violence to be funny.
- Slapstick and practical jokes can be very funny for school-age kids. They understand the difference between a good-natured practical joke and one that hurts someone physically or causes hurt feelings.
- They may deal with anxiety by laughing or making inappropriate jokes.
- They like puns, riddles and other forms of wordplay. They'll also start making fun of any deviation from what they perceive as "normal" forms of behavior or dress, and gross-out jokes related to bodily functions are a hit too.
- More subtle understandings of humor, including the ability to use wit or sarcasm and to handle adverse situations using humor.[66]

Adolescents

By the time boys are 10 years old, they are telling jokes that are very physically violent and very sexual. Girls at that age like humor that is less physical but

◀ Figure 5.17

Most of the time when we see young children playing together, they are also laughing and being silly.
Source: author photo.

more sarcastic or verbally aggressive. The jokes help define membership in a particular social group.

Boys and girls are using humor to accomplish the same goals, coming to terms with the issues of greatest concern to them. They are far too emotionally stressful for them to deal with directly. They use the jokes as an opportunity to determine cultural norms and acceptable behavior.[67]

Design Prompt: Humor

Approach your project with each of these different types of humor. How do the solutions differ?

Anecdotal
Deadpan/dry/witty
Hyperbolic
Ironic
Juvenile/sophomoric
Screwball
Situational
Slapstick
Silly

Notes

1   Pessoa, Luiz 2009
2   Weissberg, Roger 2016
3   Holloway, Ani 2016
4   Center on the Developing Child, *Executive Function & Self- Regulation*

5   Mindfulschools.org
6   Heimann, Roleen 2017
7   Navarro, Joe 2010
8   Weinschenk, Susan 2011, *100 Things Every Designer Needs to Know about People*
9   National Institute of Mental Health
10 Doheny, Kathleen
11 Kim, Jin Hyung 2017
12 Schor, Edward L. 2004
13 Kids Matter, *Motivation and Praise*
14 Hayes, Dayle 2015
15 Holohan, Meghan 2016
16 Holohan, Meghan 2016
17 Amsterdam, B. (1972)
18 Oswalt, Angela 2008
19 Oswalt, Angela, *Identity and Self-Esteem*
20 Self-concept
21 Kids Matter, *Why Culture Matters*
22 Eliot, Lise 2010
23 LoBue, Vanessa 2016
24 Genderspectrum
25 Otto, Ali 2017
26 Hines, Melissa 2013
27 Heimann, Roleen 2017
28 Johnson, Chandra 2015
29 UNICEF gender action plan
30 WebMD.com
31 Levine, Paul 2017
32 Kids Matter, *Resilience*
33 Daniels, Natasha 2016
34 Bingham Newman, Ann 2017
35 Voss, Angie
36 *Separation Anxiety* 2017
37 Passman, R.H. 1987
38 Play Therapy
39 Bingham Newman, Ann 2017
40 Benoit, Diane 2004
41 Moges, Bethel; Weber, Kristi
42 Mgbemere, Bianca; Telles, Rachel 2013
43 Kids Matter, *Learning positive friendship skills*
44 Poole, Carla; Miller, Susan A.; Booth Church, Ellen
45 Kennedy-Moore, Eileen 2012
46 Kennedy-Moore, Eileen 2013
47 Frazee, Marla 2017
48 Ryan, David B. 2017
49 Petridis, Alexis 2014
50 Bellali, Safir 2017
51 Steinberg, Laurence 2008
52 Murphy Paul, Annie 2015
53 Ryan, David B. 2017
54 Educational Development Center 2013
55 Heimann, Roleen 2017

56   Hampton, Natalie 2018
57   Madden Ellsworth, Lindsay 2017
58   Martin, Francois 2017
59   Mott, Tom 2017
60   Belton, Teresa 2016
61   Morrison, Mary Kay 2007
62   Bradley, Heather 2015
63   McGhee, Paul 2002
64   Kutner, Lawrence
65   Provine, Robert 2001
66   KidsHealth
67   Kutner, Lawrence

Part III | **Childhood Today**

Chapter 6

Play

Do you still play? When experimenting in the kitchen, do you explore the combination of ingredients to invent something new? When at the park, do you feel the physical freedom and exhilaration when floating on a swing? When designing on the computer, do you lose yourself for hours in color and image? Is design play to you? Danish play researcher Karen Feder explains:

> Play and design have a lot in common – it is imaginative, creative, explorative, iterative, meaningful, emotional and engaging. Remember how it was when you were a child, and you forgot all about time and place because you were so immersed in playing? To reach a "play state" – you have to go through the design process with an open mind, let yourself be surprised and embrace unexpected opportunities. This requires a safe and trustful environment – as play does – and an understanding of the process as an essential journey for the creation of meaningful design.[1]

Play begins in infancy, and small children spend much of their time testing hypotheses in play, helping with the mastery of skills. From an evolutionary perspective, the primary purpose of play is education. Humans are mammals, and children, like all other young mammals, play to practice all types of skills they will need throughout life, including survival skills. While young monkeys play at chasing one another, swinging from trees, human children chase each other in a playground in a game of tag. Someday, children may need to have experienced risk so that they can run to protect their bodies.[2]

In later childhood, most children play less than younger children, and today even less then they have in our most recent previous generations. This deprivation of childhood is harming many children.[3] This is one of the most significant challenges for designers, educators and everyone involved in the world of creating for play. The view that play is essential for survival and personal growth is in sharp contrast to the organized, extrinsically motivated, time-restricted activities children experience more commonly today to support growth. While the time for play is decreasing, designers are finding other opportunities to sneak in play throughout childhood. Since most kids enjoy play, almost everything designed for them could be considered and conceptualized through the filter

Play happens naturally for children and opens a kid's world because it opens the imagination.
Source: author photo.

of play. A medical device designed for kids can encourage them to engage in an active experience to heal. Teaching educational content through interactive media, a silly story, can motivate kids to learn. A playful approach to graphic design, music or toys can communicate healthy messages.

Play and Culture

Play is universal. Children's brains are wired to learn through play. All kids around the globe participate in play. The time allowed and the attitude to play activities very widely in different cultural contexts. Rena Deitz, senior education specialist at International Rescue Committee, who has worked with refugee children, explains: "In camps, there is a lot of dead time, and there is more time to do things around play. In urban areas, it is harder. There is the reality of doing chores and working. Many kids don't have time to play."[4]

Play is recognized as being so important to the development of children that the United Nations High Commission on Human Rights in Article 31 has recognized play as a fundamental right of every child, whether or not the culture in which they are socialized acknowledges, supports or sets aside time for play.[5] Through play, children learn societal roles, norms and values. Many play patterns are similar from culture to culture, although the way children play and what they play with may differ. When a natural disaster occurs, at refugee camps or in war zones, in addition to food and water, non-profits also supply soccer balls and other play tools to provide opportunities for kids. Kids need to play to feel whole.

Structural aspects of the immediate environment – time, space, availability, social environment, parenting styles – are all factors affecting the frequency, duration and nature of play activities. For example, in Asia, cultural influences tend to see play and academic activities as separate from each other, although in

Italy, there is little distinction between play and children's other activities, and they strongly emphasize social interaction in children's play. As designers, it is essential to understand the values around play and design culturally appropriate products, experiences and environments. Here are some attitudes towards play across the globe.

Play as ...
a process of learning
a source of possibilities
empowerment
creativity
children's work
fun activities

The LEGO Foundation continually develops and publishes research around play and is a great place to start when conducting research for a project.[6]

Play is fundamental to modern childhood and the way in which many children develop knowledge of their bodies, personalities, identity and social skills through exploring different play patterns. Many current theories of child development, education and parenting are based on the idea that play offers the experience for a child to construct learning rather than being the passive receiver of knowledge. What is the draw of play for kids? I have asked many kids of different ages what they like about play, and this is what they said:

"It is fun," Joshua, age 3
"It's fun," Sam, age 9
"I feel good when I play," Olivia, age 6
"I don't know; it's just fun," Carlo, age 7
"I like it," Layla, age 5

I think we can all agree that fun is the most significant appeal for children. What they are explaining is play as a state of mind. Our designs drive that mindset. Play researcher Karen Feder suggests that designers consider this question:

Why do we need to design for play? Play is the construction of our (western) world and the way we have chosen to live our life, constraining us as human beings, so we need to create a kind of fake reality to play with and upon? For instance, why do we need to design artificial playgrounds, and make them look like nature? Why don't we just prioritize wild nature in our surroundings? Maybe we can do something better or different with design, but the important thing is to constantly keep in mind, why we are designing for play.[7]

Defining Play

We all know it when we see it, but can play be defined? Even experts in the field who study play have a hard time agreeing on a definition. At a recent conference, I attended a lecture by Dr. Stuart Brown, the founder of the National Institute for Play. He defined play as a separate state of being. "In play, we identify innate talents and intrinsic motivations. It is important that we learn what our talents are and nourish them."[8] Play is an attitude. A confluence of characteristics define play. We must look at the motivations and mental attitudes towards the activity and why they are doing it, not the behavior. Two people might be throwing a ball, and one might be playing and the other not.

Play is usually defined along a continuum of more play value or less play value. The motives and attitudes can range from 0% to 100%. It is impossible to accurately quantify precisely how playful an approach is based on an experience, but we can use criteria and methods such as observational research, playtesting and interviews to determine priorities, how engaged someone is in an activity, and whether our project meets the requirements we set. A behavior that brings forth an attitude that satisfies a more significant percentage of these characteristics would be considered by most people to be more playful, while one that meets few criteria could be referred to as less playful.

1. Play is freely chosen, self-directed and spontaneous. The person playing is always free to quit. It is that expression of freedom.
2. Play must be pleasurable and enjoyable. Children must enjoy the activity, or it is not playing.
3. Play is process oriented. Children engage in play solely for the state it brings.
4. Play has structure or rules that emanate from the minds of the players. All players must freely accept the rules, either their own rules or ones created within the group.
5. Play must be intrinsically motivated. Hybrid forms of work and play such as a goal-oriented assignment by a teacher that includes active engagement and intrinsic motivation can also be play and produce playful learning.
6. Play is non-literal and imaginative. Play involves an element of conceptual constructs or make-believe.
7. Play requires active engagement on the part of the players. Children are physically and/or mentally alert and involved in the activity with a relatively relaxed or excited frame of mind.[9]

These criteria are not only useful to experts who study play but also helpful for designers to consider when developing insights during designing for play. A design that includes high play value appeals to different types of users or lengthens the life or usability of a design and would be considered by many to be a better design. A design that has more play value would most likely be stronger than a design that has less play value in the eyes of a kid. Even better if it meets all sustainability and ethical criteria, and has a positive social impact. Paul Levine, CEO of Play Science, suggests: "Designing for play is hard fun. To design intrinsically motivated, fun, scalable and marketable play is challenging. There is complexity to it. You need to think about whom you are designing for, the context, platform, form factor, structure, positioning, and distribution."[10]

Design Prompts: Play Value

Ask yourself these questions throughout the stages of the design development process.

- When you are developing your design concepts, how many of these play criteria can be achieved within a single concept?
- Can you dive deeply into one of these play criteria and add several layers to make a multi-tiered engaging experience?
- Focus on the design of the scenario of play (play experience). How would a kid play with it? What steps might they take? How might they differ depending on their personality? Interests?
- Can spontaneity or something unexpected be added to the project?
- Can play prompts be added to leave space open for the child's imagination?
- What kind of rules might a child create at different developmental stages?
- How will a social group of children differ at various stages?
- How can this project include sustainability, meet ethical criteria and have social impact?

Benefits of Play

Learning through play is a term used in both education and psychology to describe how a child makes sense of the world they are living in. Play researcher Karen Feder explains:

> Play is something we can't live without, so it is essential that we nurture and ensure to keep on creating high-quality play experiences – not only for children but all human beings. Children are not only adults in the making, they are human everyday, and deserve to be treated with respect for their value in themselves.[11]

Play meets the needs of the whole child directly and indirectly. As they play, children are improving cognitive abilities, honing communication skills, building creativity, processing and expressing emotions, developing physically and enhancing social skills. Dr. O. Fred Donaldson explains: "Children learn as they play. Most importantly, in play, children learn how to learn."[12]

Physical Development

Play fuels brain development and production of the neurotransmitters essential for growth and learning. It is multisensory and engages the child through various inputs. Children develop fine and gross motor skills. It supports strength and fitness.

Social and Emotional Development

As children learn to play in different-sized groups, they learn about appropriate behaviors within particular contexts. Play helps children figure out the complicated negotiations and compromises of social relationships. They learn which behaviors are morally or culturally acceptable and which are not.

Children learn about themselves: what they like and don't like and about different aspects of their identity. They gain emotional awareness and learn how to regulate their behavior, become independent and resilient in the face of a challenge, and build self-confidence; feelings of happiness, power over the environment, sensitivity to others, emotional strength and stability, spontaneity, humor, beliefs about self, and to manage fear, anger and frustration.

Cognitive Development

Play puts children into a present, attentive, focused, receptive, integrated state. Play provides a strong foundation for intellectual growth and creative thinking, and builds problem-solving skills and academic knowledge. Kids develop language skills, verbal judgment, reasoning and symbolic thought. Play builds the foundations for exploration, hypothesis testing and discovery for later learning. It allows children to imagine and discover. and introduces them to new situations, ways of thinking and environments. By touching this creative side, the children learn what they like to do, what they are good at and what they dream of being good at.

Consider these aspects of child development in your design. For example, say you are developing a baby toy that helps with physical development. Instead of just stopping there – elaborate. How can the toy strengthen different muscles to improve gross and fine motor skills? Which muscles have you chosen to develop, and why? How many other senses can you engage? Talk to a pediatrician. Knowing that a baby will put it in their mouth (after it has been on the floor), can it be made of an antibacterial material? Can you add a pretend play or storytelling element?

If you are developing a design that focuses on improving cognitive development, don't just stop there – keep going. What in the brain is getting wired?

Is it producing a learned skill, such as a language, or one that requires executive function development, such as the ability to control behavior? Can it do both at the same time, one as a primary objective and the other as a secondary benefit? Asking these questions will lead you to further research solutions that already exist, knowledge about the world you are designing for, and insights into what is needed to support growth. Look at academic journal publications and recently released studies. Start from where others have left off and develop new and better solutions.

Structured, Guided, Free Play

Do you learn more when you have a rigid assignment with detailed goals and deliverables, or do you prefer to make up your own? Do you like to have a mentor teaching you a new concept or skill, or would you rather figure things out through tinkering? Do you want to be a design entrepreneur, or would you prefer to work for a company? We all have our preferred methods of designing, and most likely the way we like to learn and work is rooted in the natural way we played as children.

These types of play pertain to the level of structure imposed on the play activity, ranging from a lot of structure to none at all. It is essential to focus on the quality of the play experience. In quality play, a child is engaging entirely and has a heightened ability to learn. If you have spent time with young children, you are aware that they are known to jump around from activity to activity, room to room, indoors and out, what seems to be every minute. Before you know it, there are puppets, blocks, toy cars and dolls scattered around the house and yard. Quality play has excellent benefits by expanding the child's attention span and ability to stay focused on one activity, go into more depth and stay interested because they are exploring, figuring it out and accelerating concepts. They are making connections and advancing understanding to build upon.

Structured Play

Generally speaking, when a child is engaging in structured play, learning happens through accomplishing goals by achieving pre-existing objectives. Most games with rules fall under this category: card games, board games, puzzles, classic outdoor games like tag, organized sports and model kits are structured. Rules are put in place to govern the choices of the players, identify courses of acceptable action and challenges, and give feedback.

Guided Play

Guided play is when adults or more experienced children facilitate playing. Parents and educators can offer more targeted learning experiences. The adults may set up the environment or participate in the play. The adult's role varies depending on their goals and the child's developmental level. When setting up the experience, they may provide objects, materials or an agenda with very little involvement. They may also participate as a co-player by asking questions, commenting or expanding on activities. Digital play can also be considered guided play.

Open-ended Play

Free play provides children with true autonomy to help them develop a lasting disposition to learn and freedom to explore and discover their interests, skills and talents. Controlling the course of one's learning promotes desire, motivation and mastery. It allows maximum exploration and heightened imagination. There are often unexpected twists and turns that can keep the play going for an extended period. Children engage in free play that flows without set rules or expectations.

It is much easier to design for structured play and guided play than for free play, because you can control more of the experience through rules and process-oriented tasks. When creating for free play, think through as many of the possibilities as you can and then embed prompts and features in the design to promote them.

Design Prompts: Creating Quality Play

- When designing for structured play, set up the rules of engagement. Break down the steps into tasks. What goals need to be achieved? How are they proven to be successful?
- When designing for guided play, ask: What is the role of the guide? The child? Is it a collaborative or directed experience? What kind of prompts can the guide use to promote further exploration?
- When designing for free play, try to leave as much as you can to be open-ended and up to chance. Imagine many alternative play scenarios and embed features in the design to bring those out.

Individual and Social Learning

Social play emerges over the development of the child.

Solitary Play

This is the primary play pattern for babies and young toddlers and most often seen before the age of 2 to 3 years old. The primary focus of solitary play is on exploration and learning how to do things. Playing alone gets refined and decreases throughout childhood while they start to increase in-group play activities. It is essential for all age groups, including older children, because it gives them the time they need to think, explore and create and to learn how to be alone.

There are many positive aspects to solitary play:

- Freedom to use the imagination
- Learning and practicing physical and mental skills
- Able to make their own rules
- Not having to meet anyone else's expectations
- Being fully engaged in an activity
- Exploring, creating and learning how things work

Only Is Not Lonely

In America, more children are growing up as only children without brothers and sisters as playmates and therefore playing alone at home for much of their childhood.[13]

Parents love to see their kids being independent, being creative with ways of entertaining themselves and not depending on TV or parents for engagement.

Onlooker Play

Onlooker play is when the child watches others playing but does not join in with the play.

Parallel Play

In this stage, the child gravitates towards other children, becomes more interested in making friends and plays next to them, while always maintaining their independence. What we observe is the child playing alongside another child or in the middle of a group, while having fun in solitary activity.

Associative Play

This involves social interaction between the children. Children are playing separately and may share, pay attention to others and communicate with others, but do not coordinate play objectives or interests.

Cooperative Play

Children engage in cooperative play with a common goal and work together to achieve the goal. This is common in older preschoolers or younger preschoolers who have been around a lot of older children. Throughout childhood, children choose to and need to participate in frequent social play to learn among their peers.

There are many positive aspects to peer-to-peer play:

- Builds confidence in social situations.

- Child learns to work within a group, negotiate, advocate for their ideas and deal with frustration.
- Child learns social rules such as give and take, reciprocity, cooperation, sharing and reasoning.

Adult–Children Play Interactions

Social play with adults

Adults are sometimes initiators, directors and partners promoting the play experience for children to play on their own, with peers or with them. Designing play interactions between adults and children is often favorable for bonding. Consider these different levels of involvement from the adult and how a caregiver best supports the play experience.

Onlooker: Adults observe children from nearby and may make an occasional comment, but do nothing to enhance or disrupt play. Observation can help the adult to understand children's play interests.

Play manager: Adults help children set the context for various activities and offer assistance when asked.

They may propose extensions to the play scenario, but children are free to follow or ignore the suggestions as they please.

Co-players: Co-players join in play activities. Typically adults take small supporting roles, while the child takes the lead. Co-players might extend play and often model play skills for the child, such as sharing or role-playing.

Play leaders: The adult actively guides children's play by doing the activity. Play leaders aim to enrich and extend play by suggesting new themes, props or plot twists to the current scenario. This involvement usually occurs to start or maintain play.[14]

Design Prompts: Social Play Interactions

- Imagine play scenarios that a child might experience when they are playing alone.
- How does your design promote different forms of social interaction between peers?
- How does social interaction change with co-players of various skills?
- How does the caregiver support the play experience?
- With a single concept, develop a range of play scenarios using the benefits of solitary, parallel, associative and social play.
- What will make an adult who does not want to play engage in play?

Designers as Play Advocates

These are individuals or organizations that speak, write or design to support the benefits of children's right to play. The reduction of play in children's lives today has brought about a movement in society to expand access to play in underserved communities, bring back play for children and support the benefits of play across the lifespan. We ask questions such as: How do we get parents to recognize the value of open-ended play and support families to spend time together and play? How do we bring more play to education and increase recess time in schools?

Instead of introducing something new, the easiest way to design for behavior change is to choose a behavior that already exists and include the new behavior. This might end up being a final solution or a transition solution. Fitting play into unexpected places and the down times in between other events is also a good start. When people are primed for play, it will be easier to test out a new solution.

Creating a Playground at a Nepal School

Interview with Marie-Catherine Dube, HOPE Initiative, Nepal playground

Marie-Catherine Dube is a play advocate who built a playground at a Nepal school while she was a student in college. Today she designs for water play.

What were the goals and objectives of this project?

When I first landed in Kathmandu – bright-eyed and eager to dive in – my goal was to design and build a playground, immersing myself in the culture of this beautiful nation I knew nothing about. As I spent more and more time living day to day with the Nepali people, studying the schools, how the children play, my goals shifted and expanded. I realized the impact that designing a social platform could have; each day on the construction site was a battle in literally convincing people not only how important play is for children but that they – the community – could create the same playspaces THEMSELVES once I'd leave after a few months' time. The overarching goal, therefore, became to create a social impact platform advocating and empowering the child's right to play.

What research method did you use in creating the designs?

Cultural and anthropological research: Visits to schools, playgrounds, observing children on the street, how every social class would play

Childhood development research: Understanding how play correlates to early childhood development in a third world context (versus the known; a first world context)

Design, materials and engineering research: how the locals utilized natural materials available to them, the best building methods based on the climate/materials, etc., third world playground safety (in a place where safety isn't a primary concern, how do we enforce SOME safety in the design?)

Concept testing, user testing research

How did you incorporate the research into your final design?

I wanted to put in as many opportunities for growth and development as possible while staying firmly in the realm of Free Play. I hadn't expected to add aspects of culture and social behavior into the playground but repeatedly found that these would be important in a successful final design.

Give examples of how you considered children's behavior in the design of this project?

Children's behavior in the most rural areas was surprisingly different from what I saw in the city's 'expensive'

▲ Figure 6.2

Kids on full playground.
Source: Marie-Catherine Dube (used with permission).

schools. Rough-and-tumble play was most prominent as a play type, so I had to consider ways to cater to that in the overall design. There was also 'generic' children's behavior; things I didn't necessarily see on site but through other forms of research knew were essential aspects to also design for in the final playground.

You now work designing aquatic play structures for a large company – How is your design process different from the playground you built?

Well … going from designing play spaces in a third world environment to a corporate, global company … you don't get much further apart on the spectrum! There is one sole thing that has carried over back and forth in both types of work as far as process, and that is play (how play influences the process and outcomes).

The most significant difference is that in Nepal, the process was "trial and error then learn from it and keep trying and keep failing till you get it." At my job now, the process is more rigid; we're still learning every day and developing best practices, but the creative aspect of trying and failing then learning and improving (the process I used in Nepal) isn't as strong. You need to consider, though, the two extremes: a third world country where the project is brand new versus a corporate environment that has already been researched with trials and learnings.

Splashpads are all the rage these days, replacing or being added to more traditional play environments. Can you explain from a cultural perspective why they have taken off?

Culturally for us, in North America, water is something that signifies cleanliness and life and the experience of being refreshed. In some countries in South America, for example, public water is dirty and having a communal space where everyone goes to play is not appealing. There is a huge appeal for municipalities to install Splashpads because there is typically no supervision required when there is no water depth (a pool).

Can you explain the benefits of aquatic play?

Water is such a beautiful play tool. Accessible for everyone, it adds sensory experience and development for all: from those with any disabilities to energetic teenagers all the way to grandparents accompanying a visit. It can be used to encourage high-action movement, it can be therapeutic, it draws people in and brings people together. It can teach cause and effect, it can demonstrate different textures and sounds.[15]

◄ Figure 6.3

"Water provides the perfect opportunity for multigenerational play, cooperative play, competitive play, and even individual play. Play is ALWAYS beneficial; water adds even more benefits."
Marie-Catherine Dube.
Source: Vortex Aquatics Structures (used with permission).

Play Patterns

Children have many more things to learn than other mammals and therefore, play in many different ways to learn to master them. The play patterns in this section reflect the type of activity that children typically engage in. Children usually don't follow strict rules and categorize their play. However, creators for play do, because we try to fit what we are creating into a segment of the market or produce specific experiences or learning outcomes. Often, children combine play patterns and have a few overlapping activities within an experience. For example, a child can be engaging in physical play (running) with an object (ball) or in constructive play (building) while pretending (an imagined city). Therefore, as designers, we usually prioritize these patterns and develop a design with a primary play pattern, a secondary play pattern and so forth. The primary play pattern is usually how we message or communicate the play experience to the child or caregiver, and the secondary play pattern might be embedded in and discovered by the child. However, the secondary play pattern might represent a primary play pattern for a child with different interests. Include several layers of opportunities for play in your designs and keep them open-ended so the experience can be interpreted and altered by the child.

Attunement Play

Personal connection with others brings about a feeling of joy. Sometimes that connection comes from what is said in a shared conversation, but many times what is not stated says more than the words. We begin to learn these expressions in attunement play. A newborn and a caregiver make eye contact, exchange smiles or make silly faces. With this connection, the child experiences a spontaneous surge of joy and responds. The caregiver returns the response and may add some rhythmic vocalization or baby talk. This is the grounding base of play that begins very early in an infant's life and continues into all stages of human development. Interactions such as winks, facial gestures, the tone of voice, the movement of a hand or the tip of a head all have meanings. Mastering the meanings helps individuals develop abilities in non-verbal communication. This is very important, because a large percentage of what our brains perceive in communication with others is with non-verbal signals. "It is known, through research using brain-imaging technologies, that the right cerebral cortex, which organizes emotional control, is 'attuned' in both infant and caregiver."[16]

Attunement and attachment are related. A caregiver who is responsive to a child's needs beginning in infancy establishes a sense of security within that child. The child learns that their caregiver is dependable. This creates a strong foundation on which that child can explore the world. Throughout our lives, it helps us to build and maintain our relationships with others.

Caregivers are children's first play partners.
Source: author photo.

Design Prompts: Attunement Play

- You've heard it before – A picture is worth 1000 words. Show it.
- Instead of using words to describe what you mean, convert those words into expressions, gestures, actions or icons.
- Think about the subtlety of facial gestures and the meanings they carry. How can they be used to give direction or guide the player?
- Design interactions between individuals instead of between individuals and objects.
- Can you provide opportunities for attunement in experiences?
- Can your design build on eye contact, silly faces and exchanging smiles between players?

Physical Play (Active Play)

When kids nurture their bodies physically, they also nurture their minds. Young kids don't think about the other benefits, such as health and being fit; they move for the thrill of it. It feels good. All young mammals engage in physical play. Through physical play, children learn to control their bodies and move quickly and efficiently through space to avoid and recover from falls. Walking, running, jumping, climbing, chasing and swinging are common play activities across cultures. How can physical play also be the most endangered form of play when children naturally gravitate towards it when given a chance?

Environmental and cultural influences affect physical play, such as access to rock climbing, swimming, gymnastics and bicycling.
Source: author photo.

Benefits of Playful Physical Activity for Children

Physical

- Stimulates brain development, significant organ growth and increased bone mineral content
- Improved cardiovascular fitness and aerobic endurance (heart and lungs)
- Maintenance of a healthy weight
- Improved posture, balance and coordination
- Deeper and better sleep patterns
- Building stronger bones and muscles
- Gross and fine motor skills development
- Increased flexibility
- Stimulates muscle growth and improves gross and fine muscle strength
- Helps with overall integration of muscles, nerves and brain functions

Cognitive

- Boosts cognitive abilities for learning
- Improved concentration and attention when doing academic tasks
- Innovation, flexibility, adaptability and resilience
- Greater learning is likely during and after physical play
- Organized activities such as sports require sustained attention and disciplined action
- Builds skills in risk assessment
- Regular active play and exercise enhance academic achievement

Social and Emotional

- Moderate to vigorous levels of physical activity can improve executive function
- Increased self-esteem and confidence

- Provides opportunities to make friends and enhance social skills
- Help with relaxation – children are calmer for more extended periods after physical play

Babies and Physical Play

Babies playing on their own are exploring all aspects of their environment, from the sound of their voice and the feel of their body parts to those of others. They want to gaze upon, grab, suck and rattle any object that comes their way. Babies play by engaging with an object, such as reaching to hit a ball hanging from the stroller. They also play with their bodies by wiggling their feet or putting their hands in their mouths. At a few months old, they begin tummy time.

Tummy Time

Babies need time on their tummies to develop strong neck muscles that will help them accomplish other physical milestones like sitting, standing, walking and crawling. A baby will naturally start trying to lift their head to see what is going on around them but won't be able to hold it up for long periods of time until around 3 or 4 months. Eventually, the baby will use this position to roll over, scoot and support themselves with their arms. Many toys have been developed, such as tummy time mats to promote engagement during play.

Big Body Play/Roughhousing

In big body play, children make exaggerated actions for fun. For example, a child may jump on the bed and then slam down on to the ground on all fours. It is natural and universal in children all over the world. It is rowdy, physical, and usually loud and to many parents annoying, although a vital component of children's growth and development. It supports the development of children's social awareness, builds both verbal and non-verbal communication skills and emotional thinking, and gives children sustained physical exercise. Through big body play, they become attuned to body gestures and learn how to control and regulate body movements and adapt the intensity to another person and how to compromise.

Rough-and-Tumble Play

We share play chasing and fighting with other mammals. Play fighting is different from real fighting. In a play fight, the intention is to go through the fighting motions without hurting the other person or making the person leave. This builds strength, coordination and endurance. Evidence suggests that girls are just as physically active as boys until age 4 or 5, but girls are better able to moderate and tone down their activity levels. Boys everywhere engage in more playful fighting than girls.[17] Play fighting is one way boys bond or learn restraint with other boys. From about age 11 onwards, rough play is less innocent and has an underlying theme of establishing a dominance hierarchy.[18]

Music and Movement

Dancing, musical games and movement activities also provide opportunities for physical play that enhance children's coordination, control and body awareness. Children can explore the elements of music, such as beat, rhythm and tempo, which can become embedded in their body memory.

Older Children

Many older children channel their physical play into individual or group sports activities. Often kids try out several different sports before they choose the one or a few that they like, and then they focus on the mastering of technique and the social component. Casual physical activities such as flying a kite, dancing, riding bikes, swimming, roller skating, skateboarding, parkour and jumping on a trampoline all are forms of physical play older kids enjoy. They also like an excuse to "play like a kid" or hang out in a playground.

Physical Play and Risk

Physical play is full of risk-taking, a fact that frightens many adults who are risk averse. Many experts feel that this aversion is excessive and even harmful. They point to children's natural capacity for risk assessment, which needs to be developed rather than suppressed in childhood. Exposing children to risk prepares them to meet life's challenges.[19]

The Benefits of Parkour

Interview with Caitlin Pontrella, founding partner of the Movement Creative, a nomadic studio that provides accessible fitness education and design services, focused on promoting well-being through parkour and physical play.

What does being part of the parkour culture offer kids?
Kids have intergenerational social interaction with others and the community through movement and play. It is an independent practice with several aspects of personal growth, including learning to overcome obstacles and fears. With the technical movements, kids practice the decision-making skills assessment of acceptable levels of risk. They learn how to overcome scary and difficult challenges.

How does parkour excite kids/youth at different age groups?
Kids under 12 years old are physically excited and naturally create movements in play, so they readily embrace it. From 12 to 25 years old they take it seriously and get excited to have control and power where they don't have it in other areas of their life. They enjoy the personal choice since they don't "have to do parkour." They also get involved in the broader social culture, lifestyle and social media around it. They like the interaction and acknowledgment of peers.

What does parkour offer kids to get out of their homes and interact with a city and community?

Kids get to know and gain a deeper appreciation of the place they live, can get excited by the city and go off the beaten track. The people involved with parkour are environmentally and socially conscious and community driven, supporting a healthy lifestyle, eating well, donating your time, sharing knowledge, a partnership with other people, and improving the environment.

Are there differences that boys and girls want from parkour?

There are not enough girls who practice. I believe this is reinforced by images in the media. They are adolescent male dominant, sometimes using references to the Ninja warrior, which forms our perception. Media tells girls it's more important to look beautiful than to be strong. Longer communities with older leaderships are more gender friendly; younger communities struggle with it.

What could designers be doing to be advocates for this movement?

Design spaces and details integrating play visions to make parkour possible on any outdoor site, for example, with the furniture. Study spaces with a big following. Why kids like and engage in these spaces. What makes them playful? Are they multi-tiered? Do diverse people train in walls, curves, railings, and in nature? Add ornamentation that kids could climb on. There are cheap ways to create interaction with the environment differently. For example, temporary installations, public art and using vacant lots.[20]

Body Movements for Physical Play

- Use traveling movements such as walking, running, jumping, sliding, shuffling and rolling
- Move in different directions such as forward, sideways or backward
- Move in place, bending, pushing, stretching and twisting
- Explore moving at different levels such as from low to high, or create shapes with the body
- Explore balance and control when stopping or keeping the body still to hold a pose
- Move and climb through an obstacle course (jump, hop, run, crawl, sidestep, slide, walk and balance)

Design Prompts: Physical Play

- Actions speak louder than words. What types of physical movements does your design promote?
- Can you provide a range of physical gestures, from large to small, within your design?
- How does your design support a child's natural drive to move freely?
- How does your design support a child's physically healthy lifestyle?
- What are the cognitive and emotional benefits of your project's physical play?
- Can physical play be combined and contrasted with calmer activities to enhance both experiences?
- Can you sneak physical activity into a design for the static user?
- Can music, dance and movement enhance your play experience?

Advice from Play Systems Designer Jay Beckwith and Natural Play Patterns

Interview with Jay Beckwith (play systems designer), expert in residence at Gymboree Play and Music

Jay has been designing play systems for over 50 years. He has seen the many changes in childhood and the industry over that time period. He also has worked on his own theory around natural play patterns, which I find to be particularly helpful when brainstorming a diversity of ways to move the body in play and developing play activities.

What are the most important things designers need to be focusing on related to play in the coming years?

1. Let's think about designs that reconnect kids to nature. Today most children live in cities and are less and less connected to nature. They don't get the beauty of nature or have an understanding of its importance.
2. Genuine play rises from biology. Children are triggered to play by specific environmental cues. When you get the play experience right in your design, you know it because it appears right in front of you, naturally.
3. Embed challenges. Kids are fearless, parents are fearful. Kids often play up to the point of pain. I intentionally design for small failures, what I call self-correcting errors, as this has been shown in many studies to maximize learning with little downside.
4. Inclusion: Design play settings for everyone. Include people of different abilities, ages, and from different cultures. Why are "adult" play apparatus just outdoor gym equipment? Play isn't just the domain of kids. It is for all of us. Currently, it is isolated and protected. We need to weave it into life.

Can you develop design challenges that you want designers to work on developing?

- The modern playground has become a play ghetto. They are built to get kids out of their homes and outside rather than to play in their neighborhood as a member of a community. As designers, how do we de-ghettoize play by including it in the fabric of our cities?

- The adventure playground doesn't need to be junk. How do we provide for loose parts in a public setting?
- How do we engage and inspire the community? Offer them control such as playgrounds built by the people, mainly artists, in the community.
- Think about the players and the rules or freedom of engagement. It is not just the environment. Remember, a chessboard is black and white squares until you put the players on it.

Natural play patterns developed by Jay Beckwith

Play behaviors in children arise out of fundamental biological drives and are triggered by specific elements in the environment, which, in turn, elicit predictable play behaviors in all children. For example, a tree with specific shapes will elicit climbing, a railroad track will be an opportunity to practice balance, a mud puddle is sure to be jumped in. Three characteristics identify natural play patterns:

- The play pattern begins at birth and always by 12 months
- Play patterns have clear developmental sequence
- The physical manifestation of the pattern is reflected in the areas of the brain that are being developed through the exercise of the pattern

Face: Eye contact, bond, expression, babble, words, communicate, language
Hide: Object constancy, sense of self, pretend, intimacy
Accelerate: Slide, spin, jump, eye tracking, proprioception
Walk: Roll over, crawl, cruise, toddle, balance, run, dance
Climb: Brachiate, lateralize, motor planning, spatial awareness, height
Contest: Rough and tumble, strength to weight, games, sports, fitness
Jump: Bounce, jump, fly, land
Grasp: Touch, thumb, hand, gather, collect, pattern recognition, data, science
Combine: Associate, join, make, transform, engineering
Rhythm: Listen, babble, beat, syncopate, sing, harmony[21]

Object Play

Have you ever heard someone say: "My favorite thing to do when I was a kid was banging on the pots and pans in the kitchen"? Object play includes playing with

◀ Figure 6.6

Kids see objects beyond what they are. This piece of bread became a mask and a bracelet. Source: author photo.

things: toys, sticks, rocks, art materials and all types of loose parts. Constructive play (making new objects or constructing ideas) is a type of object play. They are all examples of handling physical things in ways that promote curiosity.

Did you learn to count on your fingers before you could do it in your head? Hands-on playing with physical things helps with learning manipulative physical skills and social and emotional skills, and is also a driver for cognitive development. This influences body memory and wires kids' brains. In fact, young children learn best when manipulating objects. By freely experimenting with diverse objects, children gain information about the world and their place in it.

Inquiry is a way of questioning the world. We seek to learn something we don't yet know, looking for answers to our questions. As young children sort and arrange materials, questions arise naturally. They wonder: What will happen if I put this here? How tall will it go? What color can I make if I mix blue and yellow? What will happen if I pour water from this bucket into a funnel?

Objects, or more specifically toys, created for play are designed to promote inquiry. Leave room open in the design of the object or prompts for the unexpected interpretation by the child. The object is a means to the end, the experiential results (play) that the object (tool) produces. Think about what kind of experience you want to create, not what kind of tool you want to make. Design that experience and then tool(s) will support it.

Benefits of Object Play

Physical

- Manipulation of small objects gives children the chance to practice fine motor skills.
- Playing with larger objects or ones that promote movement enhances gross motor skills.
- Playing with balls and other objects in sports enhances skill and fitness.

Cognitive

- They gain knowledge through experimentation: They learn about the nature of objects and generalize about broad categories of similar objects.
- Creativity: Activities such as building towers and sculpting clay help develop strategies to tackle problems successfully. Objects become symbols for the imagined.
- Convergent and divergent problem-solving: Children come up with new ideas and recombine them to create novel scenarios.
- STEM–STEAM–STEAM-D: Through play blocks, puzzles, sand, balls, crayons and paper, children begin to understand logical scientific thinking, such as the concept of cause and effect. They practice experimentation, observation, measurement, quantification, classification, counting, ordering and part–whole relations.

Social and Emotional

- Through objects, children can create and express themselves.
- Social object play with adults and peers benefits children's learning to negotiate and interact with others.
- By playing with objects such as dolls or teddy bears, children can mimic or explore a range of emotions, nurturing and empathy.

Theory of Loose Parts

This theory was first proposed in the 1970s by architect Simon Nicholson, who believed that it is the loose parts in our environment that inspire creativity. Loose parts are materials that can be moved, carried, combined, redesigned, lined up, and taken apart and put back together in multiple ways. They are materials

▶ Figure 6.7

Milk crates are objects for constructive play and storytelling in a Danish playground.
Source: author photo.

with no specific set of directions that can be used alone or combined with other materials. Loose parts can be natural or synthetic. Stones, stumps, sand, gravel, fabric, twigs, wood, pallets, balls, buckets, baskets, crates, boxes, logs, rope, shells and seedpods are all examples of loose parts. Today there is a strong movement to integrate loose parts in designed objects and spaces.

Loose Parts

- encourage open-ended learning
- can be used in any way children choose
- can be adapted and manipulated or used in many different ways
- encourage creativity and imagination
- develop more skill and competence than most prescribed toys[22]

Timeline for Object Play

- Exploratory play is the first form of object play and typically begins around 5 months of age. At first, they explore the world orally. At 5–6 months, they explore objects with their hands and eyes together by looking at them, passing them from hand to hand to learn more about them. This is unique to humans.
- By the second year, children begin to combine objects in play and start to treat objects according to their intended function.
- Later within the second year, children begin to treat objects symbolically; for example, a block may represent a piece of cake.

- Over the following few years, children use objects in pretend play and the creation of increasingly complex and representationally realistic structures in construction play.
- They also use art materials to create symbolic representations of their thoughts and the world around them.
- Many objects within the same category – puzzles, games, blocks, dolls – build in complexity as a child grows.

◀ Figure 6.8

Soft loose parts where children create their own games to engage in pretend, social and physical play.
Source: Maryam Aziz (used with permission).

Manipulatives

Manipulatives are physical objects that engage a child visually and physically in learning concepts, such as coins, blocks and puzzles. The use of manipulatives is constructivist, because students are actively engaged in discovery during the learning process. Children who gain confidence manipulating objects become good at creating ideas and working with numbers and concepts.

Interview with Jed Berk/Itsabob

Jed Berk merges the fields of art and design with his balloon creations.

Can you explain how the goals of this project evolved from some of your previous work?

I was showing my work in galleries and museums around the world, and I noticed that balloons bring a sense of wonder and joy to everyone. I wanted to find a way to bring my work into people's homes and everyday lives. The feeling of letting a balloon go inspired the play experience for this product.

What research methods did you use in creating the design?

You squeeze the Itsabob to bring the balloon down, and when you let go, the balloon floats up again. I worked on many design iterations and prototypes to capture the interaction in a straightforward squeeze gesture. I developed several mechanisms and playtested many times to get the feeling right.

What were some challenges you had when you began manufacturing?

Stripping out the complexity has been the focus of bringing this product to market. Even with intense proto-typing, I spent many months working with the manufacturer to recreate my development in the prototype through their manufacturing processes.

How do you market the product?

Marketing is challenging because the product needs to be demoed, and that is difficult. Once it is seen, it sells itself. I have sold it at events and fairs, and that has been positive. Another challenge is that the product doesn't merely fit into a given market. It is part whimsical toy, part art object, which transforms balloon play. So I have to be creative with ways for people to understand and experience it.

You obtained a patent on this design. Can you explain why?

I wanted to ensure that I wasn't infringing on others and to then secure and protect my IP. The patent for Itsabob can be found on uspto.gov.

Did you have surprises in how kids used or experienced it once it was on the market?

When creating it, I always saw one or two working at a time. When 30 kids at a birthday party have them, and it is a group activity, it is an incredible experience. I love seeing expressions on people's faces at that a-ha moment when they let go.[23]

▲ Figure 6.9

The string extends by squeezing the toy and the balloon floats away. It is reeled in by squeezing again.
Source: Jed Berk (used with permission).

Design Prompt: Object Play

- How do we provide children with the tools they need for freedom?
- Can objects become a springboard to create deeper human interactions?
- How do we create play from objects that are easily accessible?
- Gather together loose parts and create three very different open-play scenarios.
- How can interactions support the concepts you would like to act out physically through manipulatives?

Constructive Play

In addition to language, a significant point of difference between humans and other mammals is the complexity and diversity of the things we make. As a species, where we lack in physical capabilities such as strength and speed, we have filled the gap and created devices to help us. Even though today most of us don't build our own homes or make our tools, transportation and communication devices, we still practice for these skills as children with blocks, sand, water, tools and chalk on the sidewalk. Constructive play can be intellectual as well as manual. Children can construct stories, songs and secret codes.

Constructive play allows children to experiment with objects and test hypotheses. They build their "what if" skills. They find out combinations that work and don't work and learn basic knowledge about stacking, building, drawing and constructing. In other cases, they replicate something that they see in real life in their culture and recreate it. For example, in hunter-gatherer societies, children make small versions of their huts, bows and arrows, and baskets in their play to practice the skills they will need to make real ones later on. In our society, they may make a spaceship from their favorite movies, possibly leading to a career in transportation design or at NASA.

Benefits of Constructive Play

Physical

- Strengthens gross and fine motor skills to manipulate and control the objects or materials

Cognitive

- Enhances creativity and problem-solving skills
- Strengthens the ability to plan the use of materials to see a design idea become a reality
- Allows children to test ideas and expand divergent thinking
- Builds a foundation for abstract thinking

In constructive play, children imagine something in their minds and then make it.
Source: author photo.

Social and Emotional

- Nurtures perseverance in the face of construction challenges
- Children learn teamwork and the ability to cooperate collaboratively and complete a task together
- Gives children a sense of accomplishment and control of their environment
- Builds confidence in accomplishment – I made that!
- Recreating scenes from their life, such as visiting the park, helps them reflect on their world
- Allows for failure
- Creativity allows children to express themselves openly and without judgment. The ability to be creative, to create something from personal feelings and experiences, can reflect and nurture children's emotional health

Timeline for Constructive Play

Constructive play starts in infancy and becomes more complex as a child grows. For young children, it involves open-ended exploration, gradually becoming more functional, then evolving into conceptual transformations.

- Babies put things in their mouth to see how they feel and taste.

- Toddlers learn what various objects can do. They build things with toys and materials.
- In the preschool years, constructive play merges with exploration and make-believe play and builds conceptual understanding.
- Has social imaginary situations, specific roles and implicit rules, and is recognizable by its persistence and tendency to become more elaborate over time.
- Grade school-aged children may turn to a potter's wheel, complex models, 3D puzzles, woodworking projects, knitting and small detailed construction toys.
- Older children will construct real objects. such as sewing their clothing or building their own skateboard.

Designing for Invention and Creativity

Interview with Cas Holman, founder of Heroes Will Rise, a toy company that encourages exploratory, unstructured play.

How do multiple play patterns in your toys work together to enhance different types of learning?

By letting children direct their play, the materials require imagination differently than if they had instructions. There is a value in practicing following instructions, but there is a much greater value in imagining and inventing something on your own – then figuring out how to make it. The figuring out is critical, and it's also the play. To try something, fail, undo it, redo it, test it, discover something you didn't expect, take it apart, learn from all of that, try again … Often there is no goal in what they are making, so there is no doing it wrong – just challenging, playful experimentation.

How does the scale of your product impact the play value? The materials you choose?

It's inspiring for children to make something bigger than themselves – to create and/or affect their environment, make it their own. There tend to be a few stages of play – first they build a thing – then they play in/on/ around the thing they built. It's uncanny how often I hear kids say "It's real!" when they approach Rigamajig. For children to feel trusted to use large, heavy materials is very empowering. That's us, adults, telling them: "You can handle this, we trust you."

▲ Figure 6.11

The scale of the blue blocks makes Imagination Playground a collaborative experience.
Source: Cas Holman and KaBOOM! (used with permission).

▲ Figure 6.12

Rigamajig provides many cues and prompts in the design for children to imagine.
Source: Cas Holman (used with permission).

Can you explain your involvement with Anji Play?
My role is to help Anji Play to standardize and "product-ify" these over 150 individual parts. I set to work standardizing the dimensions and figuring out which of the three types of wood used would be best. In talking to Ms. Cheng about her experience with each, she explained

why having one size ladder in three different types of wood is important – this is how the child learns about materiality. The bamboo ladder is lighter, the rungs and rails are one entire stalk of bamboo. It's hollow so sounds different when you drop it. Pine, by comparison, is heavy, the rungs and rails are square, and has wood grain. The sound it makes is more of a thud than a drum; solid versus hollow. The ladders are very long, so children learn how to ask for help from peers and cooperate in moving and setting them up. How could I assume the function of a ladder was to climb?

What are some key learnings from Imagination Playground and Rigamajig about making a product kit of loose parts?
Storage is critical; and design them to be easy to put away, ideally by the children. In Anji Play, children move seamlessly from playing into clean-up. An observer could easily not notice they are cleaning up, save the music that signals clean-up time. There is a lot of sorting, organizing, stacking, cooperating that happens in putting loose parts away, and kids love all those things. Anji is proof that there is value in children contributing to their community. They want the materials to be put away correctly.[24]

Pretend (Imaginative Play)

Pretending involves creating alternative realities to the real world and building on the capacities of the imagination. They think about things that are not immediately present, much as we do in the brainstorming process or as scientists do when developing theories. Children enact different people, places, times or scenarios from their imagination by establishing propositions about the nature of their pretend world, and then act them out logically. Children learn to abstract, to try out new roles and events, and to experiment with language and emotions, enact various situations and think flexibly. In an ever-more technological society, practice with all forms of abstraction – time, place, symbols, words and ideas – is essential. Play futurist Yesim Kunter suggests considering the playability quotient in designing play tools (toys):

> They need to fit easily to the child's imaginative storytelling, first his inner speech second his social speech. If it doesn't live in the imagination, then it will result in two seconds fun, and it is not well designed. It needs to have this magic spark that can evolve as the child plays![25]

▲ Figure 6.13

Through pretend play, children can examine and understand various types of interpersonal interactions and social situations.
Source: author photo.

Symbolic Play and Socio-dramatic Play

Symbolic play mentally transforms objects to represent pretend entities – for example, a child pretending a banana is a telephone.[26]

Socio-dramatic play is the social version of pretend play, where children within a social group pretend and cooperate to take on different roles, for example, "let's play superheroes." Then one child becomes Batman, another Wonder Woman and one Spiderman. As they enact roles, they are exercising their ability to behave in accordance with the shared understanding of what is or is not appropriate and practice negotiation and self-control. Kids learn very quickly that if we work to get along with people and keep our friends happy, we too will be happier and the play will last longer.

Benefits of Pretend Play for Children

Physical

- Learning by doing. Physical expression and learning through the physical gestures and body.

Cognitive

- Linked to divergent thinking, creativity, and flexible and creative problem-solving.
- Exercising decision-making as the play progresses.
- Pretending contributes to children's understanding of language and literacy symbols and their meanings. They learn to build and remember narratives. Children practice language usage in pretend play more than other forms of play.

Social and Emotional

- Navigating interpersonal reactions: Negotiating and cooperation with others and jointly engaging in the rules of engagement.
- Socialization: Children can experiment with skills and a variety of roles and learn about realities and expectations.
- Social understanding: They gain better understanding of themselves and others, and that other people have intentions and desires that are different.
- Coping and emotion regulation: Fantasy gives children skills to regulate their thoughts, feelings and behaviors.
- Offers children an opportunity to achieve mastery of their environment and control the experience through their imagination.
- Helps develop each child's unique perspective and individual style of creative expression.

Timeline for Pretend Play

(0–2 years) Babies and toddlers will use what they remember in pretend play. In their second year, children begin to engage in symbolic play. A teddy bear will represent a baby, and blocks will become a tower. At around 18 months, children begin to transform and animate objects such as a stuffed animal in their environment by giving them life. Shortly after, they give objects different functions – for example, pretending a box is a house.

(2–3 years) More imaginative fantasy gradually emerges in the pretend in two-year-olds. They sometimes lose sight of the line between fantasy and reality – even in their own pretending.

They can understand pretend actions by others and respond appropriately within the pretend context. For example, an adult or another child can pretend a lump of mud is a cake, and the child may pretend to eat it.

From toddlerhood to preschool age, the pretend becomes increasingly decontextualized, play can occur with or without objects, and children can create imaginary characters and situations.

(3–4 years) Preschoolers relate to their real-life experiences. In peer play, one child often assigns roles to the others, who may or may not make the play their own. Imaginary friends often appear around the age of 3 – and can play a significant role in a child's play. They have reached the peak time for pretend play, more realistic and detailed.

(5–6 years) They have a stronger grasp of the difference between real and pretend. Their increased attention spans and awareness of detail enable them to stick with and extend play themes for long periods of time. Pretending is social, and interacting together is fun.

(6–11 years) In the early years of elementary school, children continue to pretend. It is used in academic programs to teach literacy and other school subjects and nurtures a child's abilities in creative writing.

(12+ years) In middle school, pretend play wanes as they become more interested in organized games; however, they still engage in private fantasies or pretending in the context of digital media.

Design Prompts: Pretend Play

- Pretend a single object represents five completely different real-life things. Then build a story around it that integrates all five objects.
- Build three sock puppets and give each a name, a personality and diverse interests. Build a story that includes them each interacting with one another in a play scenario.
- Give a pretend function to a real object. Give a pretend object a real function.
- Play with an imaginary friend. Act out various emotional situations.
- Put together a costume box of scarves, purses, hats and clothes, and dress up as different characters.

Storytelling (Narrative Play)

Storytelling play is the play of learning and language, starting out with a parent telling a story or reading aloud to a child, and leading to a child retelling the story in their own words. Storytelling play often merges with pretend play. For example, a child may start recreating a scene from a book that they read and use the existing story as a springboard for a new one. Or they might start out with a story from their life and then retell it with elaborate fantasy details. It is an expressive form of play.

Stories help kids understand themselves, others, family history, morals and the human condition. Through stories, children see and hear the building of plots, characterization, climax, conflict and conclusion, and learn to organize experience structurally and communicate with others.

Timeline for Storytelling

- (2.5 years old) A child may begin to retell essential stories about themselves that they have heard.
- (3–4 years old) They can tell many kinds of stories: autobiography, fiction and reports they have overheard. They can tell stories with other people, and to other people.

Media characters strongly influence children's stories.
Source: author photo.

- (5–11 years old) They spend a lot of time learning through stories and creating their own stories.
- (12+) Adolescents are formal and conversational about the stories they tell and enjoy critiques and debate.

Benefits of Storytelling Play

Physical

- Acting out a story through body language and facial gestures helps kids learn about non-verbal communication.

Cognitive

- Through listening to compelling and fascinating speech, children build language skills and vocabulary and learn how to make a story interesting, exciting or sad.
- When children recreate stories, they have to remember key points of the plot and character names. They build memories best through stories.
- Storytelling opens children's minds to new worlds, other cultures and life philosophies. Storytelling is also a way to bring history alive and inspire further exploration of historical events.
- Children tap into their imaginative minds and provide their versions of the story.

Social and Emotional

- Stories often show us about the human condition and that we are all the same.
- Sometimes it is easier to communicate serious topics than in serious conversations.

Design Prompts: Storytelling Play

- Pick a topic for a child's social and emotional development and think of an existing story that addresses that topic. Build on it.
- Choose an existing folk tale and recreate the storyline with new characters in a contemporary environment.
- Recite a historical story in two different ways just by changing your voice and expressions. First make it dramatic, then make it funny.
- Creating a diorama or building a 3D model allows a user to interact with a story and the characters.
- Put on a costume of a known character and act out building on their identity.

Creative Play

Think about the creative potential a cardboard box holds for a kid. It could be a spaceship, a robot, a house or a car, depending on the day, mood or interests. Creative play is when one expands on what is known in the current state, to create something new. Creativity allows children to express their uniqueness openly and without judgment. The ability to be creative, to create something from personal feelings and experiences, can reflect and nurture children's emotional growth, providing opportunities for trying out new ideas, and new ways of thinking and problem-solving.

The experiences children have during their first years of life can significantly enhance or squash the development of their creativity. Any creative act can help children express and cope with their feelings and can help others to learn more about what the child may be thinking or feeling.

Creative play allows a child to develop divergent thinking: the ability to see things not for what they are but for what they could be. What is taking place in their head is that the brain scanned the objects around them, assessed their physical and aesthetic properties, and mapped those features onto alternative symbolic identities. Developing divergent thinking skills and imagination through pretend play, constructive play or creative play influences creativity as an adult.

If creativity is not a defined genetic trait but a skill and process practiced over time, then what can we do now to inspire and foster creative development from a young age? As children spend more and more time taking standardized tests and engaging in highly structured play (organized sports, video games, etc.), open-ended and imaginative play is being marginalized. Many designers and educators believe it is our responsibility to promote creative play to inspire children and build the "what if" skills that will be critical to their creative success.

If this magical age of creativity found between the ages of 3 and 5 is to be maintained and developed rather than stifled, then great care is needed to develop its evolving process.[27]

Benefits of Creative Play

Social and emotional

- Helps develop each child's unique perspective and individual style of creative expression.
- Expresses the child's personal, unique responses to the environment.

Cognitive

- Allows a child to develop the critical cognitive skill known as divergent thinking.
- Offers children an opportunity to achieve mastery of their environment. They control the experience through their imagination, and they exercise their powers of choice and decision-making as the play progresses.

Games with Rules

People need to abide by rules, laws and social agreements made between themselves and others to have peaceful communities. Social skills are practiced in formal games. In other forms of play, there are rules, but they are usually implicit; they are understood but usually not explicitly stated. In formal games, the rules are explicit. In competitive games, the rules are stated to be fair, and there are often winners and losers. There are also cooperative games with rules, such as jump rope. Games with rules are often characterized by logic and order, and as children grow older, they can begin to develop strategy and planning in their game playing. Board games, digital games, apps, party games, card games and sports can only function because everyone is adhering to the same set of rules.

▶ Figure 6.15

Students addressed the day-to-day challenges and aspirations of at-risk teenagers and set out to design an art park to foster safe, artistic expression.
Source: design by Anycia Lee, Evian Olivares; photo by John Dlugolecki (used with permission).

Games require practice to master the challenges.
Source: author photo.

Often children create their own games with rules, either by modifying an existing game or by creating a new one. As children develop the concept of their game, they need to negotiate with each other to make the game enjoyable for all players with various skill levels. Adapting the rules to make the play fair for everyone makes the game more fun and extends the play time.

Requirements of a Game

- Games require a definite beginning and end.
- Rules are the essence of the game and must be honored to engage successfully.

- Social games require players to take turns and follow specific procedures to complete the game.

Benefits of Games with Rules

Physical

- Active games with movement help develop fitness and gross motor skills.
- Dexterity games help practice with manual dexterity.

Cognitive

- Games for small children often have educational benefits: number, color, shape, letter recognition, grouping, counting, reading, visual perception and eye–hand coordination.
- Board games such as chess and checkers strengthen logical thinking, problem-solving, rationalization and strategy abilities, since they must consider both offensive and defensive moves at the same time to succeed.
- Memory, math, planning and deductive reasoning skills are exercised when playing many board games.

Social and Emotional

- Teaches children an important concept: The game of life has rules (laws) that we all must follow to function productively.
- Social games enhance social skills and strengthen self-regulation, and kids learn to be winners and losers.
- Team sports require and teach cooperative play, competition and teamwork.
- Games expose children to real-life situations, stories and narratives that expand their knowledge of the world.

Timeline for Games with Rules

Developmentally, most children progress from an egocentric view of the world to an understanding of the importance of social contracts and rules.

Preschool children participate in simple games with rules such as follow the leader, tag, musical chairs, matching games, and board games with spinners. Digital games are continually being developed to reach a younger and younger audience, including toddlers.

Games, both physical and digital, allow children to imagine in a fantasy world as they play through the game.

Before the age of 6 or 7, competing and reaching a goal, strategy and logic aren't important to many kids, because they don't understand what it takes to win. As kids get older, the competition, challenge, feedback, winning and losing, and social aspect of board games become more essential.

Board Games

The in-person social component to games is a welcome reprise from interacting with or through a digital device. We are seeing not only an increase in board game playing but also a diversity of games kids are playing.

Board games include games of chance (outcome is strongly influenced by a randomizing device (cards, dice) and games of skill (determined by the mental or physical skill of the players). They often combine both chance and skill. They may be themed with a clear narrative, such as Monopoly, or have no theme, like checkers. Many board games also have digital versions that reinterpret or extend the gameplay and storytelling.

I have experienced that developing a novel, useful and non-obvious game interaction and mechanic that could be patented is important in the game industry. Including novelty features that enhance the interaction between the game and among players and visual development of the board game enhance gameplay as well as branding the game.

Genres of Board Games

Strategy games
Cooperative games
Party games

Dexterity games
Adventure games
Card games
Collectable card games
Role-playing games

Design Prompts: Games

- Can your game appeal to different age groups, personalities or interests? If so, how?
- How does the style of the artwork contribute to the game experience?
- Do your users care a lot, or are they indifferent to the rules, winning or losing the game? How might this change the game strategy for different users?
- Can you develop a unique interaction among players?
- How will you playtest your game throughout the process?
- What types of social dynamics might there be in the group of players?
- How do these areas – strategy, socialization and physical skills – translate for your audience throughout the game?

Community Play, the Board Game Café

Interview with Robert Cron, co-owner of GameHaus, a board game café in Glendale, California, with over 1,400 board games

How do you see the GameHaus fulfilling the role of a third space and provide a sense of community?
We're all here to "nerd out" over the same thing, in varying degrees. The fact that you're *having fun* is tantamount to the experience we provide. Enjoying a game like Diplomacy doesn't override fun playing a game of Chutes & Ladders.

How did you decide on which games to include in your library?
We have a good variety to appeal to a broad range of tastes – some people only want to play familiar family favorites or old standbys, so there's Battleship, Clue, Stratego, Mastermind, and all those titles that have been around for years. The downtime of learning rules is minimal. But we also try to appeal to the more experienced and hobby gamers, so we have games like Settlers of Catan, Ticket to Ride and Pandemic on the shelves. We

even have a small section of old hex and tile wargames (because in a sense we're completionists), though they rarely get played. You wouldn't go into a major bookstore and only find romance novels, or horror novels, or biographies.

Do different types of games appeal to different age groups, personalities or interests? Why?
For diehard gamers, in particular, this is an extremely focused hobby. Websites and YouTube channels are dedicated to the newest and latest games coming out, forums are filled with back-and-forth discussion of whether or not the area-control mechanic of this game is utilized to the fullest, or if the components and map artwork does justice to the gameplay, or whether or not the theme is "pasted on" and not relevant to the experience. To some people, an experience like what we provide might be a novelty. To many of the more hardcore hobbyists, it's an entertainment focus, so it's something in which they invest a lot of time and effort and money and focus. It's serious and something they enjoy immensely.[28]

"On any given day, you'll find the family who want to introduce their kids to the games they played when they were young, high school kids hanging out playing party games, and a dedicated selection of twenty-to-thirty somethings. The space is great for first dates, too – it melts the ice a little bit." Robert Cron.
Source: Marcus Anthony Photography (Used with permission).

Digital Play

Computer games can be a valuable tool in playful learning. Digital gaming has also been accused of inhibiting the development of social skills, increasing rates of obesity and promoting violent behavior. In a lecture I attended at the Play Conference, Dr. Stuart Brown shared an insightful comment. In his opinion, "It is not about if digital games are good for you or not – if the state of play in a digital game is beneficial then it is good, if the state is about addiction then it is not so good."[29] It comes down to the content that the kids are consuming, the amount of time they are spending with digital devices and the state that it brings them to.

As designers, we are looking for ways to integrate the benefits of what the physical and digital worlds have to offer. The boundaries are increasingly blurring to phygital, and concepts merging the two will be increasingly developed in the next decade. We see that Gen Z seamlessly integrated technology into their lives, and Gen Alpha already don't differentiate between the digital and physical realms. Virtual reality (VR), augmented reality (AR) and mixed reality (MR) are just reality. Developers of digital products are looking for ways to enter the physical world, while developers of physical products are looking for ways to integrate digital products and experiences.

Kids and youth today have the tools and knowledge to create their own content. Coding is becoming a skill that tech toys and many schools are providing at an early age. As the graphics increase in sophistication, content created by children may look very similar to that created by professionals. This, in addition to the advances in 3D printing devices and materials and the simplicity of the software, means that younger and younger children who are interested will be completely capable of designing and making their own games and toys. Tom Mott, interactive producer and learning designer, explains:

> One thing I find fascinating is that digital play experiences can engage kids in a fantastic way. Kids spend hours figuring things out

on their own; not getting frustrated when something goes wrong; learning complicated things and then remembering them later. These are fantastic executive function skills. But what's interesting is that the data is mixed on whether there's a crossover effect outside of the game.[30]

For genres on digital games, see p. 429. For gamification in everyday life, see p. 431 in Chapter 9.

Technologies to Look For in Digital Play

Connected toys	Voice recognition (Siri, Alexa)
Artificial intelligence	Non-screen connectivity
VR, AR, MR	GPS
Robotics	3D printing
Smart environments	

Benefits of Digital Play

Cognitive

- Games promote playful education, especially for older kids, when other forms of play have diminished.
- Games challenge kids to make decisions through graduated levels of complexity and advanced knowledge and skills.
- Games motivate children to play, teach them to stick to tasks longer, and increase visual-spatial abilities, memory, critical thinking, problem-solving, executive function, qualitative thinking, exploration, experimentation and creativity.
- The multiple modalities (visual, tactile, auditory) can serve the needs of different types of learners.

◀ Figure 6.18

In this digital interactive immersive aquarium environment, kids can draw their own fish and add it to the digital display.
Source: author photo.

- When applied to educational content, it has the ability to increase the retention of the material. Content that promotes call-and-response interaction promotes more learning and engagement than static consumption.

Social and emotional

- While electronic play and gaming can be used for a solitary activity, it can also increase group social play and community and develop social skills.
- Cooperation and teamwork are needed for multiplayer games.

Physical

- Practice with hand–eye coordination, spatial skills and fine motor skills

Design Prompts: Digital Play

What part of your project is best experienced in the physical world? The digital world?

What should the players interact with on screen, on the device, in the real-world environment?

Will the devices talk to each other, kids and real-world objects? What will they say?

What mental models do users in your audience currently have?

Technology and Play

Interview with Dan Winger, senior innovation designer, LEGO group
Industrial designer Dan Winger develops new experiences that bridge physical and digital play.

Which technologies do you see as having the most promise in the future of play?
In the past several years, we've experienced a technological revolution like no other in history. There have been major advances in the areas of computer vision, artificial intelligence, machine learning, big data, blockchain, 3D printing, connected devices, virtual reality, augmented reality and mixed reality. These will soon be disrupting the entertainment industry (along with many other industries as well) and enable all types of new experiences. I'm most excited about mixed reality because it can layer playful entertainment and elements of gamification within our interactions with the real world. Many technologies are driving us deeper into digital experiences, but MR has the power to drive us to back to physical play and interactions.

Can you list three advantages that digital play can offer in the activity of play that the physical world cannot?
Some play experiences are better in the physical world and some better in the digital, but in many cases, the two working together can enhance both. The list of advantages for both sides can be quite long, but here are a few examples.

a. Ease and access to creative material. For example, when building in Minecraft or LEGO Worlds, you have an infinite supply of bricks to create with, you won't have to sift through your entire collection looking for that element, and there's no mess to clean up when you're finished.

b. Competing, sharing and social interaction on a global scale. Many digital experiences allow you to play with your friends remotely, meet new like-minded friends, and collaborate or compete with others around the globe. Also, there are many online creative communities – sharing galleries of user-generated content.

Mark Hatch, the author of *The Maker Movement Manifesto*, says, "You cannot make and not share." The ability to share your work can be one of the most influential motivators to create. Creatives motivate each other, teach and learn, share ideas and projects, and make friends with others that share similar passions for creating.

c. Greater levels of wish fulfillment and escapism. In video games, you may have superpowers to save the world, compete to be the strongest warrior, explore fantasy environments and alien planets, tend to a farm, collect and train exotic pets, or many other fantastic experiences.

How do you envision the seamlessness from physical to digital play evolving?

Many existing experiences merging physical and digital play – such as the Toys to Life category – often involve a physical toy or creation being unlocked within a digital game. A shallow physical experience (or even just a physical product) leads to a rich digital game, in which the material and digital experiences never truly happen at the same time. Shortly, with technologies such as the Internet of Things, computer vision or mixed reality, you can have all the rich digital content occurring as you play with your physical toys.

How do you decide what should remain a tactile experience versus a digital one?

My work has been focused on creating new hands-on play experiences. This is often achieved by leveraging digital technologies to adapt to today's evolving play patterns. So whether it's weighted towards the digital or physical, the distinction can be quite clear by asking a few questions.

- Does the digital experience extend, enhance and create more engagement with the physical product?
- Does the digital content inspire more hands-on creative building and play?
- Does it enable new types of play that could not be achieved otherwise?

If the digital features and experiences address these and similar questions, then it adds value to the physical experience.

How can technology help kids with these different types of play?

Constructive Play

Technology can open up new ways of playing with your creations and strengthen emotional engagement. Minecraft gave kids their own *entire world* to build, customize, interact with and explore with friends – something that could only be achieved digitally. Similarly, products allow users to build and share their own games, and leveraging computer vision technologies, you can even bring your physical constructions into your digital experiences.

Active Play

Over the last several years there has been a lot of development in this area, mainly leveraging technology to gamify fitness and active play. Data tracking of your runs and bike rides opens up asynchronous competition with other members of the community. Athletes can compete with others in real time without leaving the house in a simulated video game experience. The hugely successful craze of PokemonGo simply got users walking around their city to catch fantasy monsters in an augmented-reality, geolocation game. Many of these services have targeted older early adopters, but we're beginning to see these experiences being aged down for kids as the technologies become more accessible.

Pretend Play

Many video games, such as World of Warcraft, offer an immersive escape into a fantasy life and world, but it is one of the more challenging play types to integrate physical and digital play together. It's difficult to balance and align the digital content with the stories and fantasies happening within a child's own mind. I believe core value is that digital content can remove the 'blank canvas,' and provide kids with inspirational triggers to spark their imagination. This can be in the form of missions or story starters to get them acting out a variety of scenarios. This area will become incredibly exciting when near-eye mixed reality devices become mainstream as fantasy content, features and stories can be overlaid on your physical toys. Also, voice-activated AI platforms could even trigger creative storytelling and role-play through audio content alone.[31]

Toys

Toys guide play in many different ways. A simple square of fabric used as a cape will guide play differently than a Batman cape. The toy designer's role is to keep the experience as the central focus and design the tools as a catalyst for children to use, construct worlds, hypothesize and imagine. In my conversation with Dan

Children around the world play with toys that reflect their culture.
Source: author photo.

Winger, senior innovation designer at the LEGO group, he recommended for designers to consider:

> The product is a tool for the experience. Think beyond the toy and physical features, and focus the playful experience it can enable. What stories will they begin to role-play? How can they play with friends? What triggers their interest to engage each day, etc.?

In aboriginal cultures, they make toys from found materials and replicate their lives in play. In consumer societies, toys are purchased. Handmade and the craft of toys exist all over the world. Exceptions I have found are in "toy deserts," where underserved communities have little access, and households that withhold toys as a statement against consumerism and a means to give their children natural development.

Toys Research

In our research, toy designers are most interested in learning how children play. How children play can reveal a lot about who they are and how they see the world. How does the toy enhance or create an experience that further develops a child? What are the individual benefits? The social benefits?

- Through observational research, we try to uncover what children do when they play. What are the prompts that will bring about an engaging play experience?
- When creating new designs, study analogous experiences through classic toys that are out there to understand the overall patterns of behavior.
- In playtesting with multiple children in different age groups with different personalities and interests, we can map out how different children respond.

- Use scenarios to think through possible play experiences. We can never really foresee all the ways a child will use a toy, but we can embed features that open up opportunities for the imagination.
- We look at education through psychology studies on play to get to the root value of the play experience we are creating.
- Talk to psychologists and educators who continually observe and evaluate many different children and can offer insights from a professional perspective.
- Review the Toy Industry of America trend reports every year to see how megatrends in culture are affecting the toy industry.

Suggestions for designing toys from play futurist Yesim Kunter:

- Understand the audience: Who is going to play with it? What is the age group it will be aimed at? What are the needs of this demography? Understand the audience's needs and motivations.
- What type of play does it provide: Is it provoking solitary play or social play? Object play (functions, mechanisms), social play (imaginary-role-play), active play (fine–gross motor), games (play with rules and objectives).
- Focus on the "fun" element of the design.
- How is all of this captured in the final design: fun, function, color, form, and subtly communicated to the customer within seconds.[32]

Toys and their Role in Learning

The framework for the TIMPANI (Toys that Inspire Mindful Play and Nurture Imagination) study can be helpful for designers to build assessment techniques that can be used in playtesting and design refinement. It can also be used to assess a toy once it is on the market to see whether it is working as intended. The study codes children's use of the toys to produce quality play in different areas, using a coding instrument that they have developed:

Verbalization: Talking to others, singing
Thinking/learning/problem-solving: Figuring things out, working through a problem
Creativity: Unexpected ways to play with the toy. Transforming the toy
Social interaction: Parallel, associative, cooperative

A few things they have learned from the study over the years:

- Basic is better
- Construction toys score very well in multiple categories – lacking gender, different socioeconomic backgrounds
- Children have different background knowledge. That creates different impacts on children[33]

Use this method as a framework to build your own assessment tool for play value.

Learning through Play

Interview with Tom Mott, senior producer, and interactive learning designer

Tom Mott is a designer who works in many areas, including project management, user experience design, learning design, game design, gamification, research and writing. He creates learning products for companies such as LeapFrog. He has a background in Fine Art.

What are the most important things for toy and game designers to know about child development?

Physical: Kids need to move! Little kids don't have exceptional motor skills yet.

Social/Emotional: Kids have different temperaments. A timed game can be fun for one kid and stressful for another kid.

Cognitive: You always need to think about appropriate cognitive load: At what age can they follow two-step or three-step directions instead of just one?

Play: Think about repetition: If your toy or game has something that people love, they'll want to do it again and again.

How much do you consider children's behavior throughout the experience in the design of a product?

I consider it constantly. When working on interactive products, I strive for a balance between "free play" and "guided play."

What are the most important things for designers to know about the culture of childhood today?

For 2-6-year-olds: I think a lot of the old paradigms still apply. Parents don't want just to stick their kids in front of screens. They want developmentally appropriate toys and games. If you do happen to be designing screen-based products (web or apps) for that age range, there will typically be an expectation that (a) you're not advertising to the kids or continuously upselling them; and (b) there are educational benefits, whether they're explicit or implied.

For kids 7 years old and up: once you get into this age range, you're competing with school, after-school activities, sports, homework, and the aspirational culture of smartphones, tablets and video games. More and more of these kids want to be playing. If you're a designer of learning games, you're going to be designing for a niche market. If you're a board game designer, you need to think about gameplay that's appealing to an entire family. One big challenge is that your target audience (the kids) are now in school for 6 hours a day. They're learning all day. The parents say "We don't need to keep buying all the learning products."

What types of projects do you find best experienced physically? digitally?

Developmentally, kids still need to play with things, just because your brain does different things when you're manipulating and touching actual objects instead of just processing audio-visual information.

Digital: One thing digital provides, of course, is solo gameplay: you can play chess with a video game without anyone else. You can listen to audio stories even when there's not an adult available to read to you. I think social-emotional skills are best offered through direct interaction with other people, not through an app. That may change of course.

Physical: Kids need to burn off energy. And, like it or not, sports ability is a primary way that elementary school kids establish social bonds and relationships, so it's essential to get the kids out of the house and moving around, not just playing video games.

Do you have any suggestions on how to best merge physical and digital play?

Many of the products I've worked on do this. They were physical products (books, maps, board games) that also have digital content in the form of interactive audio: tap the page with the stylus and hear audio responses. A couple of insights … with the caveat that the suggestions would change depending on the target age and type of product you're making.

1. When possible, the physical item should function even without the digital aspect. A talking stuffed animal should merely function the way a stuffed animal does typically.
2. I love when the digital play supports or suggests the physical play. The physical movement reinforces the audio/visual learning. When we designed the Animal Adventure Board Game, some of the activities we included involved getting kids to move around the room like an animal (flap your "wings" like an owl!) which kids love.
3. So far, the best merging I've seen takes an existing play pattern (building with blocks, controlling a remote-control car, electric trains) and instead of merely *replicating* it digitally, you *provide something they couldn't otherwise do*. Building blocks that come to life. A remote-control car that lets you explore from the car's point-of-view and film videos. The ability to place a 360-degree camera on top of an electric train and film immersive videos that you can watch later. That's so cool!

As a game writer, what are some of the things that you consider in story and character development and how that relates to the play experience?

Most of my products have many mini-games or shorter interactive experiences built into a more extensive product. For example, with an interactive book, the book itself (say, *Cat in the Hat*) is telling the story, and then the games and interactive audio that I write are supporting reading comprehension, engagement with the text, or just providing an additional way to interact with the characters in the story.

Two ways I think about a story and character development:

1. Ideally, these types of mini-games should either relate to the story or the characters. Not just a random "Let's play a guessing game!" or "Let's race!" but extend what's happening on the page. If the Cat in the Hat is making rhymes … let's make rhymes with him. If Fish is getting angry at the Cat in the Hat, let's explore *why* he's getting angry.
2. Fun games often have some conflict and resolution. At a minimum, that just means helping a character on the page. ("Let's help the Cat in the Hat clean up the mess!" is much more engaging than "Let's find things on the page.") ("Oh no! Let's set the table before the guests come!" is stronger than "Find the words on the plates.")

What does the future of play look like to you?

I think there will always be social and physical aspects of playing: go to the park and play with friends. Run around. Play hide-and-seek. Build a clubhouse out of cardboard boxes.

I think that for 2–6-year-olds, a lot of traditional things will stay the same: blowing bubbles, imaginary play, dolls, action figures, hide-and-seek, developmentally appropriate toys, physical board books and picture books. Older kids (7 and up) are dying to get onto screens, whether that's smartphones, video games, tablets or the TV, so we'll continue to see video games, apps and phygital products for that audience. For adults, I see an increased interest in "gamification." I think any teaching methods that make the material more engaging and help with retention are always positive innovations.[34]

Toys and Technology

New tools in the coming years will enable designers to conceptualize, model, present and fabricate our visions. They will also influence the toys we make. Technologies have always been embedded in toys. Increasingly, toy companies are integrating technology, an upgraded experience factor that is being delivered by digital devices. The question for designers is when it is appropriate to use technology, and how we use technology to increase and produce a better-quality play experience to further development. The challenge for designers will be to maintain tactile engagement and the value of learning in the physical world.

Play futurist Yesim Kunter explains:

Any technology that becomes a norm for the daily life of the culture will translate into toys. Technologies that are here to stay that we can't imagine life before them will have the most significant impact on the future of play. Some of the emerging technologies such as AI, IoT, Advanced Sensors, Robotics, Drones will change the way we interact with things/ourselves and objects, which means toys will adapt into these new lifestyles. At the same time the backlash of screen time and need of being outdoors and in nature, soft skills (empathy, intuition, understanding of our emotions) and creativity will also become exceptionally important in the development of our future humanity, which means some toys will follow this wave.[35]

In many toys, specifically those that use lights and sounds, the technology is just there to sell it off the shelf or for a limited out-of-box play experience. A rocketship that has sounds and lights is exciting for one play session, but has no more interest once the child has experienced that it doesn't lead them anywhere. Technology can be limiting when building an experience that needs to be programmed with a beginning, middle and end. It is not open-ended. The scripting takes away from a child's ability to pretend and imagine. It is someone else's story.

The best use of technology is when it is flexible and adaptive; when it is used to motivate kids in play, to learn about technology and to use it to create in a way that they could not without it.

Smart toys, with their own intelligence through embedded sensors, respond through speech recognition and use accelerometers and other electronics for the child to engage with the toy. Some technologies, when combined, can animate the toy and give it a lifelike persona or personalize the experience or play for the child. Adaptation of new play technologies in education, health and science is a burgeoning field. The problems occur when the toy does too much for the child, not when the child is playing with the toy.

Making Custom Toys

We are all familiar with the process of designers creating toys and kids playing with them. Today, through the DIY and maker movements, the interest in sustainability, the value of artisan handcrafted toys, and the proliferation of toy innovations on social media, parents and children are encouraged to make their own toys. Customization and 3D printing allow kids to design their own toys, not just for themselves but for broader distribution. Making the toys is part of the play.

Manufacturing techniques in the toy industry are slowly changing too. Online tools for toy configuration, crowd-sourced products and on-demand manufacturing provide opportunities for mass-customization. Toys can be tailored to kids' specific needs, and they can be treated as individuals. The co-design allows kids to take an active role and is an opportunity to create value in the toy. Mass-customization also offers great promise to reduce the waste due to unwanted products.

Age Compression

Today kids are getting older, younger. Children are moving through play stages faster than they did in the past. Kids are growing up and transitioning away from toys and traditional unstructured play to digital experiences and gaming at earlier and earlier ages. They also progress to more complex toys at a faster pace. This is a big challenge for today's toy industry, when they are losing children by the age of 7 to digital gadgets.[36]

Sustainability in Toy Design

What we choose to make for children needs to have a direct impact on progress. The question at the outset of every project lies in the trade-offs between a product existing and the immediate benefits it produces. Professor Heidrun Mumper-Drumm suggests:

> Put a high standard on the design and manufacture of children's products. We don't care about them enough to invest in them. Children are designed "down to." It is as if someone said, "Let's just put some glitter on it to cover up the seam, or a sticker because they won't notice it anyway." Working from a the approach of sustainability gives us a positive context to explore.

Toy packaging is a significant problem on its own; Professor Heidrun Mumper-Drumm added: "Packaging is designed to sell products. Toy purchases are often impulsive, and packaging is designed to encourage 'I want it, and must have it.' The packaging displays all of the pieces, promising more than it delivers when it is opened." As designers, we can surely develop better solutions beyond shelf-appeal.

Inspiring Kids to Learn and Discover with Toys

Interview with Eric Poesch, senior vice president, Design and Development, at Uncle Milton Toys
As the leader of the design team, he is responsible for the conceptualization, development and design of products.

As a product designer, what are the most frequent topics that you address when designing toys for kids?

Play value and engagement: Great toys provide open-ended play that encourages engaging experiences. They allow kids to use their imaginations, creativity and invent their own ways to play.

Kid appeal: The product theme and styling need to connect with today's kids.

Ease of use: Features and function have to make sense. Designers always have to have in mind the target age range that you are designing for.

Ergonomics: If it is held, looked through, strapped on a wrist, fits on the head, there is always that challenge to design for the child, but often accommodate an adult.

Cost of manufacturing: How do you deliver an excellent experience for the right price?

How do different design disciplines contribute to toy design?

Illustrators: Theme, style, color and detail.

Brand managers: First, we need to ask and decide: Who the brand is genuinely targeting, the child or the parent/grandparent or both? In the science and learning space, we are often targeting the parent/grandparent.

Graphic designers: Simplicity and impact of visual and written communication. Make it kid friendly – and don't presume that anyone is going to read. Make it as visual as possible. Put into words only what cannot be visually depicted. Include just enough written content to bring more significant meaning to the visuals.

A brand logotype is critical, but more critical is establishing a strong, cohesive branded look that works across all of the products in a line to make a unified statement at retail.

Packaging designers: Strong visuals and clear hierarchy (order of the visual read). If you try to say too much, you end up saying nothing. Packaging structure plays a role as well in the communication. An open box shows the value and quality of the product, while a closed box often means little to no blisters (clear plastic packaging) and less packaging cost and provides you a greater principal display panel for imagery that can bring the product to life and show it in action. How can we as designers create less packaging waste and use more sustainable materials?

What cultural shifts will affect the things you make in the coming years?
I foresee a split in culture, between those who have the means and adopt every new technology into their lives

---

and those who may or may not have the means but choose to limit or exclude specific technologies.

The effect of technology on developing minds and the potential redefinition of play is most concerning. Unstructured, imaginative, inventive play is critical to the healthy development of future generations of kids. In the coming years, I know technology will enter into children's products and toys in ways we can't imagine today. I will always opt for products that encourage unstructured play and exploration while incorporating technologies that support real-world discovery and experiences.

What do you see as the role of a designer in the future of play?

Traditional, core play patterns are in danger of becoming extinct and being replaced by digital play experiences. Ironically, this is precisely what some toy makers are encouraging with the introduction of LCD screens and other advanced technologies into infant and preschool toys. Critical roles designers must play in the future are to uphold and promote core play patterns and unstructured play.

As a Dad, what are essential recommendations you have for designers to consider when designing for children?

Visit kids' museums with interactive exhibits. Here you can observe kids at play and explore each of the exhibits. Note which exhibits get the most attention. Note the age range versus appeal per display. Which ones make the kids laugh or produce greater emotional responses and "wows" and why. Note what percentage of adults and/or kids are reading the signage and written explanations of each exhibit. Note how they interact with the displays and if the interaction appears to be as the designers intended or are the kids creating their own ways of playing and exploring the space.

And of course, be a kid yourself, play and explore and spend time at both the most popular exhibits and the least popular exhibits to try to determine the difference in appeal. Is it the subject matter? More hands-on versus less? Fixed interaction, versus more open-ended exploration and experimentation? Characters and/or colors? …

Define the experience you are trying to provide the child. Is it purely fun and whimsy or is there learning? If learning is part of the core experience, it still must be fun first, or it will not hold the interest of the child. The best experiences are those that provide learning as a result of play. Humor and emotional engagement along with engagement of as many of the senses as possible are best to connect with all children. Become familiar with child development. I have seen with my two children; each child is born a unique individual with innate aptitudes.[37]

The Future of Play: 25 Topics to Know about Play in the Coming Years

This is my list for looking at what some of the leading topics and technologies around play are today. Use this list as a starting point for inspiration and to create your own vision of the future of play.

1. Personalized play: Users will build a cognitive profile, and products and experiences will adapt and respond to kids' individual interests and abilities. Feedback loops will guide the behavior of toys and games over their lifetime.
2. Growing social dimension to play: There will be new ways to interact with others physically and digitally. There will be a shift of importance in child development areas from cognitive and physical development into more toys, games and experiences about social and emotional development.
3. Create and share: Kids will create their own toys, games and digital experiences at home, in community spaces and with mass-customization

manufacturing. Apps, robotics, 3D printers and back-end manufacturing systems will transform to provide flexible products and experiences. Brand engagement includes crowd-sourced product development and changes and updates throughout the life of products. Kids will share and distribute their creations with users worldwide.

4. Modeling different play behaviors: Traditional play patterns will remain the same (especially for the younger ages), but they will branch out and shift in priorities, for example, from constructive to pretend play. How play is acted out will change based on cultural shifts, technology and material resources.

5. More play in education: Technology and new media will address diverse learning methods in classrooms and in online education. Focus on using play for 21st-century skill building (problem-solving, thinking, collaboration and creativity) for all ages from kindergarten through college.

6. Play everywhere: Every time and place is an opportunity for playification and gamification. Play expands across industry sectors, including home, community, retail and healthcare. It happens during the bursts of free time in between other things.

7. Play outside: There is a renewed interest in playing outdoors. Tactile experiences will include playing with loose parts and nature, risk in active play, and exploration in natural wilderness and with biological systems. Technology will enhance natural experiences in context.

8. Open-ended play: Decrease in fixed predesigned object play and increase in open-ended toys, games and play experiences.

9. D is added to STEAM: Creativity and divergent thinking and STEM and STEAM deepen and are elaborated into STEAM-D by adding design.

10. Seamless play tools: Technology will be seamless and off the screen. It will be embedded in the landscape, wearables, networked toys and experiences. Greater levels of wish fulfillment and heightened fantasy play will be achieved with new technologies, including VR, AR, MR, AI, robotics and more.

11. Universal design in play: Inclusivity/accessibility for special needs is no longer for compliance but to best benefit all users. It greatly expands beyond physical challenges to include more cognitive challenges.

12. Narrative interactions: A shift from things (toys) to experiences and immersive storytelling environments will allow kids to interact directly with their favorite characters in real time.

13. Design ethics: Designers expand into leadership positions and address research methods for play, gender stereotyping, violence, cross-cultural play and sustainability.

14. Play for social impact: Small non-profits model change around play, and larger companies build active foundations to support children's rights.

15. Play throughout the lifespan: Kidult market, play in the workplace, play to a better self, and senior market expands.

16. Global play: Communities of like minds play online and across the globe.

17. Emerging markets influence play: What companies learn about play from research and distribution in China, India, Brazil, etc. will create a cultural mash-up of toys and characters with cross-cultural influences. Designers are challenged with global and regional design strategies.

18. Play for behavior change: The geography of play makes it easy for each of us to be our better selves. We learn to allow more time for play.
19. New design processes: The designer considers the benefits of the child in all areas of product development, including research, design, manufacturing, marketing, distribution and post-consumption.
20. Toys, games and advertising merge: As corporations continue to finance educational play experiences, branding becomes more integrated into everyday life through interactive advertising and new social marketing methods.
21. Social connectivity throughout phases of the play: Currently, people share completed products. Soon, they will share the creation process and play experiences with others through real-time data.
22. Kids and parents are more involved in what toys companies make: Crowd-sourced product development and manufacturing and new distribution models take hold and break down the current monopoly structure in the toy and game industry.
23. Democratic play, locally and globally: Bringing physical and digital play to small spaces throughout cities in underserved communities narrows the digital divide.
24. "Nutrition label" for play: A playability scale will standardize the measurement of skill-building properties, including fine and gross motor, cognitive, social and emotional, and creative. Each product's play value will be transparent with a labeling system.
25. Best practices in sustainable design: Design for longevity is brought to the play market. Materials, packaging, processes and business models are readdressed for a closed loop system. Where and how toys are made and business practices will be more environmentally and socially conscious as consumers demand sustainability increases.

▶ Figure 6.20

Using autonomous intelligence, this interactive retail environment encourages customized hands-on social play.
Source: Jeremy Dambrosio, Harmonie Tsai, Jack Xu (Used with permission).

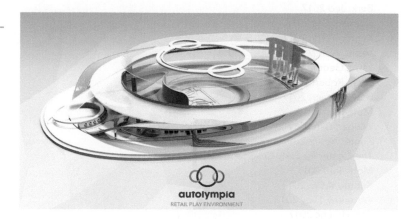

Design Prompts: Future of Play

What does the future of play look like to you?
How do your interests tie in with different types of play?
How do you see some of the current play theories evolving in the future?
Will technology improve and extend your design? If so, how?

Notes

1   Feder, Karen 2017
2   Gray, Peter 2014
3   Gray, Peter 2010
4   Deitz, Rena 2017
5   United Nations Human Rights Office of the High Commissioner 1989
6   The LEGO Foundation
7   Feder, Karen 2017
8   Brown, Stuart 2017
9   White, Rachael E. (2012)
10 Levine, Paul 2017
11 Feder, Karen 2017
12 Donaldson, O. Fred
13 Hanson, Cynthia
14 White, Rachael E. 2012
15 Dube, Marie-Catherine 2017
16 Brown, Stuart 2017
17 Edwards, Carolyn P.; Knoche, Lisa; Kumru, Asiye 2001
18 Kennedy-Moore, Eileen 2015
19 Gill, Tim 2010
20 Pontrella, Caitlin 2017
21 Beckwith, Jay 2017
22 Kable, Jenny 2010
23 Berk, Jed 2017
24 Holman, Cas 2017
25 Kunter, Yesim 2017
26 Leslie, Alan M 1987
27 Land, George 2011
28 Cron, Robert 2017
29 Brown, Stuart 2017
30 Mott, Tom 2017
31 Winger, Dan 2017
32 Kunter, Yesim 2017
33 The Center for Early Childhood Education 2017
34 Mott, Tom 2017
35 Kunter, Yesim 2017
36 Waterlow, Lucy 2013
37 Poesch, Eric 2017

Chapter 7

Education

What motivated you to learn when you were growing up? Did you have a favorite subject that you couldn't wait to listen to the teacher share more about? Did you enjoy participating in team projects and giving presentations on what you learned? Did you enjoy reading books and writing reports? Education can be delivered in many different ways, from systematic formal processes to informal engagement. Learning can happen in a classroom, under a tree, with a toy, watching TV, visiting a museum or playing. It can be at home, in the community or while traveling. It can be intrinsically or extrinsically motivated. This chapter outlines different approaches to childhood education and its importance in development and building a foundation for future learning. Everything kids experience or interact with in their lives, from their classwork to entertainment, is an opportunity for learning.

Designers who work on projects that are intended explicitly for learning, such as educational toys, games and programs, think about the subject, the best way to teach it, and the desired learning outcomes. Some build in assessment methods to follow through and find out if the kids are learning. We consider different learning styles, educational philosophies and how to give feedback to the user to test or showcase their learning. We think about how kids learn at different ages, so we understand and include the best methods for communicating the educational content. We also consider the best context for learning specific content. For example, should a teacher, a peer or a mentor teach it, or do children acquire the learning on their own through experiences?

When developing a project for learning, it is helpful to research different educational approaches. For a preschool project, are you going to use the tenets of Whole Child, Montessori or Reggio Emilia as a foundation? Does each have something to contribute to your thinking? Are you developing a project for a public school, a gifted, charter, private, or homeschooling program? How does that impact your use of government educational standards? How might the project differ in a formal or an informal educational environment?

Many companies that produce products, services and environments for learning have experts such as child development specialists, educators and learning designers on board to collaborate on the process. For advice on your project development, get educators on board early on so they can highlight

▲ Figure 7.1

Kids enjoying their friendships at school.
Source: Marie-Catherine Dube (used with permission).

approaches, considerations, methods of assessment or the benefits of your project to develop. Hopefully, this will strengthen the learning of children.

What Does It Mean to Learn?

There are many ways that we can define learning. Most of the time, what we mean is that we have retained the information. It is a cognitive process whereby we have memorized the information, stored it in long-term memory and can recall it. We also learn methods to find answers to solve problems. We can take in information and apply it to the real world. We see something and recognize it. We can imitate something we have watched and replicate it. We can label it and verbalize the meaning. Repetition and using information are often the best ways to learn and retain it. Since we learn our whole life, and there is no beginning or end to learning, one of the most important things for kids to do throughout childhood is to learn how to learn.

To enable learning, we look at several factors:

   (What) is the content or skill to be learned?
   (Why) is it essential to learn this content?

(Who) At what developmental stage is the learner? What intelligences do they prioritize?
(How) is it learned? What is the best method to showcase the content?
(Where) What is the best format to distribute the content?

Developmental Ways of Learning

Although children learn in many ways at the same time, the brain prioritizes different ways that kids pick up cognitive skills at each developmental stage.

Babies and toddlers learn through the senses: They explore the world and take in information through all of their senses and their body. Through repetition, the brain strengthens the neural pathways and helps make the activities easier.

Between 2 and 7 years old, kids learn through language: Kids first build vocabulary and then begin to read and write. Kids develop neural pathways making communication of concepts through language easier.

From 7 to 12 years, kids learn through logic: They can make connections between topics and ideas and reflect on them.

Teens learn through reasoning: They can think abstractly, with more complexity, and consider possible outcomes. They can think through complex abstract concepts in math and reading.[1]

Contexts for Learning: Formal, Non-formal, Informal

Children learn in the home, at school or in the community. Each context affects the content that is taught and methods of distributing the content. Although schooling is essential, what kids do with their free time can have a significant impact on their learning and development. The quality of kids' time learning outside of formal education is tied to their parents' skills and exposure to different experiences. During informal time, kids have the chance to pursue what interests them for its own sake. Self-directed activities let children develop the ability to focus their attention and engage deeply in what they are doing.

Formal Learning

This is intentional, structured, often guided by a formal educator, curriculum and defined assessment methods, and may lead to recognized credentials. Today, there is a growing interest on developing new assessment methods along with an increase in informality to learning and more places and tools to learn from.

Non-formal Learning

This is loosely organized and may or may not be guided by a formal curriculum. Led by a teacher or mentor with more experience, it builds on children's existing skills and capacities. It is engaging, since interest is a driving force behind the child's participation. Girl and boy scouts, summer camps and after-school programs are all examples of non-formal learning for children.

Informal Learning

This is learning that happens casually at those in-between times when education is usually not scheduled or planned: a visit to a science museum, camping in a forest, a visit to a library or playing with a toy. There is no formal curriculum or credits earned, although there may be learning objectives and assessment methods planned in the development of the product/experience. There may or may not be an educator, and the design can serve as an educator. For example, an exhibit design can teach about play, or an app can show concepts of cause and effect.[2]

◄ Figure 7.2

Field trips extend formal learning from the classroom to the community.
Source: author photo.

◄ Figure 7.3

Interacting with technology can deepen and personalize the content for the user.
Source: author photo.

Reading stories you write and illustrate to kids helps to understand how they interpret it.
Source: Viviah Shih (used with permission).

Scaffolding

Scaffolding is a term used to describe sequences of learning in education. Scaffolding of concepts from previous learning builds on prior knowledge with a support that is slowly taken away.

- Learning is cumulative. Early experiences shape their motivation and approaches to learning.
- Learning and development follow sequences towards greater complexity and symbolic capacities.

- Development and learning proceed at varying rates and result from an interaction of maturation and experience.
- Multiple social and cultural contexts influence development and learning.
- Development and learning advance when children are challenged.[3]

The Uni Project, Bringing Learning Experiences to Public Spaces

Interview with Leslie Davol, co-founder/executive director
These custom-designed installations pop up in parks, plazas and other public areas to offer reading, drawing and hands-on experiences for New Yorkers of all ages.

Can you briefly describe the Uni Project? How did you determine the need? Goals and objectives? Different parts of the program included? How did you decide on the final solution?
What people on the street see is attractive, architect-designed carts, filled with books, art, puzzles and blocks, dependent on the program (READ, DRAW, SOLVE and BUILD). Portable benches provide a place to sit, and boards offer drawing and building surfaces. Prompts contributed by artists and educators provide self-guided activities and daily challenges. Low-income residents and other vulnerable populations mainly benefit from the access to culturally responsive materials, free activities and safe space that our reading rooms provide.

Define informal learning specifically as it relates to your project.
Learning at the Uni is informal because it is entirely voluntary, unstructured and self-guided. We put out materials and supportive staff, and kids can come and go at will. Kids describe reading at the Uni as something you "get" to do, "could" do – in stark contrast to the way they described reading and learning in other, more traditional environments as something you "have" to, or get "forced" to do.

Can you explain how the Uni Project uses the community to support playing, learning and growing?
Neighborhood reputation is an integral part of kids' identity that we think has been too long overlooked in the effort to change educational outcomes. Communities need help in finding ways to express and promote the value of learning outside of schools. Finding a safe place is a daily challenge in some neighborhoods, and finding

a safe place to read and learn outside of school is even more challenging.

How does intergenerational engagement play a role in the informal learning that takes place at the Uni Project?
We're not just a "kids' thing." I can't think of a better way to turn kids OFF from an activity than to say
OK kids, learning is something you have to do, and we've got these pretty dull buildings where we lock you inside, separating you from the rest of the world, while most adults you see in media or on the street don't seem to be learning at all and seem to be doing all sorts of grown-up stuff that looks more fun.

It's disheartening to discover how many kids view learning in this light.

Do you see different behaviors when the Uni Project is brought to different neighborhoods? If so, can you describe a few differences based on location?
In some neighborhoods, parents are present, involved and take pictures. In others, adults are not around at all. This makes a big difference to how things play out on the ground. Sometimes kids just want to talk about some tough things that happen to them. Our collective inability to provide some of the basics to children is not only a massive obstacle to their education – it can rob them of their childhood.

What value do you add by bringing the library or other materials to the children versus them finding other related opportunities in their communities?
Some neighborhoods where we go have been called "book deserts." But even in those that do have educational and cultural opportunities, we put this stuff in public space and make the act of learning visible to the whole community. Placing learning on display also explains why we focus only on hands-on learning opportunities and don't use screens. Yes, you have access to more information on your iPhone than the Uni collection on the street, but do you have a pop-up book on that device? We've got a whole shelf of them.

What type of training do you have for the staff that works at the cart? How does this support the experience?
Staff is the most critical aspect of our installation. Much of what you see put in public space is unstaffed because this is most convenient and (some would argue)

cost-effective. When you add terrific staff, you can do so much more in public space. We do have a process for training our staff as well as a method for picking people with instincts for this kind of work in the first place.

How does the design of the cart and the benches support the experience?
It makes a place that is attractive and beautiful. It makes people feel comfortable while they are there. The design also creates a place to gather around something, in both a physical and symbolic sense. Most of our installations are in the round.

What can formal educators who develop learning outcomes, process and assessment criteria learn from your process and experience?
I think they should look at us as an exciting new angle to evaluate educational activities. What happens when you make a specific learning experience entirely voluntary, letting kids walk away anytime they want? There is no classroom door you can shut out there on the street. It forces you to think in new ways about reaching people. How many activities that happen in the classroom would survive that kind of test? It might be exciting and fun to find out.

What type of editing criteria do you use to reinforce learning in choosing the books and art supplies included in the cart?
Materials we provide have to be eye-catching from a visual standpoint, and "high quality." Ideally with lots of pictures. We want some elements that reflect New Yorkers' experience (in all of its diversity), and we want some that merely delight or give people a little bit of escape and the experience of wonder. The supplies support creativity and self-expression by providing people the opportunity to draw!

Your project is an excellent example of design for social impact. Are you documenting the expected and unexpected learning of these experiences? How do you see it having a long-term social impact?
We have a fairly rigorous approach to documenting our work, so we can improve and also share what we learn with other groups via our blog or conferences, etc. I hope we have a long-term social impact. I don't think we're going to transform New York City into a Scandinavian educational utopia, but I do think we will leave the city a better place than we found it by the time we're done.[4]

The Uni-Read program includes high-quality books that reflect the community.
Source: Uni Project (used with permission).

21st Century Skills Movement

The National Education Association established the Partnership for 21st Century Skills and believes that the four Cs – critical thinking, communication, collaboration, creativity – are the skills students need today. You may notice that they are the same skills required to be a master of the design process. Educators are currently incorporating the four Cs in education and play.

Critical Thinking

- Address real-world problems and arrive at solutions using the appropriate tools.
- Manage information presented in a variety of media and synthesize it to make it useful.
- Use an interdisciplinary approach and are engaged in project-based learning.

Communication

- Apply speaking, listening, writing and reading in varied contexts.
- Clear, engaging dialogue presented in a way that is meaningful to the individual and audience.
- Understand how to integrate appropriate and a variety of media.
- Engage in self and peer review and are taught information, media and digital fluency.

Collaboration

- Engaged in the community of learners. Includes partnership and teamwork, leadership and assistance, and alliances that serve to benefit the whole.
- Ability to self and peer assess, have leadership skills and can use collaborative media and technologies.

Creativity

- Creating something new and adding value to the knowledge that they have.

- Using novel thinking as an avenue for expression or to move a project forward.
- Develop qualities like entrepreneurship and innovation.
- Ability to imagine, incorporate the design process, integrate function, and use interdisciplinary and STEAM approaches.[5]

Other skills that are considered necessary for kids to acquire today are: global competence, culture, multicultural literacy, the arts, technology, character education, citizenship, confidence to fail and iterate, knowledge of content, reasoning, evidence collection, civic, ethical and social-justice literacy, health and wellness literacy, conservation and environmental literacy, scientific literacy, reasoning, economic and financial literacy, entrepreneurialism, global awareness, and humanitarianism.

Engagement in Learning Begins with Attention

- Attention is a variable commodity
- Active learning engenders attention
- Novelty and change get attention
- Physical movement fuels the brain
- Seat location affects attention
- Environment influences thinking
- Learning has a rhythm[6]

Global Competence

Today's kids are closer to a worldwide community through digital spaces, and they will need to build attitudes, knowledge and skills in global competence to create a sustainable future. Most educational frameworks for global competence include:

- Attitudes: Openness, respect and appreciation for diversity; valuing multiple perspectives, empathy and social responsibility.
- Knowledge: Understanding of global issues, interdependence, globalization and its effects on economic and social inequities locally and globally; world history; culture; and geography.
- Skills: The ability to speak, listen, read and write in more than one language; collaborate with people of diverse cultural, racial, linguistic and socio-economic backgrounds; think critically and analytically; problem-solve, and take action on issues of global importance.[7]

Learning Outcomes, Tactics and Assessment

When designing learning products, it may be necessary, or at least helpful, to develop the goals and desired outcomes of the content. Including assessment methods in your product, experience or program can also help to determine if the children are learning.

- Develop goals and learning outcomes. What do I want the students to learn?
- Develop tactics (methods, materials) for how to "teach" the topics of the project.
- Develop ways by which students can demonstrate that they have met the learning goals.
- Create assessments that are flexible and can be customized and adjusted for individual needs.

Tom Mott, senior producer and interactive learning designer, explains:

> We always have a learning designer as part of the core team throughout the production cycle. At the beginning of a project, they provide guidelines and input. Learning efficacy is tested through qualitative data (e.g., parents and teachers providing feedback) and quantitative data that's collected.

Step 1: Develop Learning Outcomes

Developing clear goals on knowledge, skills, attitudes, competencies and behaviors that you want the child to learn from your project, at the beginning, will help to focus on priorities in the design. Make sure to keep the statements focused on what the kids learn rather than what the project will teach. Use words that could be easily assessed.

What will the students learn?
What will the students be able to do?
How will they apply that skill or knowledge?

For example, learning outcomes could be identified in these groups:

- "Hard skills"
- "Soft skills"
- Higher-order thinking skills
- Skills defined in state and national standards
- Specialized skills (for example, STE(A)M subjects, 21st century skills)

Bloom's Taxonomy 2001

Designers can benefit from using frameworks to organize objectives because it helps to clarify objectives for themselves and for students, plan and deliver appropriate instruction, design valid assessment tasks and strategies, and ensure that instruction and assessment are aligned with the objectives. Bloom's Taxonomy (1956) for teaching, learning and assessment was updated in 2001 to be more dynamic:

Remember (Recognizing, Recalling)
Understand (Interpreting, Exemplifying, Classifying, Summarizing, Inferring, Comparing, Explaining)
Apply (Executing, Implementing)
Analyze (Differentiating, Organizing, Attributing)
Evaluate (Checking, Critiquing)
Create (Generating, Planning, Producing)[8]

Step 2: Teaching Methods

Activities such as lectures, experimentation, field trips, class projects, using media, reading assignments, interacting with software or objects, and experiencing are all different ways to teach content.

What methods will you use to teach the content in an engaging way?
What types of activities can you include for different kinds of learners?
Which content is best as instructor-led or uncovered by the child through the experience?

Step 3: Assessment

Testing, critiques, presentations, demonstrations, show and tell, and observing are all valid methods of assessing learning. A rubric lists outcome to evaluate or rate performance. They are helpful to show students ahead of time, so they understand what is expected of them and at the level they are performing.

Design Prompts

Can you clearly define your learning objectives?
Which teaching and assessment methods will you use?
How do you know the child learned what you intended?
Can they perform a task or showcase knowledge that they were unable to before the experience?

School Segments

If you would like to distribute your educational product, service, software or experience to a school, there are many different types of schools to choose from. Each school segment has goals and curricular objectives. It is up to us to determine the school segment that aligns with our objectives. Special education (advanced and delayed) can be integrated into the regular school system or may be a specialized program.

Many countries have educational systems that define academic standards of what needs to be taught and learned in a particular time frame. I have always considered educational standards a good starting point to think about a project's content. This inspires us to think about how a project can connect to the school curriculum and where it fits into a child's overall learning experience.

- Preschool (2 to 4 years) School readiness is the focus of much of early childhood education, either at home or in a preschool environment.

- Elementary school (kindergarten through 5th or 6th grade) Children gather a broad range of academic learning and socialization skills they need to succeed in further schooling and life.
- Middle school (6th or 7th to 8th or 9th grade) In these transitional years, students are given more independence, move to different classrooms and teachers, and choose some of their class subjects to prepare for the structure of high school.
- High school (9th or 10th through 12th grade) Graduating from high school has become increasingly important and is viewed as a minimum require-ment for success. An ongoing debate in education has been whether high schools should become college prep for the masses or an avenue to career readiness.[9]

Types of School

There are many different types of school, and it is helpful to research which kind of program or school environment would be best for your project. Teachers and children have rituals, such as the use of space and the daily program. Each school has a culture, and each class has one too. Understanding the cultures of different schools helps us to create better experiences for kids throughout their day.

Public schools
Charter schools
Magnet public
Vocational and alternative high schools
Private schools
Parochial schools (church-related schools)
Independent schools
Proprietary schools
Homeschooling

Technology in the Classroom

When does a live feed of what is going on in the world today surpass listening to a teacher talk about it or sitting and looking at it in print? When does it make sense to include interactive whiteboards or use paper and pencil? What do kids who are immersed in technology at home need from schools to learn at their best? How can technology bring more sensory information to accommodate all learners?

When used as educational tools in classrooms, for homework or edutain-ment, media and technology offer exciting, diverse ways to learn. Educational content, whether on TV, in video games or online, provides stimulation and engagement for countless youth who could not gain that exposure through any other means. Although new media has given them more access to information, it has hurt them when it comes to digesting this information and knowing what to do with it when they do get it.[10]

Waldorf schools discourage any form of digital technology for children under 12 years of age. Beverly Amico, leader of outreach and development at the Association of Waldorf Schools of North America, explained time-tested truths about how children learn best.

> Teachers encourage students to learn curriculum subjects by expressing themselves through artistic activities, such as painting and drawing, rather than consuming information downloaded onto a tablet. Lessons are delivered by a human being that not only cares about the child's education but also about them as individuals.[11]

However, Don Marinelli of Carnegie Mellon is in support of technology in the classroom:

> We are losing the kids' interest. How do we teach better? Sitting in a school on creaking furniture, looking at the clock and old textbooks is not engaging. Give them the information through the media skills they have. Instead of just looking at a shield in a photo bring the world to them with technology.[12]

Eston Melton, an assistant principal at West Potomac High School in Alexandria, VA, says students can better internalize their lessons when they're doing them on their smartphones or tablets. "My education becomes something I walk around with in my pocket," he says.[13] Maggie Hendrie, chair of the interaction design program at ArtCenter College of Design and a consultant in UX Strategy and Interaction Design, stated:

> Designers should consider micro connections. Connectivity over distances. There are strong benefits to kids playing with children in other countries or with people of different points of view. Connect with others around play. They can explore new modes of education. What can digital technology do to enhance the relationship of the child, the family, and the teacher and how they interact?[14]

Design Prompts

How can schools and cultural institutions expand their reach and audience to connect children and youth with content outside of the classroom?
How can digital networks better connect learners and those with knowledge?
What forms will school take? Can it be self-organized?
How can mastery be demonstrated with new methods of learning?
What do new learning communities look like?
How will learning technologies address resource constraints?

Methods of Teaching

Instructor-Led

Information is being imparted to one or more people by another. It is interactive and in real time. Students can answer and respond to questions and participate in activities like group discussion and demos. The retention rate has the potential to be very high if the level of engagement is high.

Computer-Led

This type of teaching includes interacting with a computer to learn. For example, it could be through an digital game, an online course or videos.

Flipping the Classroom

Educators are leveraging technology to create a different role for themselves in their classrooms. Instead of using class time to verbalize information, technology is helping them use their time with students more efficiently. In this combined model, teachers record lectures on video and post for the students to access online. Students can watch the lectures when it's most convenient for them. In the classroom, building on the material they studied, students work on a project with teachers at their side, coaching, answering questions. Students can seek one-on-one help from their teacher when they have a question and learn in an environment conducive to education. Andrew Kim, a Steelcase education researcher, says:

> More and more, classrooms are becoming places where knowledge is created versus consumed by students ... As students start to have more control over what they use to help them learn, you need to have spaces that support more creative or generative activities. Teacher–student interaction and peer-to-peer interaction within the school is fundamental to learning.[15]

Educational Methods, Models, Approaches

Constructivist Learning Theory

Educational programs promote specific learning methods throughout the school years. Some of these programs are based on constructivist learning theory. John Dewey, Lev Vygotsky and Jean Piaget had strong influences on the development of this theory. One of the primary goals of using constructivist teaching is that students learn how to learn and take the initiative for their own learning experiences. Examples of constructivist activities include experimentation, research projects, field trips and media. Class discussions and reflections are the most important distinctions of constructivist teaching methods.

- The active learner (Dewey's term) uses sensory inputs and needs to engage physically.
- People learn to learn as they learn: Learning consists of constructing both meaning and systems of meaning.
- Learning involves language: The language we use influences learning.
- Learning is a social activity: Our learning is associated with our connection with other human beings, our teachers, our peers, our family and acquaintances.
- The activities are interactive and student-centered. The teacher facilitates a process of learning in which students are encouraged to be responsible and autonomous.
- Learning is contextual: We learn in relationship to what we know, what we believe, our prejudices and our fears.
- One needs the knowledge to learn: It is not possible to assimilate new knowledge without having some structure developed from previous experience to build on.
- It takes time to learn: For significant learning, we need to revisit ideas, ponder them, try them out, play with them and use them.

Motivation is a critical component in learning. Unless we know "the reasons why," we may not be very involved in using the knowledge.[16]

Furniture and Toys for Progressive and Constructivist Approaches

Interview with Dana Wiser, design manager, Community Playthings

Designed for the early childhood classroom, the approach to the designs appeals to Reggio, Montessori, Waldorf and High/Scope, and is compatible with some of the commercially available curricula.

What are some similarities and differences in designing for kids rather than a general user?
Good design is sound design, and many general principles apply across the board. Also, as we design children's products, we know those teachers' and parents' needs obtain as well, so it's never only about kids.

> Similarities: Aesthetics, human factors, universal design (as per the spectrum of abilities) – these apply as elements of good design.
> Differences: Products for young children, 0–8 years, must be developmentally appropriate, requiring some expertise in child development and psychology. Specifically in today's society, this suggests freedom from screen technology and a commitment to open-ended design that engages children's' imagination.

How have you seen the early childhood market for furniture and toys change over the years?
The early childhood market changes with society. Significant trends that influence our designs are:

a. Daycare. Mothers joining the workforce changed child care. Daily duration went up to 10 hours, and centers became surrogate homes. Children's needs in the daycare environment include softness, security and cozy spaces.
b. Infant care. Another effect of mothers working was to reduce the age of children in care.
c. Developmentally appropriate practice. As standards for child care evolved, the role of furnishings grew. Learning centers, for dramatic play, block play, literacy, the science area, all needed boundaries defined by storage of respective materials; they needed work and play surfaces. These became design and market opportunities.
d. STEM and nature. A resurgence of interest in science, nature and the out-of-doors led us into what is emerging as a whole new product line for outdoor play
e. Technology. This substantial societal change is a main detrimental to young children. Developmentally appropriate practice recognizes the need for young children to deal with the real world, not the virtual. We won't even design a computer table for this reason, or support screen technology in other ways.

How do these different types of play patterns influence your product designs?

Pretend play: The play kitchen is iconic – no early childhood classroom can be without it, along with small tables and seating. We leave details to the children's imagination with open-ended materials. Support for dramatic productions includes puppet theaters and staging.

Constructive blocks allow children to work through issues of real life by building their own worlds.

Creative play: Children become engrossed with sand and water play, preeminent open-ended materials, whether indoors or out, where the line between creative and science blurs.

What types of details do you use in your designs to make them safe and child-friendly?

Freedom *from* details contributes to safety because it supports durability. We treat wooden parts with heavy radii and pride ourselves in machining bare wood to a high finish. Product failure lets children down; durability, our hallmark, contributes not merely to physical safety but to psychological well-being – I am worthy of this well-built plaything. A green chemistry policy adds much to safety.

What is the most important thing to know about the difference between designing for indoors and out?

For play equipment, durability is critical, and outside all issues are heightened. We would expect to support rougher play, especially in the first half hour of release from indoors, but after that, and given sufficient time in the daily schedule, the play dependably settles into the basic exploration of the world and relationships through interaction with the materials. In or out, this should involve access to loose parts in open-ended play; also to child-managed risk.

Are there particular considerations when designing each of these products?

Toys: Freedom from detail. Open-endedness. Durability. How to empower the child's imagination.

Storage: Shelves' backs are exposed to contribute to the interactive environment. The height of shelving affects room visualization and supervision. Shelves also present materials for children's choice of where to work and play. Consider storage of teacher-only materials and child-proof security. Shelves and furniture both must be flexible as to room layout according to constantly changing needs.

Furniture: Anthropometrics and ergonomics are essential for small, growing children. We recognize that room furnishings are a foundation for teachers' individualizing and localizing of their classrooms. We design prompts for such interior design into our furniture.

Play equipment: Children's needs are deep and wide, so equipment capable of being many things depending on today's narrative is essential. Versatility, flexibility. Large muscle play needs supporting alongside dramatic and constructive.[17]

This section outlines a few philosophies of education. Some are specific to early childhood; others expand into elementary and high school. In our design process, we can use these models for inspiration. One might be in tune with your views towards education, and you may decide to include the principles in your project or to expand on the current methods. Or you might find approaches that you like in each of them to influence the approach to your project.

Montessori Method

Dr. Maria Montessori developed this educational approach, which has been tested for over 100 years in diverse cultures throughout the world.

Five basic principles represent how educators implement the Montessori method:

- Respect for the child
- The absorbent mind
- Sensitive periods
- The prepared environment
- Auto education

◀ Figure 7.6

Montessori learning tools are each designed to teach specific concepts. These trays teach about different bodies of water. Source: author photo.

In a Montessori environment, children learn by exploring and manipulating specially designed materials for the curriculum. Each material teaches one concept or skill at a time and lays a foundation from which students can comprehend increasingly abstract ideas. Children work with materials at their own pace, repeating an exercise until it is mastered. Throughout the classroom, curriculum areas contain a sequential array of lessons to be learned. As students work through the sequence, they build and expand on materials and lessons already mastered, developing qualities with which they'll approach every future challenge: autonomy, creative thinking, and satisfaction of accomplishment. The teacher may gently guide the process, but the goal is to inspire rather than instruct.[18]

Whole Child Approach

This initiative looks at the development of the whole child as an individual within nature and the community. It promotes these areas of learning:

- Intellectually active
- Physically, verbally, socially and academically competent
- Empathetic, kind, caring and fair
- Creative and curious, disciplined, self-directed and goal-oriented
- Free
- A critical thinker
- Confident
- Cared for and valued

The Whole Child approach identifies five kinds of learning that each child is exposed to, every day if possible. They are:

- Cognitive-intellectual activities
- Creative-intuitive activities (the arts)
- Structured physical movement and unstructured, self-directed play
- Handwork – making things that can be useful
- Engagement with nature and community[19]

Waldorf Method

Waldorf education is based on the insights, teachings and principles of education outlined by the world-renowned artist, and scientist, Rudolf Steiner. This method integrates the arts in all academic disciplines for children from pre-school through 12th grade to enhance and enrich learning.

In the lower grades. *artistic* elements in different forms (rhythm, movement, color, form, recitation, song, music), as a means to learn to understand and relate to the world, building an understanding for different subjects out of what is beautiful in the world.

In the upper grades and high school, this leads to the conscious cultivation of observing, reflecting and experimental *scientific* attitude to the world, focusing on building an understanding of what is true based on personal experience, thinking and judgment.[20]

Anji Play

This somewhat new philosophy and approach to early learning has been developed and tested by educator Cheng Xueqin in Anji County, China. It is beginning to achieve international recognition, and we are starting to see some programs based on Anji Play in the US. Through sophisticated practices, site-specific environments, unique materials and integrated technology, the Anji Play ecology focuses on self-determined play in an environment defined by love, risk, joy, engagement and reflection. Play designer Cas Holman explains:

> Anji Play is a revolution and is already changing everything. Ms. Cheng Xueqin developed the model from the ground up with play at its core. She made no assumptions about *what* the children needed to learn. They discover that. She made no assumptions about *how* they needed to learn. They explore that. This puts the process at the center, rather than the outcome.[21]

▶ Figure 7.7

Exploring risk is a tenet of Anji Play.
Source: Anji Early Childhood Education (used with permission).

Reggio Emilia Approach

The Reggio Emilia approach values the child as strong, capable and resilient. This approach originated in the town of Reggio Emilia in Italy out of a movement towards progressive and cooperative early childhood education. Every school adjusts the philosophy to reflect the community's needs.

Fundamental Principles

- The Hundred Languages of Children: Children use many methods (drawing, sculpting, dance, movement, modeling, music) to show their understanding and express thoughts and creativity.
- Children search out knowledge through their investigations. They are motivated by their interests to understand and learn more.
- Social collaboration and playing and reflecting in groups, having their thoughts and questions valued is a focus.
- Communication is a process, a way of discovering things, by asking questions, using sound, music and language as a way to explore and to reflect on experiences.
- The environment is a teacher. A space filled with natural light, order and beauty encourages collaboration, communication and exploration.
- Mentor and guide: The adult's role is to observe children, listen to their questions and their stories, find what interests them and then provide them with opportunities to explore.
- Documenting and displaying visual representations of children's thoughts and learning and making them visible.[22]

◀ Figure 7.8

Listening to stories and reflecting with the teacher is an important part of the Reggio Emilia Approach. Source: author photo.

A Reggio classroom.
Source: New School West (used
with permission).

The Reggio Approach

Interview with Roleen Heimann, co-founder and co-director of The New School West, a preschool that uses the Reggio Approach

Can you describe how the Reggio Approach is used in your preschool?
Everything is intentional. We focus on the children, families, teachers and the school with building community. Through observation, we look at the children as collaborative thinkers and ask, "What do they know?" Then we ask, "What do they want to know?" and we develop our curriculum to support their questions and theories. We use open-ended materials in a co-constructive process, and the teachers and children become researchers together. One group may be experimenting with space and materials. One group is writing a play. Through group meetings, the work is shared and interconnected via more dialogue and questions. This is where the children can think about their thinking, a metacognitive process that makes the meaning that is important in connecting thoughts and extended ideas. The parents are partners and bring what they know to the community. We are genuine learners together.

What are the most important things designers can learn about children by studying the Reggio approach to bring to other products, environments or programs for children they are designing?
Think about designing for process and the development of critical thinking skills. How do you get past a product as a stationary object? How can children use open-ended materials in relationship to multiple ideas by using them in many ways? How can the time and space be flexible?

Can you explain specifics about how your environment is set up to nurture the Reggio approach? How about the materials and products in the environment?
A lot of times designers see what they like in a design and copy it for what they want to do in their new design. Instead, challenge yourself to observe without a pre-set idea and look at what the children are doing. Be creative with them. Where are their interests? Is there more than one way to do something? The environment is essential in setting the tone and availability for an extended time to make a relationship with the materials. Children (and we) have a right to have time to think and rethink, time for both individual and collaborative thinking, and the time for flexibility. What do you think will happen here?

Can you elaborate on why these are important for child development?
Sensory manipulation of materials: Children learn through their senses. In urban areas, children don't often get to feel and touch nature as much as they should. Natural materials are an essential part of learning. At the beginning of the school year, we have these big boxes of clay. Children don't just start sculpting with it. We slowly introduce it to them. First, we put the blocks of clay on the floor, and they jump off it, poke it, touch it and start to make forms. They move into sculpting with their hands and tools later on. With kinesthetic learning, we also play with sound, light and tactility.

Technologies are materials too: Overhead projectors add a dimensional element to the space. The children see things in different ways. We use digital cameras that support metacognition (thinking about their thinking). Teachers use video and photos to record what's happening so as not to miss any important processes. There are many times, the teachers play back the recordings with the children so they can reflect on what they were thinking.

Documentation makes learning visible: With teachers recording and documenting elements of what the children are doing at any point of the day, the focus is no longer on the end product but, instead, on the process of thinking and connecting with others and materials that were involved. Parents forget that children are so connected to them that they can't imagine that their parents don't know what they did. By making the interactions, connections and learning visible, the parents (and the children and teachers) can see it.

Hundred languages of children as a means of expression: There are different ways to express the same ideas and children have the right to have access to a variety of materials for expressing these ideas. Each of us learns differently so including materials as part of the learning process supports the context of learning. Young children must touch, feel, see and hear in many ways to construct this knowledge. It's not just about workbooks, pencils and paper and one-dimensional work.

Using language as play: The children and teachers are supporting communication all day. They write down memories, leave notes to parents and send messages to each other in their personal folders. We have small group discussions. We have a communication table.

Do you have any ideas on how society can further support children of today?

What struck us most as visiting teachers in Reggio Emilia, Italy, is that children are very much a part of their community and they are seen as viable contributors to their society. They are capable, curious, strong communicators and they have a view of the world that is worth seeing and listening to. Here in the States, we see our role as protectors of children. We even have phrases like "They are our future" and "They are empty vessels ready to be filled." We are so busy trying to "teach" them what we think they should know that we fail to see and hear what they are telling us. How can we support them with their questions?

They are citizens and present now. We need to get past the teacher mode and look at what are they offering as individuals and a valuable part of the community.

Where can designers be most helpful using their creative skills to help child development?

1. Create materials to be used in many ways, not just one. Keep it open and have them develop it.

2. Cost is a big problem. There is a divide between who has access to products, environments and experiences. Some don't get access to the opportunities others do. Are there alternative solutions?

3. Some parents feel guilty because they don't spend enough time with their children and they replace that with buying things for them. The best "toys"/ "activities" are ones that involve challenging thinking which could engage both children and adults. In the child development books, it is said that that most valuable material is one that has moveable parts, that can be used in many ways.

4. Designing for someone else, you must explore yourself. To be creative designers for children, you must stop and take the time to observe, watch children, and spend time with them. You will see things that you won't think of otherwise. You can't design and intrude your ideas on them.

I like how the Reggio Approach includes creative development and expression of ideas as a core aspect of child development. What do you think are the most important things for designers to know about each of these topics?

Cognitive development: Metacognition is essential. We spend time thinking about thinking. Co-construction is the social aspect to constructivism. Children working in a group will have the ability to see things differently with many different perspectives. It's very different than working alone.

Social development: From the moment they are born, children want a relationship, and it is our responsibility to support the interconnections. It's not something we "teach." It is something we "model." Classroom environments have a lot to do with enhancing relationships.

We respect one another. We take care of each other, see other perspectives and learn to live in and to respond to the world. This is a community, and there are acceptable behaviors that have to be modeled. The children learn to think about who they are as part of a group and listen to others. How do I express myself in socially acceptable ways?

Physical development: Children need areas to explore both small and gross motor skills. People place a lot of importance on small motor skills, and yet it's most important in the early years to offer many more gross motor skill development. Balance and coordination are a part of supporting small motor development, as well.

Creative/expressive development: There are many ways in which we take in information and having the materials, environment and time to express ourselves and our ideas is something of value and an intentional part of both our indoor and outdoor classrooms.

Having strength in all these five domains contributes to having healthy self-esteem.

What are your thoughts on media play?
There is a lot of research that tells us that limiting the amount of media is essential. When children get stuck in play that is copied by the media, they are less creative in supporting others' ideas and taking on roles that are not predefined. Defined story versus collaborative storytelling. (ex: Once we had a child with glasses and all the children wanted him to be a specific character with glasses, but he didn't want to be that character. To be a part of the role he wanted to play, he hid his glasses.) Children must feel powerful. Sometimes a character offers a way for a child to enter a group to play.

What are the most important things adults can learn from children?

- Mistakes are valued. Don't rush to fix or make rules
- We need time to theorize, reflect and review
- Co-constructivism has us learning together
- Free ourselves from judgment
- Slow down and observe[23]

STEM + Art = STEAM

Innovation remains tightly coupled with science, technology, engineering and math, and therefore in recent years it has become a focus in education, toys, and programs and experiences for children. STEAM education incorporates the "A" for the arts – recognizing that to be successful in technical fields, individuals must also be creative and use critical thinking skills, which are best developed through exposure to the arts. Ana Dziengel, a designer and creativity blogger noted:

> I look at STEAM as the design process. It's important for kids to learn to problem solve and make connections, which is what the design process is all about. More parents are becoming aware of the Process Art Movement and valuing the process over product. Process art, STEAM, and design all emphasize the same things: exploration, making mistakes, and problem-solving. If you want to be an innovator you need to think differently.[24]

I spoke with Tom Mott, senior producer and interactive learning designer, about his thoughts concerning STEM and STEAM. His response was:

> I have mixed feelings. On the one hand, I love the process behind STEM: identify a problem, brainstorm to propose a solution, try something out, iterate, and arrive at a goal. I love that! Those are amazing skills to support. I love that maker spaces, and maker fairs are popping up. It's good to see kids MAKING things. But … there's a faddish aspect to STEM/STEAM that I am wary of. An excellent STEM product will find an audience and impact a child's life. And if that's *your* passion too, then you're going to feel connected to making products like that.

He also had some suggestions when working on a product with the primary goals of teaching STEM skills or early math skills: "I strive to strengthen executive functions as well, like flexibility, short-term memory, following multi-step directions, etc."[25]

◄ Figure 7.10

STEAM Carnival attendees wrestling with the Hexicade.
Source: Two Bit Circus (used with permission).

NASA JPL, Merging the Arts and Sciences

Interview with Jessie Kawata, former industrial design lead and creative strategist at NASA JPL

Why is it important for children and youth to have a strong foundation in both the arts and sciences?
The arts and sciences offer an understanding of how the world works through direct interaction. Design bridges the arts and sciences. It also preps them for interdisciplinary interaction throughout their childhood and adult lives.

Artist skills to acquire: Creation making and visual communication are vital for innovation. Artistic expression allows kids to articulate ideas.

Science skills to acquire: The ability to understand how the environment works as a holistic system. The merging of contexts helps to understand implications. We are a part of something bigger and our actions matter.

What is your take on the interest and growth of STEM and STEAM in education?
I believe that STEAM should be STEAM-D, adding in a "D" for "Design." There is a preconceived notion in the science and tech industry that the "arts" are meant for public outreach or marketing. It is perceived as the "decoration" and not part of the "cake batter" itself. Designers build, invent, architect, research and operate just like engineers and scientists do. So the question for educators is how to mitigate these biases early on so students can understand the value of working with creative minds?

When teams of artists and scientists work together, what are the strengths of the different backgrounds? How do they influence the project?
Scientists often look at the micro: artists or designers often look at the macro. Both disciplines ask "why." However, when scientists try and understand, artists and designers ask more questions. The artist or designer might ask: Why do people care? How is this relevant?

How are the thinking process and skills you learned as a designer used in the science world to create innovative solutions?
Designers can help make science applicable, usable and relevant to society. By using processes such as ethnography, systems design and design thinking, design can help formulate usability requirements for science data. These approaches have helped me to work easier with engineers and technologists to create blue-sky solutions that are more impactful and elegant.

What makes space so fascinating to kids?
The mysteries of space foster curiosity and the imagination. I'm sure science fiction and sci-fi movies contribute to the fascination. I've always wondered what it would be like if astronomy and earth science was taught in school as an essential skill such as reading. Would

children grow up having a closer connection to the cosmic universe?

What are the three things that you would like people interested in designing for kids from an earth science social impact perspective to work on in the coming years?

- Climate change: Understanding climate change is an excellent way to teach kids the perspective of cause versus effect and pattern change over time.
- Biogeochemical cycles (water, clouds, rain, ocean): Natural cycles are sometimes hard to understand. Consider creating a narrative where they can find personal interest through societal implications. Examples are how the water cycle affects farming or how cars affect the carbon cycle.
- Biodiversity: Each organism, whether it be a plant or animal, has a role in an ecosystem and therefore plays a vital role in the health of our planet. Creating a product or solutions about this subject allows kids to understand the importance of diversity of life.

What role do you think designers can play in education and influencing children's interest in the arts and sciences?

The environment is at a tipping point right now, and in the next coming decades, the intersection of technology, science, design and art will be more critical than ever. Engineers will need to collaborate with designers to make sustainable solutions that fit seamlessly into people's lives, and scientists will need to collaborate with designers or artists to communicate the complexity of scientific data. To influence this intersection at an early stage, designers can create educational tools and toys that foster multi-disciplinary creation and collaboration. They can create children's products that help to solve real-life problems, work to bring art and science workshops to schools, improve and develop curriculums, or create experiences for children that integrate science. Another impactful way that designers can help drive children's interest in art and science is to work at a science organization to influence the way they do public outreach. Many science institutions do not have core artists or designers integral to the way they operate.[26]

▶ Figure 7.11

Industrial design team members at JPL brainstorm about a mission to the caves of Mars in Left Field.
Source: Jessie Kawata (used with permission).

Project-Based Learning

This approach to learning focuses on developing a product or creation. The project may or may not be student-centered, problem-based or inquiry-based. Students create written, oral, visual or multimedia projects with a real-world audience and purpose. Project-based learning is usually done in English, social studies or foreign language class. According to John Larmer, director of product development, and John R. Mergendoller, the executive director at the Buck Institute for Education, the difference between making projects such as building a volcano or a diorama and project-based learning is that meaningful cognitive engagement – rather than the resulting product – distinguishes projects from busywork. Project-based learning includes:

- A Need to Know: An "entry event" that engages interest and initiates questioning.
- A Driving Question: A good driving question captures the heart of the project in clear, compelling language, which gives students a sense of purpose and challenge.
- Student Voice and Choice: Regarding making a project feel meaningful to students, the more voice, and choice, the better.
- 21st Century Skills: Students investigate questions, synthesize information they gathered and use it.
- Inquiry and Innovation: As they find answers, they raise and examine new questions. Students integrate the information they gathered to create their team's product related to that question.
- Feedback and Revision: A process for feedback and revision during a project makes learning meaningful because it emphasizes that creating high-quality products and performance is an essential purpose of the endeavor.
- A Publicly Presented Product: Students answer questions and reflect on how they completed the project, next steps they might take, and what they gained regarding knowledge and skills – and pride.[27]

Inquiry-based learning and design-based learning are all forms of project-based learning.

Problem-Based Learning

This approach to learning that focuses on the process of solving a problem and acquiring knowledge. The approach is also inquiry-based when students are active in creating the problem. It often takes place in math and science class. It doesn't necessarily include a project at the end, so it doesn't always take as long as project-based learning.

Inquiry-Based Learning

In this approach students explore a question in depth and ask further questions to gather knowledge. This student-centered, active learning approach focuses on questioning, critical thinking and problem-solving. It's associated with the idea "involve me and I understand."[28]

Design-Based Learning

DBL "sneaks up on learning" by engaging students' innate curiosity and creating a fun, interactive environment that develops higher-level reasoning skills in the

Design-based learning projects.
Source: Paula Goodman (used with permission).

context of the standard K–12 curriculum. A teacher using DBL challenges students to develop never-before-seen solutions to specific problems. For example, a 6th grade class studying ancient Egypt, China, Greece and Mesopotamia might begin by working in teams to build a 3D scale model of a place in that ancient world. Doing so compels them to answer questions such as "What types of dwellings will people live in?" and "What values will they live by?" Armed with their solutions, students investigate topics about these societies in required texts.[29] Paula Goodwin, the Director of the K-12 programs at ArtCenter College of Design, explains the difference between project-based learning and DBL:

> DBL focuses on how to frame the challenge – words matter. It doesn't say what the solution could be in the question and is open-ended. For example, other forms of learning might frame the problem as: Design a container for water. Design-based learning would frame it as a Never-Before-Seen Design Challenge: Create a way to transport water. The framing is intended to bring about more innovative solutions. If you say use blue – you use blue. DBL philosophy is not about what to teach but how to teach it. Design challenges can be tailored to the process. In preschool, it can be used for the alphabet. In HS it can be used to build the garden that deals with sin after reading *The Scarlet Letter*. It doesn't matter what the design looks like. It is the big idea. What are the concepts?[30]

Personalized Learning

In his best-selling book *The Element*, Ken Robinson writes

> the key to [educational] transformation is not to standardize education, but to personalize it, to build achievement on discovering the individual talents of each child, to put students in an environment where they want to learn and where they can naturally discover their true passions.[31]

Sometimes called "teach to one" or "adaptive technology," personalized learning uses interactive software to tailor lessons and assignments to individual students, to reflect their strengths and weaknesses, at the pace at which they learn. Soon, with technologies, children will have a diverse learning ecosystem in which learning adapts to each child instead of each child trying to adjust to school. Learners and their families will create individualized learning playlists reflecting their particular interests, goals and values. Those learning playlists might include public schools but could also include a wide variety of digitally mediated or place-based learning experiences.[32]

Learning through Play and Gamification

Educators are increasingly seeing the value of using play and games in learning.

Ben Mardell, of Project Zero, Harvard Graduate School of Education, explains: "learning through play in school involves promoting students' feelings of Choice, Curiosity and Delight."[33]

Game-based learning is active learning and engages kids to interact with subject matter and make solving problems enjoyable. The theories of game strategy that make them alluring, motivating and sustain attention are being brought into schools in many creative forms. Appropriate, informative, educational and interactive content showcasing stories with engaging characters is a strong learning tool.

- As early as kindergarten, kids are playing with games in the classroom to learn essential reading and math.
- Middle and high schoolers are being tested using competitive game show formats to compete against each other to show the retention of knowledge of their lessons compared with traditional testing methods.
- Board games that align with state and national curriculum standards are now part of school library collections.
- Game-based learning systems are being used to teach courses and have been proven to increase intelligences.[34]
- Instead of starting with an A and going down from there, kids begin with a zero and level-up achievements through quests.

Successful educational games must identify the ability level of the player and provide rapid feedback. The games must be slightly more challenging than the learner's skill. Paul Levine, CEO of Play Science, explains:

> It is important to know what play type is appropriate to drive the learning. Research the science. What is the pedagogy? Framework? Where is it on the play spectrum? Should it be directed? Should you use scaffolding? Or should it be open-ended? Is there playful learning? Are they playing as individuals or in groups? Is there Coplay within one or in different age groups 0–3, 3–5, 6–9? What does the

experience look like? Design so there is just enough rules and framework and not too much to inhibit creative and pretend play. STEM learning happens best in a project-based approach because kids reason through a problem and there is more than one way to do it. Many game mechanics use scaffolding. Use playification versus gamification – choose intrinsic motivation versus extrinsic motivation – you always want to shoot for driving play and intrinsic motivation.[35]

Paul Levine's suggestions for designers:

- Work with learning designers and curricular framework.
- Give kids agency in their learning. Don't force parents to be part of the process. Education relies on intrinsic motivation.
- Experiential is the most impactful kind of learning.
- Understand the culture's attitude and values around education. Don't assume it is the same as yours.

Summer Camp

"Shrinking summer" is the term used to describe the trend towards earlier school start dates or year-round schooling. Summer camp has evolved to include a series of crash courses of child enrichment.[36] Camps include themes such as Chinese Language Immersion Camp, STEM Technology Camp and Broadway Music Camp. This provides a lot of choices for kids to stay on top of their education with activities they enjoy. Many parents feel pressured for their kids to take advantage of this time to advance learning.

Camps offer something different from other community experiences:

- Kids learn how to adapt quickly to a community that forms for a short time and has a finite conclusion.
- They interact with positive role models outside of their family and explore their talents, interests and values.
- They learn to work together, make choices, take responsibility, develop creative skills, build independence and self-reliance, and gain confidence.

Old-fashioned nature camp (with s'mores and campfires) is recognized as valuable in helping children mature socially, emotionally, intellectually, morally and physically.

Benefits include:

- Reconnecting with nature
- Spending their day being physically active
- Unplugging from technology
- They have free time for unstructured play
- They can grow more independent
- They can experience success and become more confident

- They can gain resiliency
- They can develop life-long skills and wilderness training
- They can learn social skills and make friendships[37]

Cultural Differences throughout the World in Education

Culture impacts the way children participate in education. There are different expectations from parents, children and teachers in different cultures and environments. As designers, we must understand ideologies regarding education and learning and the cultural patterns and beliefs of groups as well as including variations for individual children.

Individualist and Collectivist Cultural Perspectives on Education

Individualist Perspective

Students work independently; helping others may be cheating.
Students engage in discussion and argument to learn to think critically.
Property belongs to individuals, and others must ask to borrow it.
Teacher manages the school environment indirectly and encourages student self-control.

Collectivist Perspective

Students work with peers and assist when needed.
Students are quiet and respectful in class to learn more efficiently.
Property is communal.
The teacher is the primary authority, but peers guide each other's behavior.
Parents yield to teacher's expertise to provide academic instruction and guidance.[38]

What Can We Learn from Finland, Denmark, Sweden, Norway?

Finland is highly regarded for its educational philosophies. As the Harvard education professor Howard Gardner once advised Americans, "Learn from Finland, which has the most effective schools and which does just about the opposite of what we are doing in the United States."[39] You would expect to find the best high-tech gear and learning programs in Finnish schools, but instead, as Kelly Day, a math teacher and Fulbright scholar researching in Finland, put it,

They truly believe and live by the mentality of less is more. Conversely, in the US we truly believe

"more is more" and we continuously desire and pursue more in all areas of our lives including education. We can't even stick to one philosophy of education long enough to see if it works. We are constantly trying new methods, ideas and initiatives.[40]

Today, the Scandinavian and Nordic countries have extreme regard for children and are modeling many ways of combing play with education successfully. In later years, they value equality more than excellence.

Educational Equality

When all children have access to quality education, it gives them the tools for success and has a ripple effect on the family, community and economy for future generations. Educational equality is one of the most significant challenges our world faces. Many of the world's most disadvantaged children remain excluded from school, and many of the children in school are enrolled but are not successfully learning. How can we use our skills as designers as a way to improve the inequality gap in education? Through products, services and systems, what kinds of tools and support can be developed to enhance learning? These are some of the most significant problems to solve. To create global impact, UNICEF encourages us to work on:

- Enrollment and achievement
- Education during and after conflict
- Equality in education
- Child-friendly schooling[41]

Improving Education, a Global Perspective

Interview with Rena Deitz, senior education specialist at the International Rescue Committee
I had the pleasure of interviewing Rena Deitz about her humanitarian work with children. Her focus is on social and emotional learning. She shared with me some of the challenges of improving the education for children who have experienced trauma or are growing up in difficult circumstances. Her experience in working alongside design researchers offers insights into where they can provide value in the process.

What are the most important things for designers to know about children's issues related to education from a global perspective?
We tend to think about broad humanitarian issues or culture, but we need to look at each child. Each has a different background and story. If we just look at the overarching similarities, we overlook the individual needs. With humanitarian work children have been through the loss of a parent, attacks, some were child soldiers. The background of the child and the trauma needs to be addressed first. Then we can work on a general approach to literacy and numeracy.

What do you see as the most significant challenges at the International Rescue Committee when working on projects for kids/youth?
The individual needs as just mentioned. The other is navigating between kids and the local governments' approach. Everything mixes with political. For example, in Lebanon, the government is the provider of education

for Syrian refugees, which is a great long-term solution. In the meantime, while they are still unable to provide every refugee with quality education, they are only allowing NGOs to support refugee children *who are enrolled in school* with "remedial education." As a result, the children who are not able to access formal education miss out altogether.

You have worked with design researchers in the past. Do you see benefits to design researchers and educators collaborating?
It is good to have a separate time and space allocated for this research. It is a parallel process to what we already do – but designers bring a fresh lens. Designers tend to look at broader issues. It all comes down to the experience of the design researcher. There is a considerable range of rigor and skill in the field. You have to have the knowledge and background, and be willing to learn from the experts in the subject area you are designing for. One of the major risks in designing for humanitarian education is inadvertently doing harm. It is essential, also, for designers to be aware of the humanitarian "Do No Harm" principle and follow it, and any guidelines provided for them by the experts.

In what ways do you hope to see new technologies affect international education in the social impact field in the coming years?
I hope that tech will become a priority and even out the playing field. Right now access to tech does the opposite. It creates a greater divide. I hope to see

technology get to more people. I worked on an Arabic literacy app for Syrian refugees that had a social and emotional component. You don't just drop US tech (iPads, mobile devices) into a different context. Design to the tech that exists on the ground. Most people have phones, so create text-based communication. Not many people have access to educational software, so focus on the development of curriculum and storytelling. Content alone works. Technology alone doesn't. Come with the tech and teach both the tech and content.

Do you think about play's role in child development and learning in your approach? If so, how?

Play is in every area of IRC's education work, literacy and numeracy. There is a playful approach to the activities. In early childhood, formal play and structured play are at the core of what we do. Informal play is also an important mechanism for children to develop and practice skills they are learning elsewhere.

Are there specific ethical issues you encounter when working on projects for kids/youth? If so how do you address them?

Ethics is the most significant challenge with design. IRC staff use a child safeguarding policy. When we use outside designers, they are briefed on the policy, but they do not always adhere to it. It is a bit of a challenge. Taking pictures is a significant controversy. We need to safeguard and get formal consent, although the context is complicated and it may be difficult to obtain written permission. Maybe there isn't a parent or if there is they can't sign consent. If we use pictures, we change names to protect the identity of the child. They may be at risk if people can find them – that is why they are displaced. Designers are sometimes hesitant to change names, etc.

Can you develop questions that you want to ask designers to think about when working on projects in these areas?

- How, as designers, can you build on what already exists?

Build on what we have learned from other projects. For example, schools in the box, libraries without borders. Designers should be able to take what is there and make it better instead of starting over.

- How can you include the people that are already doing the work?

Don't work outside the sector. Integrate design into what the people are doing every day already instead of something that is stand-alone. Get at the real needs of the beneficiary/client and the implementing staff.

- How can you complement areas of expertise?

Rely on your expertise as a designer and the process that you use. Designers are good at looking outside of a small box. Usually, in education we are looking at a smaller piece, often focused on traditional formal or non-formal education. Designers can include a broader perspective, and pull in insights from other sectors.[42]

Notes

1 Stanford Children's Health
2 Eton, Sarah Elaine 2010
3 *Principles of Child Development and Learning*
4 Davol, Leslie 2017
5 National Education Association
6 Steelcase Report 360° The Education Edition
7 Tichnor-Wagner, Ariel 2016
8 Armstrong, Patricia
9 Weber, Steven
10 Common Sense Media 2012
11 Jenkin, Matthew 2015
12 Marinelli, Don 2010
13 Shane, Brian 2012
14 Hendrie, Maggie 2017
15 Steelcase Report 360° The Education Edition

16 Hein, George E. 1991

17 Wiser, Dana 2017

18 American Montessori Society

19 ASCD

20 Association of Waldorf Schools of North America

21 Holman, Cas 2017

22 The Reggio Approach

23 Heimann, Roleen 2017

24 Dziengel, Ana 2017

25 Mott, Tom 2017

26 Kawata, Jessie 2017

27 Larmer, John; Mergendoller, John R. 2010

28 Teacher Tap

29 ArtCenter

30 Goodwin, Paula 2017

31 Robinson, Ken 2009

32 Sweetland Edwards, Haley 2015

33 Mardell, Ben 2018

34 del Moral Pérez, M. Esther; Guzmán Duque, Alba P.; Fernández García, L. Carlota 2018

35 Levine, Paul 2017

36 Senior, Jennifer 2015

37 Rockbrook Camp for Girls

38 The Education Alliance

39 Doyle, William 2016

40 Day, Kelly 2015

41 Unicef.org

42 Deitz, Rena 2017

Chapter 8

Children's Spaces

Think about your favorite place as a child. Was it a soft chair you cozied up in to read in your bedroom? Maybe it was the sidewalk in front of your apartment where all the kids gathered on summer evenings to catch fireflies? Was it your grandparents' backyard that extended into the wilderness? You probably remember vivid details of the smells, colors, textures, voices and feelings of this child-friendly and inviting place. Our ability to recall details about these special places so many years later speaks to their importance in our lives.

Places are the center of experiences, socialization, education and play. They are places for families and friends to interact. They bring the community together. The home becomes the foundation of identity, who they are and where they are from. Children spend a significant amount of time in school, and it is the central focus for education and friendships. The third space, that transition place between school and home, places like the YMCA, the church, the mall and skate parks, engage kids in the community they are from and nurture their individual interests. Children's hospitals are environments that heal children both physically and emotionally. Entertainment spaces create experiences that can physically bring children to their fantasies and the world of make-believe. Through digital technology, children today can connect with children in other places around the world. Designers provide a diverse network of spaces, indoors and out, in which young people are free to engage in the world in a range of activities that will nurture them creatively, physically and emotionally.

Spaces are Centers of Experiences

Formal sites are those planned spaces designed for use by young people, such as schools, playgrounds, skate parks, kids' museums and youth centers. Informal spaces are not particularly sanctioned as children's spaces but are used by children (the block, alleyways, sidewalks, shopping malls). Kids also prefer to carve out their own spaces: a hollow space between the bushes, a fort built out of blankets, or a parking lot as a hang out spot. They use spaces to act out their imagination, carry out their adventures and socialize with peers. Determining the functional needs of space, health, safety and security for kids helps us to design better spaces.

► Figure 8.1

Kids carve out their own spaces.
Source: Michael Lyn (used with permission).

Range of Environments

Children need a range of settings to explore to learn about the world and express themselves. Babies, toddlers and preschool children spend much of their time in the home, with a focus on playing, learning and safety. They tactilely experience the environment to understand it and reinvent it. As children grow, they have a broader range that they can explore. Younger school-age children have a stronger network of spaces, playgroups and organized activities. Kids move towards third-space experiences, first supervised, then unsupervised. Teens socialize and do things together in spaces.

Spaces Kids Like

- Places of excitement, challenge and thrills
- Places that support fantasy and engage the imagination
- Spaces created for them, although they also want to make places closed intimate spaces of their own
- Many kids enjoy the outdoors for the freedom it offers

- Forbidden places, especially tweens and teens as they separate from their parents
- Spaces where they can spend time with their peers and socialize

---

Design Prompts: Determining Needs

As you get started on determining the developmental, functional, and health and safety needs of your space, here are some questions to build on.

Developmental Needs of the Children Who Use the Space

- What are children capable of doing and their interests at a particular age?
- How do they interact together and alone?
- What do they consider comforting or threatening in the environment?

Functional Needs of the Space

- What experiences need to be delivered at a performance level?
- What is the range of small, large, individual and group activities?
- What is in the built environment? What is delivered in props, objects?
- What are the ergonomic and inclusivity requirements?
- How do we design systems technology, power, water, storage needs and restrooms?

Health, Safety and Security

- What is the most efficient and sustainable heating, cooling, air quality and electrical system?
- How do you eliminate sharp edges and reduce sound levels and ambient noises?
- Have you considered furniture and props, safe cleaning formulas and processes?
- What are the open lines of sight for parents and supervisors?
- What is the security access to space?

Sustainability in Environmental Design

- Location: Encourage spaces for all, community and programming, surrounding context, previous uses of the site, and traffic patterns. Use only as much space as needed to complete the job.
- Climate: Consider macro- and microclimate conditions, wind direction, temperatures and topography.
- Low-impact and renewable materials and finishes: Use locally acquired natural, salvaged, recycled and reclaimed materials, non-toxic and sustainably produced, that require little energy to process. Choose longer-lasting and better-functioning materials and products.
- Energy efficiency: Use and control natural light, heating and cooling systems and structures as much as possible. Use manufacturing processes and products that require less energy. Limit the unnecessary use of water. Use permeable pavers, native and drought-tolerant plants.
- Healthy buildings: Indoor environmental quality, especially air quality intake as well as environmental contaminants such as mold, allergens, and chemicals found in cleaning products and building materials.
- Reusability: Create multi-use, pop-up, portable space exhibits and installations for maximum use.
- Noise exposure: Use appropriate materials and acoustic buffers to reduce exposure to high levels of internal and external ambient noise.[1]

The Home

Children's early experiences of home profoundly affect development. It is where they will form bonds with family and gain a feeling of security, and where they have many joyful and learning experiences throughout childhood. Deb White, formerly head of store's training and development for Pottery Barn Kids, explained to me: "The home is where they learn social behaviors, model family life. They figure out how to play safely, and norms for interacting with others."[2]

Children's needs from a home:

- A sense of security from family
- Physical room and freedom to move, play and grow
- A place to sleep, eat and perform personal care
- Opportunities for creative freedom and exploration
- A stimulating environment

Rooms in the Home

The spaces in the home for children change to suit their needs. Deb White noted:

> As kids evolve they have different needs for space. When they are younger, they want to be with the family all the time. As kids get older, they need private space. Parents would come in the store wanting to design a bedroom space. They should consider the entire house. How will they interact in the kitchen? Do you need a small table, a high stool to make cookies and pizza dough? Get into every room and think about how it will be used.[3]

Pottery Barn Kids offers an in-store design experience as well as online tools for parents and kids to be creative and design their own spaces. Deb White suggested:

> Get the kids involved in the design and purchasing decisions at any age. They will feel like they own it and care about it more. Look at a child's favorite art project and look at the colors they use. Start the conversation there. What do they love? It also is a learning opportunity. While kids are designing the space they are learning space planning, using grid paper and pencils or technology on the computer.[4]

I was curious about her thoughts on the differences in the purchasing experience online versus brick and mortar retail. She noted: "Interacting with home products and accessories is a sensory experience – for example, different people have different reactions to the weight of fabrics. For specific types of products you need to feel it."[5]

Bedrooms

Babies

Atmosphere: Tranquility and comfort are essential for kids and parents in a nursery.

Sensory: Since babies spend a lot of time looking up, their view from the crib should be stimulating: a dimmer on lighting, chimes for sound and a mobile for movement.

Function: The crib is a restful environment for sleep. A comfortable chair for feeding, and storage for clothes, supplies and a changing table.

Safety: Short curtains, no window cords, socket covers, high shelves attached to walls. Be sure that all of the furniture follows the consumer safety guidelines and has been duly safety tested.

Aesthetics: At this age, the décor is based on what the parents like and their hope for their child. Keep it simple and stimulating, so it is easy to add to and swap out later on.

Toddler to Later Elementary School

During about 10 years, this room will go through many transformations. It should be designed with flexibility, grow and shift from day to night to day and over time as the child grows. Deb White offered:

> Design projects for younger kids usually begin with a theme. As kids get older, the room becomes more neutral. They express themselves and their identity in their spaces with their own creations. Their room reflects what they have learned and are doing in school. They are proud of what they made, want to share that and show that off. Their creations comfort them.[6]

Atmosphere: Parents and kids have a different view about what is essential in the space.

◀ Figure 8.2

Adults see toys everywhere as messy, while kids see it as excitement and flexibility for play.
Source: author photo.

Sleep: Children will usually transition out of the crib between 18 months and 2½ years. Some children will move into a toddler bed, while other children will transition into a twin bed. A mattress with excellent support and a safe bed for jumping, climbing and rowdy play. A bunk for children over 5 or raised bed frees up floor space.

Play: A versatile layout with lots of floor space to spread out with toys and transform in the child's imagination. A creativity space to build and make things with storage can nurture their artistic side. As children get older, they spend more time in their room: relaxing, playing with friends, using media and the computer, reading, playing video games, listening to music and enjoying quiet time. They play games and develop their personal lives and identity. The room becomes a space where friends are brought.

Study spaces: Older children need a place to do their homework and study. A desk that is ergonomically correct and adjusts to the child promotes learning.

Storage: Everything needs a place. They need space to spread out and engage in activities. Storage can hold the stuff they accumulate, toys and collections that rotate out. Everything should be at the child's level so they can pull out what they like to play when they are inspired.

Safety: Furnishing should be robust, and all details should be considered. Consider rounded corners, wall anchors, joinery hardware, varnishes and paint.

Aesthetics: As children grow, they try on different roles and gain and lose interests quickly. Provide flexibility to change out the décor and displays frequently to try on different identities, themes and personalities. Deb White noted: "The décor changes based on what they are interested in at the moment. What passions do they have today? Bedding, artwork and accessories can also inspire them for their future."[7]

Shared Spaces

Being next to a sibling could be comforting, although to make shared bedrooms conflict-free, create a neutral backdrop. Divide the territory and define each child's space.

Teens

Atmosphere: Adolescents are busy coming and going and need a comforting sanctuary when they are at home.

Function: Their room is placed away from the world to feel at ease. They like to lounge and hang out to entertain and socialize with friends. Create privacy and soundproof. Deb White said: "When they invite friends over and other people into their space they want cool things. Beanbags, rugs, and accessories. This is how they want people to see them."[8]

Play: With the increased use of technology, bedrooms are being transformed into digital bedrooms. With television watching, video games and mobile tech, children are physically isolated from peer groups but socially connected virtually through social networks.

Sleep: Control lighting to adjust to sleep patterns. Teens go through growth spurts, go to bed late and sleep a lot later in the morning.

Study spaces: They have a lot of schoolwork and sit at the desk for a long time. Standing at a desk is currently known as the healthiest way to work.[9]

Storage: They have multiple activities they are involved in and need to organize their stuff.

Safety: Media safety is a significant concern for many parents. Some parents choose not to have access to media in their bedroom but in a more public space in the home.

Aesthetics: The space should represent their independence and interests.

Living Room/Den/Playroom

This is where media and play activities, family bonding and memories take place. Adults usually organize these more controlled spaces. This room should have floor space to spread out toys and move around and furniture that is child-friendly.

Kitchen

Preparing and eating food together provides opportunities to model healthy behaviors. The family dinner socializes kids to learn about their day and can teach about healthy eating and nutrition.

Cooking together

- Strengthens family traditions
- Provides an opportunity for learning about healthy eating
- Provides experience for children to practice math and reading skills
- Boosts children's self-esteem
- Helps them explore different cultures and cuisines

When families have dinners together, they have

- Better academic performance
- Higher self-esteem and sense of resilience
- Lower risk of substance abuse, teen pregnancy and depression
- Lower likelihood of developing eating disorders and rates of obesity

The Family Dinner

According to the non-profit organization The Family Dinner Project, sharing a fun family meal is good for the spirit, brain and health of all family members. Recent studies link regular family meals with grade-point averages, resilience and self-esteem. Additionally, family meals are connected to lower rates of substance abuse, teen pregnancy, eating disorders and depression.[10]

Outdoors

An outdoor space offers a place for kids to be free to explore and engage in different types of play, explore nature, run around, get dirty, relax, soak in the sun and quietly read a book. If there is room and light for a garden, there is the opportunity to teach kids to grow their own foods and flowers.

Outdoor features could include:

- Playhouse, sandbox, mud
- Open space to run around
- Secret play places
- Table and chairs
- Play equipment climbing structures
- Gardening and engaging in landscaping

Children's Furniture

Aesthetic: Furniture is usually either a feature in the room or a neutral piece. Furniture is an investment, so consider longevity. Accessories are much easier to swap out aesthetics then furniture.

Sizes: The biggest challenge in designing furniture for kids is their changing size while trying to keep the longevity of the furniture. Do you need to make the same product in different heights, or will it grow with the child? For example, play tables are sold for under 3 and 3–6 age ranges, and each has different user requirements.

Safety: Anything fragile, breakable and potentially dangerous has no place in a home with young children. Think about the weight, or the furniture, especially tables and chairs that slip easily and can easily fall over. At the same time, you will want it to be portable and easily move around. Avoid sharp corners or edges. Keep in mind that kids will climb into open drawers and use bookshelves as ladders, so don't make them too high, and make them firm and weighty with a broad base. Anchor to the wall.

Materials: Sturdy, smooth plastic or polished wood is ideal. Metal furniture, like beds or chairs, is too hard and can hurt if there is a fall during rough play. Use non-toxic finishes.

Distribution: Furniture is expensive to ship so consider ready to assemble options but ensure safety and durability.

Playborhood, a Movement

Interview with Michael Lanza, entrepreneur and author of *Playborhood*

Michael Lanza was so concerned about the diminished standard for a good childhood compared with how he grew up that he transformed his home into a space that promotes risk for his kids and the community. His manifesto captures many of the current concerns around childhood today. It is useful for designers to use as a framework.

Playborhood manifesto

I want my kids to ...

- Play outside with other neighborhood kids every day
- Create their games and rules
- Play big complex games with large groups of kids, and simpler games one-on-one with a best friend
- Decide for themselves what to play, where to play and with whom

- Settle their own disputes with their friends
- Create their own private clubs with secret rules
- Make lasting physical artifacts that show the world that this is their place
- Laugh and run and think ... everyday

Michael Lanza offered suggestions for designers to consider when developing play places for kids.

- Many spaces designed for kids are culturally contrived and preprogrammed by adults. They limit kids. Designers can provide opportunities and give them a foundation to own the place.
- Designers need to think about all of the things competing for kids' attention, technology and other things kids want to do. The activities and environment need to be attractive to them and resonate with them.

- The same ideas behind placemaking hold true to create an enriching environment for play at home. There needs to be shade, a place to sit, and food. We have indoor and outdoor spaces but emphasize the outdoors. The trampoline being at ground level is a prominent feature. The playhouse is big, so it is a private, protected space but suitable for older kids, groups as well as small gatherings. Kids love to draw, graffiti and deface the space and then come back at another time to see what they did.
- It takes time for a place to develop character. When creating spaces think about to how kids will manipulate it to make it their own. Do they feel free there? What kinds of parameters are put in place? Is it only a short time? Supervised? Can they create their own culture in this place?[11]

◀ Figure 8.3

Friends watch as a boy jumps from a ladder to an in-ground trampoline.
Source: Playborhood (used with permission).

Community Spaces

Why is it that when we meet someone, we ask them where they are from? It says a lot about who we are. The community of people and places where children grow up is central to how they experience their daily lives, identity and play. Play and socialization can happen anytime, any place, in supermarkets, at train stations and at laundromats. Third spaces between home and school, such as public spaces, markets, rock climbing gyms, public swimming pools, church and the mall, although not designed specifically for children, are places children frequent and opportunities for play and socializing. Digital places provide new opportunities for social spaces and new types of communities.

As we move into a generation of families who are squeezing play and family time into our lives, how do we as designers promote places for play?

► Figure 8.4

Neighborhood little libraries build community and promote reading by fostering book exchanges.
Source: author photo.

Leisure time is now organized time with structure and kids are supervised at all times, and some children are reacting to this with increased signs of stress and anxiety.[12] The increase of privatization of organized activities versus investment in public parks and community places widens the gap between the middle class and the working class and access to healthy, enriching spaces for all children.

Placemaking

A space is a physical description of an environment, whereas a "place" connotes an emotional attachment to it. Placemaking utilizes a user-centered approach to the planning, designing and managing of public spaces, although the term can be used to represent any environment. It involves a collaborative design process to discover needs and aspirations of the users to create a shared vision. The space design, a strategy for providing programming opportunities to build character, planned activities, uses and meaning in the community, is as important for designers to consider as physical design.[13]

The role of Community in Children's Lives

Interview with Mark Atkinson, markets development officer of Portobello Road and Golborne Road in West London
Mark is responsible for ensuring the long-term viability of the markets and market streets. He received a BID (Industrial Design) from Rhode Island School of Design, and an MSc in City Design and Social Science from the London School of Economics and Political Science.

How do you integrate your background in industrial design, graphic design and placemaking in creating events and experiences at the market? Are there similarities in your innovation process?
The environment I work in is extraordinarily messy and complex: spatially, socially, economically, organizationally and politically. I deal with a level of multi-dimensional complexity far beyond anything my design training equipped me to manage. I do keep a deck of IDEO Method Cards on my desk to remind me that design thinking is a useful framework. The most relevant concept from my design background is rapid iterative prototyping.

What do you see as the role of cities and community organizations in providing experiences for kids playing, learning and growing?
In the US it seems relatively mainstream that cities aren't the preferred places for kids. When I was growing up in the suburbs of NYC, getting to the museums, parks and neighborhoods I loved was expensive and took a long time. There weren't many interesting places I could walk

or ride my bike. Seeing a movie required convincing a grown-up to drive. I believe cities are ideally suited for kids because they enable greater autonomy and expose them to more diversity of people, environments, experiences, cultures and lifestyles.

What are the most important things cities and non-profit community organizations could do for kids/youth over the next 10 years?
It is vital for cities to maintain and enhance their openness, accessibility and diversity. It is also essential for them to ensure a safe and clean environment. Designers can work on ways for people to engage and interact with one another in ways that contribute to human flourishing and well-being.

Are you currently thinking of ways of combining physical community networks and digital spaces (social networks) specifically for kids and youth?
It's on my radar, but the best people to think of these things are the kids themselves. I recently sat on an enterprise day judging panel at a local school. Teams of students were tasked with getting young people to engage more with the public street market. All of them came up with gamification apps that merged digital and physical spaces, networks and communities.

What is your take on the roles designers/innovators play in tactical urbanism, informal education, pop-up play and the development of third spaces?
These terms are all responses to constrained public budgets, a pressure to maximize return on real estate

assets, and growing city populations leading to increasingly contested urban space. Designers and innovators can apply design thinking and rapid iterative prototyping at a systems level (often called service design, social design or process design). The basics of devising ways to address unmet needs through careful observations of users' behavior are as relevant to urban environments, community systems and place-based economies as they are to products.

Can you briefly describe the Food Explorers program?
As part of the 150th anniversary of the market we collaborated along with another local organization, the Hammersmith & Fulham Urban Studies Centre, who created a children's map and guide and commissioned a short film.[14] I also hoped that exposing children to fresh fruit and vegetables in a fun and entertaining way could help to foster curiosity about different foods and encourage healthy eating and encourage them to shop locally. At the event they took a market tour, followed by a mini international food festival that showed them the local history as reflected by the cuisines of different cultural groups that are part of the markets and market streets.[15]

◀ Figure 8.5

Children from local primary schools visited the market to meet stallholders and local shopkeepers, trying different fresh and cooked foods.
Source: Food Explorers (used with permission).

Mobility

The way to move between sites – home, park, school, sidewalks, street – and modes of transportation are always a challenge for children. In urban environments, many adults find the street traffic too unsafe for children. In rural environments, children are often isolated due to the distance between places. Michael Lanza, author of *Playborhood*, explained the importance of independence for children: "Consider proximity and where the place is. Once you drive a kid somewhere an adult is probably also there overseeing and controlling the experience."[16] Jay Beckwith, a long-time play systems designer, explained:

> Cars have a huge influence on urban planning and, hence, on play in neighborhoods. We have seen in many places that play blossoms when you take the cars away. Instead of following the old thinking about planning we need to think about what will happen when we have self-driving cars?[17]

Automobility LA Design and Developer Challenge

Interview with Martin Sanders, director of innovation-partner projects at the LEGO Group
LEGO partnered with Honda, Trigger Interactive and children to envision a future Los Angeles where autonomous driving changes the way people travel, the infrastructure, and the design of cities and lifestyles.

What were the goals when deciding to involve the kids in the process?
Getting children onboard to primarily play with the idea, we could join them on that journey and be inspired by them – to develop and turn their stories, thoughts and imaginations into realities.

How was the brainstorming proposed to the kids?
I'm a firm believer in beginning any process of inventing play at LEGO with the question: What wish do we want to fulfill? With the children, it was simply about that – getting to know and understand what they wished for, at the same time recognizing some natural needs. Since they are the builders of tomorrow, it would be their imaginations and wishes that shape the future.

We set out some basic ideas and frames of reference to start with and let the children build with LEGO, tell stories, act out and role-play situations together. It was just a good fun play session guided by some underlying questions. In that way, I feel we tapped into all our inner-children and began to trust those intuitive free-thinking parts of our imaginations.

What exciting ideas did they develop around self-driving cars?
The children were fascinated by the idea that AI could also become a friend or companion. They started to imagine travel not as a necessity but as a time to look forward to in the future. With AI they could continue playing and use the travel time for entertainment.

There was also a play exercise where the children built and visualized a future Los Angeles made from LEGO bricks and elements. Their stories were very considerate of nature and the need to preserve green spaces and parks. We loved that these builders of tomorrow had high regard for the planet and the harmony of bringing advanced technology and nature together. The Honda design team particularly resonated with this notion in the subsequent concept design and development phases.

Were any taken further into the design?
The team considered everything and blended some of the key ideas into a set of concepts. In particular, a modular approach to transport emerged as a consistent theme and became a foundation for the whole project. In keeping with LEGO philosophies and play, the modular approach also allowed the design team at Honda to incorporate and cater to the broad variety of scenarios described and played by the children.

What were the benefits for you to have them involved? What were the benefits for them?

For the design team, it was a refreshing reminder of the playful state of mind from which most great ideas often come. Releasing oneself from the constraints of over-analysis and moderation can remove the barrier to great ideas that many design teams can often face. Children became our teachers, showing us how to reacquaint with that inventive and playful state of mind once again.

The children felt a great deal of pride and joy from the apparent responsibility of designing the future of Los Angeles. I also saw the kids enjoy interacting with each other – sharing and being inspired by hearing other groups tell the story of their designs. They naturally built on each other's thoughts and became united through this single creative mission to design the future. And because the future doesn't exist yet, the children were excited by just the pure possibility.

A few weeks later, we unveiled the 3D printed concepts and AR walkthrough experiences to the children at Honda Advanced Design studios. They were thrilled to see how their ideas and play had turned into reality. I think we also engaged and influenced some of the children, who said they now wanted to be designers. For children who never previously thought that play could lead to real-world designs and solutions, it was a huge realization for them.

What was learned from this experience?

In many ways, it proved a long-standing hypothesis that the best ideas come when you are playing and essentially "take your guard down." It also shows that intelligence, insight and creative intuition is not something that naturally comes with age. It's more to do with approach and attitude. Developing ideas into reality does require specific knowledge, skill, and experience, but great ideas come from anyone. In fact, I might say that the best ideas come from children … of all ages.[18]

◀ Figure 8.6

Children's ideas around the future of Los Angeles were taken further into 3D printed models and renderings.
Source: LEGO–Honda–Trigger (used with permission).

Playgrounds

No city or suburb in the world could provide enough space for play. Play can happen everywhere in a community, on the sidewalks, on a rock or by a tree, at the mall, but playgrounds serve an important role in the community as a place where kids, families and neighbors gather together while interacting and learning with their peers and family.

A well thought-out playground

- Provides stimulation and challenge
- Has a strong emphasis on physical play; kids move and strengthen their bodies
- Has areas for relaxation, activity, surprise, encounter, isolation
- Becomes a meeting place for kids to gather and socialize

Formal playgrounds may be the only experience many kids have with the outdoors. Many playgrounds are located in public parks where there is a mix of interests that are served for all members of the community, with community centers, barbeques, water features and amphitheaters.

Developmental Benefits of Playgrounds

Physical Development

Spinning, stretching, swinging, climbing and moving around are good for overall health and growth. Many organizations and people from the public health sector believe that playgrounds with well-designed equipment will aid in the solution to the childhood obesity crisis on the US if we can change families' behaviors and get kids there to use them.

Cognitive Development

Appropriately designed play areas allow children to explore new ideas and test their abilities, risk and resilience, leading to improved memory, considerably more attention and concentration, and reduced disruptive behavior.

Social Development

Good play areas offer children opportunities to play alone or with other children. The playground should allow younger children to manipulate items easily, explore spaces, and begin to interact with others. Play areas for school-aged children encourage social growth and cooperation.[19]

In Playground Design, Include a Diversity of Spaces

Active spaces: Encourage physical play, varying topography, changes in height, gross motor skills development. Promote fitness and health.

Experimental spaces: Discovery, exploration, hypothesizing, sociable, flexible, messy, building, testing, creativity, idea generation, loose parts, mud, sand and water.

Individual spaces: Small and protected spaces away from the noise that are cozy and enclosed. Support private time, accommodate one or two children, observation and listening.

Gathering spaces: For large or small groups, they encourage interaction and offer comfort, seating, shade, and soft and hard features. Could function as a stage for planned events or spontaneous creativity.

Ecological spaces: Natural live vegetation, trees and shrubs that contain ecosystems that attract birds, insects, worms and butterflies. Access to water, soil and plants. Invite exploration, creative thinking and observation. Provoke inquiry, evoke an emotional response, nurture a sense of responsibility and offer time for reflection.[20]

Balancing Safety and Challenge in Playground Design

Matthew Urbanski, principal at Michael Van Valkenburg Associates, thinks the most obvious challenge is: "How do we incorporate risky things in the landscape that are accepted?"[21] In play environments that are too safe, children will often find ways to take risks and find challenges that are sometimes hazardous. A quality play environment is challenging but also safe.

Design Considerations for Safety

- Supervision security and line of sight
- Dangers of traffic, distance from street
- Entries and exits
- Access to sun and plenty of shade
- Lighting
- Keeping kids away from hazards
- Ground materials and fall zone surfacing
- Crowding – parties and special events
- Mix of pedestrians, bikes, scooters, skating, traffic
- Sanitation and toxicity
- Equipment design
- Age-appropriate play spaces and equipment

There Are Two National Standards for the Design of Play Equipment

1. "Playground Equipment Safety Guidelines" by the US Consumer Product Safety Commission (CPSC)

2. "Consumer Safety Performance Specifications for Playground Equipment for Public Use" by the American Society for Testing and Materials (ASTM)

Playground Safety

Interview with Kenneth S. Kutska, executive director of the International Playground Safety Institute
Ken and I discussed several factors designers should be aware of when creating safe playgrounds for kids.

What are three key factors that designers should consider when designing a playground environment?
Play Value (Taking Risks)
With increased play value and graduated challenges comes an increased risk of injury. Does the developmental value of teaching a child new skills such as developing their own risk assessment process outweigh the potential increase of a few more fractures or even a mild concussion?

Age Appropriateness (Developmental Needs of the Whole Child)
Children do not develop at the same pace. Some say children will self-regulate their activities by developing their own risk assessment skills. These skills develop through their experimentation and failures in the play environment. This is not true for all children, and therefore serious accidents will occur. Whose fault is that?

Safety Concerns (Eliminating Known Hazards)
One could argue that supervision should be in there somewhere in the top three, but children will do what they

do when they want to do it, supervision or not. What is good supervision? The definition escapes most people.

Where are the most prominent mistakes made in playground design?

Designers have hundreds of opportunities to make mistakes along the way. One area that is often overlooked is the failure to know who the intended users are and understand that other unintended users will also be in the environment. Too often the owners and their designers are governed by what funds are available for the project. While the end result might be a compliant design, it may not constitute good design.

Overall Design

The two most significant concerns for playground location are surface runoff as well as subsurface drainage, which will affect the play equipment and the concrete footers and the life of the surfacing materials regardless of whether it is a unitary rubber surface system or some loose-fill system. Neither like a lot of moisture and can influence the surface's longevity and performance.

Equipment

The standards were written considering swings, slides, climbers, merry-go-rounds, horizontal ladders and springing/rocking equipment. Times have changed, and designers and manufacturers have pushed the envelope with all these traditional types of equipment and have developed hybrids by combining types of equipment requiring the designer and manufacturer to conduct their own hazard assessment of the new product that does not neatly fit within the current standard.

Surfacing

Surfacing seems to be the next area of concern once the manufacturers and designers have done their job of addressing all points of transition when a child is moving about the play equipment. What they fall on and where they fall then becomes the issue.

Maintenance

Everyone involved needs some basic training on injury prevention and what is required to maintain both the equipment and surfacing.

▲ Figure 8.7

Lack of maintenance is the cause of many playground accidents. Source: author photo.

What do you think are the most important topics that designers should be working on in public spaces for play?

I believe there is room for improvement in designing the public playground as the public commons. A thriving community shows it cares about youth in the types of parks and education facilities they provide. Public common areas bring the community as a whole together, and relationships can flourish by providing shelters, benches, tables and shade through greening of the square.

What are the best resources that designers should look at for safety standards?

The first document I would look to is the US Consumer Product Safety Commission's *2010 Handbook on Public Playground Safety* (www.cpsc.gov/s3fs-public/325.pdf).

After that, I would get copies of all the related public playground safety standards from the ASTM. I would also suggest that all designers attend one of the many NRPA [National Recreation Park Association] Certified Playground Safety Inspector courses that are run all over the country. They would learn a lot of useful information about these two documents that may not be easily understood by just reading them.[22]

Skate Parks

There are two types of places where skaters practice: skate spots and skate parks. Skate spots that hack current public spaces are what give skaters a bad boy image in communities. Skate parks are built by the community or private developers for many beneficial reasons to the community. Safir Bellali, the global innovation lead at Vans, explains how skate parks promote community:

> With a skate park a neighborhood can transform from nothing to do to something to do. We have worked on programs such as Summer Nights with GRYD and City of LA Mayor's office to include

skate parks for gang reduction and youth development. We have events with portable skate parks and keep parks open to midnight. People come together and participate in activities such as skateboarding, Capoeira (a Brazilian martial art), creative projects and team sports instead of getting into trouble. The skate competitions were hugely successful. The sense of community is really palpable when everyone is out at night in a positive atmosphere.[23]

City Play Spaces

Matthew Urbanski, principal, Michael Van Valkenburg Associates Landscape Architects
MVVA landscape architects have created many parks and play spaces to build healthy communities and quality of life, specifically in cities. Matthew Urbanski shared what he thinks about and his design process when working on projects for children and the communities where they live.

How do you decide on your approach to a park/playground design?
We start out by defining the missing experiences in the community. What is in the immediate vicinity? Are a lot of apartments or single-family homes? Is there little access to nature, a creek, a beach? Can kids get dirty anywhere?

We also look at the specific site under consideration. What could it do well based on what the community needs? There may be a need for a soccer field, but it doesn't fit, so you develop a place to kick the ball around?

As a designer, you get yourself in the mind of a kid, and you think about the play capacity of everything you are making from a child's perspective. The benches, the pathways, and how they move through space. We think about interaction with nature and the natural setting for spontaneous play.

Can you explain your approaches to this variability in the play experience throughout the environment?
Landscape as a medium has the capacity for exciting juxtapositions. Different rooms can happen in close proximity. In Chelsea Grove there is a garden, pier, skate park, carousel, the boardwalk – you would say they don't go with each other – but yes they do. There are different audiences – a variety of people; families, and teenagers.

A well-thought-out play space

- Provides stimulation
- Has an area for relaxation, activity, surprise, encounter, isolation and interaction
- Becomes a meeting place for kids and families to gather and socialize
- Addresses the community needs

- Has a sense of eluding the obvious
- Not didactic or boring – awards seekers
- Has a sense of mystery

Can you explain how vegetation such as grasses, trees and flowers is chosen to enhance play throughout the seasons?

- Hydrangeas are summer shrubs. After the spring flowers go, it is nice to still have flowers. They are like big balls and not too precious.
- Rose of Sharon and smoke tree add color and are fanciful.
- Golden rain tree is a yellow flowering tree in summer – it is unexpected.
- Kids find fruit pods and paper lanterns fun to play with. They open them and see the seeds inside.

What are some other features that work well for a mix of interests for older kids (over 12) to get out and play?
Older kids like playing with younger kids and want an excuse to do it without being embarrassed. They are not in a rush to grow up – they feel pressure to do it.

They also use these spaces to socialize with their peers – hang out on or inside the play equipment. They lie down at night to talk and look at the stars.

There are the old-fashioned games – basketball – handball and less aggressive sports – shuffleboard, ping-pong. Everyone uses Exercise for Seniors – kids climb on it. The challenging outdoorsy sports are really in today – biking, BMX, rock climbing and kayaking.

A good playground allows a child to engage in the space and use their imagination. Can you name a feature in one of your parks that is an experimental space that does this?
Ambiguous creative spaces invoke various interpretations. In Brooklyn Bridge Park, there are rock outcroppings made of salvaged stone from a bridge. They are vague and look sort of like ruins, a fortress and coves. Their forms evoke various narratives. Kids say what is this? We can play hide and seek, make it into a battlefield or create a theatrical setting.[24]

Custom play structure at Maggie Daly
Park, Chicago.
Source: author photo.

Inclusive Playgrounds

While playgrounds are required by the Americans with Disabilities Act (ADA) to
be accessible and not to discriminate, some designers in the industry are moving
towards inclusion, benefiting people of all ages and abilities. Safety, comfort
and social participation for all types of abilities are now being considered within
shared play experiences. Playgrounds can also be used for therapy. Some
occupational therapists prefer a natural setting for rich multisensory settings of
playgrounds, and play therapists take kids on walks and to the playground and
out into nature to make it easier for them to express their feelings. The combination of these three elements – *Play Experience*, *Environment* and *Variability* –
allows every child to choose how they want to engage in the play space.[25] The
US Department of Justice provides technical assistance regarding policies,
programs and more (www.ada.gov).

Wilderness Playgrounds

In wilderness playgrounds, a tree becomes home base, a meeting place, or a
place for climbing and hiding. Found natural materials including twigs, seedpods
and grass serve as loose parts for play. Some may include manufactured
equipment, but the designers also respond to the local ecology, natural vegetation and seasonal potential for the play setting and play materials. Play gardens are places where children can freely play and reclaim the ability to learn
in a natural environment through exploration, discovery and the power of their
own imaginations. Past trends towards the loss of wilderness are shifting to
rebuilding planned wilderness.[26]

At a Danish adventure playground, children run free and play with animals, bonfires and junk materials.
Source: author photo.

Adventure Playgrounds

Adventure playgrounds offer unrestricted play in either a natural or a junk environment, and kids invent with found materials. The first junk playgrounds were based on the ideas of Carl Theodor Sorenson, a Danish landscape architect who realized that kids played everywhere but in the playgrounds he designed in the 1930s. Inspired by kids playing on a construction site, he created a junk playground. In the 1940s, Marjory Allen, an English landscape architect and child welfare advocate, suggested that they use their bombed building sites with scattered junk that kids were playing in all day to create places similar to what she saw in Denmark. Play supervision (play workers) was introduced, and some of these sites became adventure playgrounds.

Today, there are many public adventure playgrounds in Europe, and there are a few here in the US. Europeans more readily embraced spaces for children to engage in what developmental psychologists like to call "managed risk." "It's central that kids can take their natural and intense play impulses and act on them," says Dr. Stuart Brown, a psychologist and the founding director of the National Institute of Play. Children need an environment with "the opportunity to engage in open, free play where they're allowed to self-organize," he adds. "It's a central part of being human and developing into competent adulthood." Wild play helps shape who we become, he says, and it should be embraced, not feared.[27]

Patty Donald, the Berkeley, California, adventure playground's longtime coordinator, says: "I find there are a lot of adults who don't know how to play with their kids. The cell phone probably is the biggest problem we have. The parents are physically here, but they're not really present."[28]

Social Equity and Play

There is an economic divide in the access to play spaces in the United States. I have experienced many beautiful parks in upper-middle-class wealthy suburbs. These kids spend their time engaged in after-school programs and weekend activities and have little time for the playground. Lower-income neighborhoods have fewer opportunities for enrichment and playgrounds, and many children are kept away due to the derelict population. Kids need play spaces and nature nearby, especially in the most urban neighborhoods.

PlayBuild, Bringing Design and Play to a Community in New Orleans, Louisiana

Interview with Angela Kyle, co-founder of PlayBuild

Define "informal learning" and "design education" specifically as it relates to your project.
Informal learning is accidental learning. We embed info and knowledge about architecture, design and urban planning in our programming. There is very little diversity in design education and design professions. Our hope is, with a broad approach to critical thinking, to propel kids into this thinking and these careers. Many of the kids that have come to PlayBuild are design aware.

Can you explain how PlayBuild uses community to support playing, learning and growing?
By creating the space for kids and families to come together, we become a community together. We create these family moments. We need to be present in the space. When nobody is there, then we become part of the problem – blight – underutilization. Parents sign a release so the kids can come and play on their own. It is a place for kids to go.

What type of editing criteria do you use to promote design and engagement in choosing toys, construction materials and other play tools?
There are many toy deserts in low-income areas. Kids have very few toys and if they do, they are from the dollar store. We have toys these kids don't have access to and promote constructive play. We engage kids in design-based craft activities with prompts in context and subject matter. We provide a space where kids make cool stuff. Parents actually shut down creativity. We bring history of New Orleans design. For example, we have a project that is a shotgun birdhouse. Working with student volunteers from Tulane's School of Architecture and the Taylor Center for Social Innovation & Design Thinking, PlayBuild produced laser-cut wooden birdhouses in the style of the iconic New Orleans shotgun house and hosted a special event for neighborhood kids to assemble and decorate their creations.

PlayBuild is a good example of design for social impact. How do you see it having long-term social impact?
Our framework does not just solve one problem. Our first goal is place making and moving the needle in the neighborhood. Our next is education. STEM, STEAM are the hard skills, while soft skills are the social and emotional.[29]

▶ Figure 8.10

The goal of PlayBuild is to engage underserved kids aged 4 to 12 years old through creativity and active learning. The program stimulates curiosity, enthusiasm, and an interest in exploring careers in design, architecture, city-planning and related disciplines. Reading materials and programs have an emphasis on design.
Source: PlayBuild/Gigsy (used with permission).

Pop-Up Play

Since it is expensive and challenging to get playgrounds built, time to play needs to be integrated into kids' daily lives in the in-between spaces. Today, small companies and non-profits are filling the gaps that are missing in play to create places for communities to share in play experiences. They supply the space, materials and mentors for play through pop-up events on the city streets, under freeway overpasses, in open areas, parks and empty storefronts. Each event is different, but loose parts, creativity supplies and constructive play are usually involved. Diverse materials stimulate the children's creativity into making objects, installations and their own play environments. Kids go wild and build cardboard cities, obstacle courses and whatever their imaginations create.

Tactile urbanism projects are quick, often temporary, cheap projects that aim to make a small part of a city more lively or enjoyable. Many of these projects and events are open source, DIY and easy to replicate, and have a more significant political message. An example of tactile urbanism is PARK(ing) Day, an event where artists, designers and citizens transform metered parking spots into temporary public parks.

KaBOOM! Inspiring Play in America

Interview with James Siegal, CEO, and Amy S. Levner, vice president, Marketing Communications of KaBOOM!

KaBOOM! is the national non-profit dedicated to bringing balanced and active play into the daily lives of all kids, particularly those growing up in poverty in America.

If you had a wish list of project topics for designers to work on, what would they be?

Inequity is the common thread in everything we do.

1. Play everywhere: How does the entire city become a place for play? How do we claim space in the built environment not explicitly designed for kids' play and include kid-friendly opportunities and infrastructure?
2. Reimaging the town center: Playgrounds become a gathering space for kids and families. How do we create environments that are welcoming to kids and families, and incorporate elements for kids?
3. Playspaces as content delivery mechanisms: How can dynamic content and the use of technology trigger interaction and prompt learning in the built environment? For example, can a play structure be programmed to ask why the sky is blue when a child climbs to the top? Can kids hear the sound of water to prompt pretend play when going down a slide?

What type of research did you conduct to develop your outline for the decision-making process around play, specifically around getting to a playground?

We researched play through the behavioral science lens and worked with the behavioral scientists from ideas42 about why we have had this generational decline in play. What we found was that proximity matters. Kids need to live near quality play spaces to go out and play. We found that low-income children have fewer alternatives for childcare, therefore, spend 25% of their time with their caregivers doing chores. The need for proximity highlights the importance to integrate play into every day and translate it into the built environment that usually does not consider play. There are very few natural prompts, and without prompts, people are not forced to think about it. There are also a lot of hassle factors to go and play such as not having the gear or time, and it is too easy to say no. By integrating prompts into everyday spaces and making it as easy as possible, we can solve the hassles. We are following up on some of our projects and are starting to see that our simple theory is proving to be working.

A couple of design challenges suggested by Amy:

How do you make play the natural choice?
How do you make joyful moments with a community-driven approach to the built environment?

We work mostly in the built environment. There is a lack of physical spaces out there. This provides many opportunities for designers. There are other organizations like safe routes to school that is programming rich. For us, it is a combination of light community-driven programming in conjunction with the built environment.

Miami city

There was a city grant to transform dead-end streets using hyperlocal artists and a lot of paint to create play areas. The community drives the artwork and the programming. They put in hopscotch, and the kids did not know how to play hopscotch, so they organized light programming (events) where the grandparents came out to show the kids how to play. Young adults use the space to have picnics and dates.

Can you explain the importance of social equity and play, and play during a time of crisis, and where designers can be most helpful using their creative skills to help advocate for play?

Social equity infuses all of our work.

- Designers need to consider the engagement of the community as part of the solution. Especially in under-resourced communities.
- Collaborate with local and government systems that touch kids' lives such as schools, parks and housing. In many underserved areas, there is not a playground at the school or in the housing development. It either has not been included, has been ripped out, or is so decrepit it offers diminished play value.

Play during a time of crisis

Play is a critical element of recovery in a time of crisis. It is very top of mind right now. Kids need a space to be kids in natural and human disasters to help create a healthy childhood. Including play as part of recovery, the design is a critical part of the solution. We work with Save the Children in relief response.

What are your most significant learnings from hosting your design competitions?

1. There is a considerable demand for this type of work – we had 50 winners but over 1000 entries. Before the competition, nobody was using tactile urbanism for kids and families.
2. Their solutions are as broad as people's imaginations. There is so much creativity. One entry used a bubble machine on a roof. There nothing was built and was minimal but created a great opportunity for play.
3. We developed principles for all designers to use for Play Everywhere.

> Wondrous: Sense of wonder
> Inviting: Feel welcoming
> Convenient: Easy to access
> Challenging: Through play, cognitive, physical, social and emotional
> Unifying: Bring people together
> Shared: It happens in real time with others – not off on your own

For more, see the playbook KaBOOM! has developed at https://kaboom.org/playbook[30]

Locations to Consider for Pop-Up Play

Built environment

Streets, sidewalks and trails
Parks and open spaces
Transit
Residential
Civic spaces
Commercial spaces

Moveable Pop-Up Play

Festivals
Events
Farmers' markets
Outreach for organizations – children's and science museums[31]

◄ Figure 8.11

BUSt! Boredom from Lexington, KY brings play to the transit system.
Source: KaBOOM! (Used with permission).

Education Spaces

Schools play a central role in the lives of children. Architect Alice Fung explains:

> The school environment is a child's second home. The spaces that we provide affect their relationship to the physical world powerfully. The spaces should come from a place of optimism, and they should give the occupant a sense of agency and belonging.[32]

Creating a non-institutional home away from home for education is about improving students' engagement in the process, attention and success in outcomes. New methods of curricular engagement, such as 21st century skills, flipping the classroom, the use of physical and digital technology, and project-based learning, require designers to rethink the current school environment.

The environment affects:

- Learning potential and effectiveness
- Relationships, a sense of inclusion or exclusion
- Quality of engagement into the broader community
- Values (mutual respect and understanding)
- Moods

Facilities Design

Facilities have a significant impact on student behavior, grades, teachers' performance and community satisfaction. Don Marinelli of Carnegie Mellon's Entertainment Technology Center states: "How do we expect students to think outside the box when they are trapped inside a box? The environment is key. Why does cool have to be separated from learning?"[33]

School should be the future; however, many of our public schools are a relic of the past. There is little money to keep up with deferred maintenance, let alone design new schools, redesign classrooms and purchase new furniture.

According to the Center for Green Schools, $542 billion would be required over the next 10 years to modernize the nation's K–12 public schools.[34] How does this affect students?

1. Student Behavior: *Broken Window Theory* states that physical disorder, such as broken windows, run-down buildings, and so on, leads to bad behavior and disorderly conduct.
2. Student Achievement: Facilities can weaken or improve the learning environment. Higher grades have been associated with the design and condition of school facilities.
3. Health – the *Sick Building Syndrome*: Student and faculty health can be impacted by the physical facilities (acoustics, air quality, temperature) they are in.[35]

Classroom Design Considerations

The primary intent of classroom design is to create an environment for active learning, maximizing attention and information retention, and to stimulate participation.

- Make the space welcoming.
- Keep the space interactive and include active participation.
- Base classroom designs on educational philosophy, lesson plans, learning goals and choosing materials to support the teaching and learning framework.
- Let students move around, keep it flexible and offer choice, and maximize interactions within the space.
- Creating centers and learning stations, spaces for informal collaboration, socialization and individual focus. Include spaces for individuals to gather as smaller groups and the whole class.
- Integrate all of the intelligences across domains and activate the senses with diverse engaging materials and activities.

How do we apply some of the knowledge we have about learning to thinking about designing schools? Steelcase Education is rethinking how they can best leverage space and technology to improve the learning process. Insights according to Steelcase 360° Education Report:

- Person-to-person connections remain essential for successful learning
- Technology is supporting richer face-to-face interactions
- Integrating technology into classrooms mandates flexibility and activity-based space planning
- Spatial boundaries are loosening
- Spaces must be designed to capture and stream information
- High-tech and low-tech will coexist[36]

Shared Spaces within the School

Common gathering spaces: A place to bring the community together and to support social learning between students and peers, in pairs and groups.

The library: Evolving, expanding role as an analog and digital content and IT service provider.

The cafeteria: As a place of relaxation, brain break as well as social learning.

Expansion and Redesign of the Sequoyah School in Pasadena, California

Interview with architect Alice Fung, Sequoyah School in Pasadena, California

Alice Fung is co-founder and president of Fung and Blatt Inc. and won several awards for the expansion and redesign of the Sequoyah School in Pasadena, California. I spoke with her about how the school's philosophy impacted the design decisions, the research and the project development process.

How did the educational mission of school and learning philosophies influence the design?

We worked with the school's philosophy of place-based learning, which sees the students' local environment – the schoolyard, the neighborhood, the community – as a primary learning tool. We wanted to provide an environment that is rich with spatial choices, that is connected to nature, that supports creative engagement and that stimulates learning. We also looked for opportunities to engage the students in conceptualizing the new building by rethinking what their learning environment can be.

Can you explain the needs of the school and the types of spaces you created to fulfill those needs? What were special considerations for the different kinds of spaces?

The school needed space to accommodate increased enrollment. The building was to accommodate three junior high classes, art and science labs, and a flexible music/theater/multi-purpose space. We were able to incorporate many outdoor gathering spaces as well into the program.

We thought a lot about the physical–cognitive connection, as in how we experience space and how space affects the way we behave. For instance, music

▲ Figure 8.12

"We created a variety of flexible and loosely programmed spaces at different scales; areas that give the occupants agency to choose which to use and how to use it." Alice Fung.
Source: Fung and Blatt (used with permission).

and theater spaces can be combined and expanded into a lecture hall. It can open up into an indoor–outdoor gathering space and further into a performance stage that commands a large outdoor audience.

The classrooms too are dividable to adapt to smaller class sizes. They have breakout areas for small groups and window lofts for solitary work. Students can choose to work on a rug, or at the outdoor learning pods.

What did you see as the benefits to designers, educators and students collaborating on this project?
We had worked on incremental improvements to the campus before the classroom building and got to know the school quite well. Collaboration afforded us deeper insight into the curriculum objectives, knowledge of how current spaces were used, as well as existing deficiencies and needs; all of which helped us as designers to fill in the holes. More importantly, collaboration and communication allowed the community to take ownership of the process; from the school's leadership to the teachers and staff, to the students and their parents. Without that collaboration, the project would not have succeeded on as many levels as it has.

How do you integrate issues around sustainability and LEED certification in your design?
We applied passive solar design principles, using concrete floors for thermal mass, south-facing glazing with deep eave overhangs, and north-facing high windows to bring in natural light and enable convective cooling. The primary roof has a south-facing slope that accommodates an extensive photovoltaic array.

We developed the rainwater distribution system as an essential feature. From the way water gets collected, drained and dispersed through rain gutters, expressed down spouts, spillways and bio-swales to subterranean rain-cages, students can follow how the water gets filtered back into the aquifer.

In your designs, how do you balance factors related to space, products, furniture and technology?
Space and light come first: creating spatial volumes that are well proportioned and uplifting, optimizing their relationship to each other, the inside to the outside.

Along with the school, we chose furniture that allows for flexible configuration to accommodate small and large groups and that invites teamwork. The presence of technology is understated. It is integrated just like any other tool that supports learning.

Can you explain how you interpreted the research into the final solution?
We researched by immersion. We were objective listeners and gatherers of information. We worked with students on their visions. We considered the community's feedback carefully as we developed design iterations. Doing so enabled us to formulate the right questions and, in the end, to provide a satisfying user experience.

Anything else you can share that we can learn from and bring to other projects?
Change is never easy, perhaps more so for children. The project was successful because the school's leadership team understood the careful way in which the process had to be unfolded.

The design is for the adults as well. The real test is for children to return after a decade or two and not to feel like they had been condescended to like children! Designed with empathy and respect, the school environment can be the entry point towards developing engaged citizens and life-long learners.[38]

Learning Labs and Maker Spaces in Public Libraries

Have you been to your local public library lately? You may be surprised to see that, in addition to books, some also check out sewing machines and power tools. At some, visitors can explore VR and other new technologies.

Many have play areas for children and maker spaces for tweens and teens. Public programming for children/teens includes story time, crafts workshops, 3D printing workshops, science shows and puppet shows. They show movies with free popcorn, have LEGO and comic book clubs and creative workshops for kids to create their own zines. The role of the library today is to serve the community's needs, and that is customized to each community.

Libraries are places where kids and youth of diverse backgrounds can connect with one another and with adult mentors to explore topics of personal interest and relevance.

Libraries allow all members of the community to explore, play and learn from new technologies.
Source: author photo.

- They provide access to media, with a mix of digital and traditional tools.
- They emphasize interest-driven and production-centered learning.
- They offer new contexts for youth to build skills and gain knowledge that connects them to future opportunities.

Teen-only: Direct involvement of teens in the planning and design process is one of the signature characteristics of Learning Labs. Inviting youth to help create the physical spaces ensures buy-in. Teens frequently talk about needing areas that are "teen only" – spaces that are safe, flexible, and provide opportunities for youth voice and public display of work. The most important features are not the size or shape of the space, but rather, the ways of thinking, behaving and working that emerge within it.[39]

Nature Wilderness

Humans are wired to be in nature. Throughout human history and pre-history, children have spent their formative years outside playing. Nature deficit disorder is a term coined by Richard Louv, author of *Last Child in the Woods*, that describes the psychological, physical and cognitive costs of human alienation from nature, particularly for children in their vulnerable developing years.

Kids Who Play Outside

- Are less likely to get sick
- Are less likely to be stressed or become aggressive
- Are more adaptable to life's unpredictable turns
- Are more likely to protect the environment

- Have fewer symptoms of ADD and ADHD
- Are more creative

I asked him about the role that designers can play in education and influencing families to spend more time outdoors in authentic nature. His response was inspiring.

Designers play a vital role in either aggravating or preventing nature-deficit disorder. Today, cutting-edge urban planners, architects and industrial designers are addressing biophilia, our genetically wired affiliation with the rest of nature, by incorporating natural elements in their designs for homes, retail outlets, schools and workplaces. Nature-smart designers can blend energy-efficient housing with nature trails, vegetable gardens and rooftop gardens. They can conserve the natural environments where they still exist, and in cities create more of them. New schools should be designed with nature in mind, and old schools retrofitted with playscapes that incorporate nature as a central design principle.

Why Aren't Kids Exposed to Nature?

- Stranger danger
- Families are too busy to search it out
- Urbanization, suburbanization, densification

- Changes in our lifestyle, sports, gym and planned activities, and electronic distractions
- Risk of nature itself

I asked Richard Louv what more designers can do to create more sustainable solutions for social equity and nature. His suggestion was:

Connection to nature should be an everyday occurrence, and if we design our cities to work in harmony with nature and to be more biodiverse, direct nature experiences could become more commonplace. Why not think of cities as incubators of biodiversity and engines of human health? As the designer William McDonough, who has done extensive work in China, would suggest, communities should be created that not only reduce our carbon footprints but create wildlife habitats, even in densely populated cities. He asks, metaphorically: What if our species not only reduced its carbon footprint but left wetlands everywhere it stepped?

Mobility and proximity are often a challenge to getting kids to a neighborhood playground, let alone natural wilderness. Richard Louv suggested:

> Pop-up play and micro-play experiences can be beneficial. Conservation is no longer enough; now we need to create nature. Creating biodiverse environments will improve our psychological and physical health, our sense of pleasure and happiness, and our ability to learn, and replenish biodiversity where it is threatened.

Richard Louv points to that moment when you are in nature listening to the leaves – turning over a rock and seeing what is there. This is spiritual growth. He explains how the child–nature connection and environmental literacy should be considered as fundamental elements of children's cognitive development, as well as their psychological and physical health. Future education reform must widen the definition of the classroom (www.childrenandnature.org/learn/newscenter/).

Richard Louv offered several recommendations to enhance the exploration and discovery of nature, specifically for older kids, 9–18 years old, who have outgrown playscapes:

> I often tell teenagers that nature is the most powerful social network. An environment defined by screens has its charms, but also tends to blunt the use of many of the senses, essentially narrowing our experience of life. But young people have an inherent interest in exploring their own powers (which is why the theme of superpowers is so present in children's and young adults' books and movies). Environments and programs can be created to help them learn to use more of these senses at the same time, to sit under a tree and consciously listen to every birdsong (and actually understand bird language), to watch, to be aware of what the body is touching, what the nose smells, what nature is broadcasting. Older kids can try wildlife photography, plant a garden, help restore butterfly migration routes, become a "natural leader," start a nature club, go on techno-fast – leaving all electronics behind for a few days. Physically challenging or extreme outdoor experiences are gaining in popularity.

In his book *The Nature Principle*, Richard Louv includes a section on what he calls "techno-naturalists." He described:

> Taking technology with us into nature isn't new. A fishing rod, a compass, binoculars are examples of technologies we've used for nature exploration. Today, the family that goes geocaching or wildlife photographing with their digital cameras is doing something as legitimate as backpacking; these gadgets offer an excuse to get outside. The attitude of young citizen naturalists towards technology is bound to be different from that of many older people – and

► Figure 8.14

Exploration and the sense of wonder is a critical developmental trait that nature offers children. Source: author photo.

that could be an advantage. However, I'm not keen on the kind of gadgets that go over a certain line, to the point where we become more aware of the gadget than of nature; apps with guided tours of natural areas, for example, offer audio information at the cost of the use of many other senses. The litmus test for some of this technology should be how long it takes for someone to look up from the screen, or forget the gadget, and actually experience nature, and to feel a sense of wonder. Another test is whether the technology is preventing other people from fully experiencing nature. Loud engines don't pass that test.

Design Prompts: Nature

As suggested by Richard Louv:

1. How can you incorporate natural elements to make your design more restorative?
2. What are some ways to transition yards and open spaces to improve biodiversity?
3. What are some of the economic benefits to children's greening a city?[40]

Edutainment and Entertainment Places

Many families plan after-school time, weekends and summers around children's entertainment offerings that adults often enjoy too. Venues bring people together to experience fun within the community or privately for parties or with friends. As childhood has evolved, kids' edutainment and entertainment have expanded. Cultural institutions such as children's museums, zoos, aquariums and botanical gardens offer full-day experiences for children and families. Youth entertainment venues include storytelling, interaction, thrills or technology. Bowling, skating, water parks, theaters, arcades and mini golf promote a safe place to hang out for "clean fun."

Children's Museums

These interactive spaces focus on informal, hands-on, multisensory learning for young children. They have themed exhibit storylines and/or collections. Many museums are geared towards children under 8, while some focus more on preschool-age children and school readiness. Many explore culture, health and wellness, literacy, performing arts, science, technology, engineering, math and visual arts. Exhibits are best when they are tactile and get kids interested in finding out more. Kids don't do things the way you intend. Material choices and durability have a significant impact on how kids react to a space.

At a recent children's museum's conference I attended, the relationship to technology was widely debated. Some insist that there is no place for technology in children's museums, since early childhood is about tactile exploration and they get too much tech at home. Other disagree and believe museums are the ideal environment to explore technologies that kids don't have access to at home.

Carol M. Tang, the executive director of the children's Creativity Museum in San Francisco, outlined guidelines for technology use in children's museums:

- Facilitated
- Mixed-age engagement
- Digital and analog
- Skill mastery
- Advances/uncovers storytelling
- Advances sharing
- Screen devices are usually "opt out" in a separate room[41]

Traveling Exhibits

These exhibits are created by an exhibit company, a museum, group of museums or independent companies and circulate to different venues. Touring is a way of sharing with like-minded institutions and of achieving economies of scale, which allow more ambitious projects to happen.

Design Prompts: Exhibit Spaces

- Is interaction with adults essential for a child's learning, or can a young child direct their own learning?
- Does the exhibit pass the toddler test?

- Do the kids pay attention to what you expect them to? They tend to ignore the most complicated and be amused with the simplest thing.
- Does the space support differences in physical scale and mental development between age groups?

Early Childhood Learning Center at Kidspace Children's Museum, Pasadena, CA

Interview with Lauren Kaye (chief officer of learning environments), Michael Shanklin (chief executive officer) and Shellie Kalmore (chief programs officer)

What were your goals for the exhibit?
The Early Childhood Center wasn't meeting a key feature of our museum – it was a fairly generic space, and over the past few years, we have been focusing on the nature theme, one aspect of our uniqueness.

Our primary goals were to:

- Encourage adult–child interactions and prompt conversation between children and their adults
- Inspire wonder, encourage exploration, and stimulate interaction through unique and natural objects and spaces
- Serve as a platform for facilitated educational programming and for use by the local Early Childhood research community

Research

- We had a charette with people of different backgrounds, including parents with young children and early childhood educators. Susan Wood from Cal Tech asked: "Why would you want to create fake nature?" That changed how we looked at the physical space. Instead of recreating nature with fiberglass logs, it became about how to inspire parents to take kids into nature.
- We had a time-lapse video where we watched visitors' behavior in the previous exhibit. They love the slides and tunnels, so we kept those and redesigned those.

Strategy
The design provides a rich sensory environment that stimulates exploration, delivers open-ended experiences, and exercises imaginations. Objects and programming are geared towards tactile exploration and interaction between the parent and the child.

Design

- The room brings out the emotions, textures, colors and shapes that you feel the wonder of nature. The joy of exploration and discovery. There is also a sense of satisfaction and pride when they accomplish something.
- With the use of soft squishy surfaces and keeping everything low to the ground and cushioned, it is safe for kids crawling or early walkers and falling a lot to explore.
- It needed to be easy to clean.
- A flexible space that can accommodate changing out its use for classes, music, story time and open-ended play.

Implementation
We developed the programming, stories, music, classes and interaction with educators in tandem with the exhibit. All books, music and songs are about the natural world. Toys and manipulatives are made of natural materials whenever possible. Classes take kids outside, and educators encourage parents to take kids into the outdoor space.

How did the specific learning objectives of the exhibit drive the final solution?
We developed a framework that includes areas of development for 0–3 years old and ideas for how it relates to the experience in the space.

Cognitive: Children play with manipulatives such as blocks and cars to test their hypothesis. We have a worksheet that outlines the educational objectives of the exhibit for acquiring the manipulatives. They explore Wonder Boxes filled with natural materials through touch. A child is read to and listens to music.

Social and emotional: An infant explores their reflection in a mirror, or cuddles in a nuzzle nest with the parent. They play alongside and observe other children playing.

Physical/motor: A new crawler moves over the bumps and textures of the shiny stream or crawls through the Twisty tunnel. An early walker climbs up steps and slides down. Programming includes sensory, snow, water, parachute play, and art with stamps and paint.[42]

Early childhood learning center.
Source: Kidspace Children's Museum (used with permission).

Cultural Museums

There is evidence that introduction to the disciplines of art, history and science at a very young age contributes to the development of a child's identity and builds a foundation for continued and increasingly more complex learning in school and in everyday life. In art museum programs in natural history, science and history museums, after looking in the galleries, the children often make works of art in a studio-like setting.[43]

Traditional museums are distinguished from other learning and recreational settings by the presence of real objects. Many focus on the display of art objects and continue their preservation. Traditional museums are suitable settings for formal learning as well as informal learning.

Encountering objects for adults and children of all ages includes

1. investigating
2. communicating (talk about it)
3. representing (creative expression, draw, play a game, sing)
4. recalling (provides a basis for later conversation)

The kids often comment about how special it was to see the "real" thing, and if they could touch the objects and artifacts and have a multisensory interaction, the experience was even more memorable.[44]

Art Museums

Children who visit museums, concerts, and dance and theater performances are introduced to the arts. They can gain benefits for many areas of cognitive development, such as critical thinking, through arts appreciation. In art appreciation, they develop more sophisticated judgment that supports the making and understanding of the meaning and purpose of the arts. They form a broader knowledge and understanding of a diverse historical and cultural heritage. As they get older, concepts of art such as realism and abstraction can be further explored. The Getty Museum Education Department suggests that children's visits to art museums can support them to:

- Create art and reflect upon what they have made.
- Seek and construct meaning through encounters with art.
- Create narratives about artworks.
- Understand the historical and cultural contexts of works of art.

Natural History and History Museums

While these museums include archaeological material, their primary responsibility is to create spaces in which children can connect to history. Natural history and history museums are especially memorable for children because of the scale of the artifacts and storylines around animals, transportation vehicles and dinosaurs, based on their intimate interactions with their toys, picture and storybooks, and popular media. Designers developing content for museums should be aware of the prior experiences, knowledge and understanding that they bring to the museum so that messages can be communicated, remembered and utilized.[45]

Suggestions to engage kids in an exhibit:

- Familiar connections and contexts: Connections with familiar life experiences are vital links to children's enculturation and subsequent learning in museum environments.
- Personal and social connections: Seeing an adult point out something that is special helps the child care and remember. Children like seeing things with their favorite colors and artifacts that they already have some knowledge of from home or school.
- Storytelling: Using storytelling presentations or allowing children to use their imagination to create their own stories.
- Dialogue: Open-ended discussions let them contribute their thoughts about what they saw, how the elements were put together, and what meaning they ascribed to the piece.
- Humor: If there is something funny or odd, they remember it.
- Preparation, spontaneous recollections and follow-up: Discussing what will be seen beforehand, drawing and discussions at the museum, and a follow-up after the visit will help with learning and remembering.[46]

Science and Technology Museums, Observatories, Space Centers

The role of these institutions is to ignite a community's passion for science, space technology and the excitement of discovery. Museums of science and technology are concerned with the development and application of scientific

◀ Figure 8.16

Today, science museums focus on interactive learning and have a high level of exploration.
Source: author photo.

ideas, scientific literacy, inquiry skills and instrumentation. Many exhibits are designed for all ages. Some include an OMNIMAX theater/planetarium.

The modern science museum has a threefold function:

- Exhibition of collections
- Sponsoring of research
- Education

Children's learning in these museums is related to topics that may include

Simple cause and effect: The child learns the physical relationship between an action and a result. Evidence that the child has noticed the impact or outcome must exist.

Small motor: The child develops or uses small motor skills.

Information/Script: The child learns about the sequence of actions or the objects used during an event.

Categories: The child learns about classes or groups to which items or concepts belong.

Factual: The child learns information that is related to neither script nor categories.

Procedure: The child learns the particular way of accomplishing something or of acting.

Conceptual cause and effect: The child learns or gains an understanding of the relationship between items or actions in a script or in a procedure.[47]

Zoos, Aquariums, Botanical Gardens

These institutions emphasize a mission of conservation, education and habitat preservation in a way that sparks curiosity, exploration and self-discovery. They

A toy zebra is compared to the live one to
help the child make the connection.
Source: Brian Boyl.

gain the critical thinking skills by asking why. They nurture the love of biological
sciences, systems and the planet.

> It is difficult to be concerned about the fate of an animal you have
> never seen. Even a video representation of an animal does not have
> anywhere near the same effect as seeing one in the flesh, hearing it,
> smelling it. The usual response to such a real-life sight – whether in
> a zoo or in the wild – is emotional.[48]

Gerald Iles points to an extra benefit of zoo animals to education:

> They are also starting to play a role in the mitigation of the effects
> of climate change. The Zoological Society of London's conducted a
> study at the London Zoo and found that after a visit to the zoo they
> had better conservation-related knowledge, concern for endangered
> species, or desire to participate in conservation efforts.[49]

Botanical gardens introduce children to the natural world and connections
between plants, places and people in their own lives. They experience sensory
learning, observational skills, adaptations of plants, animals and insects, and
habitats. They also learn about cultural traditions such as Japanese, Chinese
and European-style gardens as art forms.

Public-Private Entertainment Venues

Children's and youth entertainment often takes place within gathering places in
cities, malls, churches, public markets, libraries, and state or county fairs. Public
live entertainment is funded by the venues (or for a small fee) for a public forum,

musical performance, community storytelling, children's theater performance, kids' movie, festivals, circus, puppet show, clowns, magicians and many more.

Edutainment Learning Spaces

This sector involves edutainment, and also includes libraries and children's museums. Either free or for a fee, you will find offerings in robotics, coding and academic enrichment companies, and in libraries, city parks and recreation:

Children's discovery or edutainment centers
Movement and gymnastics
Children's discovery farms
Tutoring and academics
Children's enrichment classes
Children's activity centers

Entertainment Centers

Children's entertainment centers, amusement centers, and indoor arcades are targeted to children relying on a specific entertainment slant, type of play or environment. They offer active play, video games, food and birthday parties. There are always new concepts being tested in this market; here are some of the current industry sectors:

- Soft-contained-play centers
- Restaurant and entertainment (eater-tainment)
- Inflatable centers
- Birthday party centers
- Indoor trampoline centers/parks
- Children's adventure centers[50]

◄ Figure 8.18

An playful interactive system where people build their food order out of Legos at Lego House, Denmark. Source: author photo.

Retail – Entertainment – Brand Engagement

Entertainment experiences are being added to the retail mix to attract consumers. As the modern consumer spends more time shopping online, and less money on things and more on experiences, these have now become a priority for many companies that previously only sold products. The expectation that shopping delivers more than just the sale of products by retail will be an exciting place for change and innovation in the coming years as new technologies are introduced.

Theme Parks

Theme parks create an idealized and self-contained world and an atmosphere of another place and time. Designers use artistic inspiration, advanced techniques, materials and technology to reproduce landscape and buildings to create a fantasy world and thrills. They design both on-stage (what you see) and off-stage experiences (operational infrastructure).

Companies with a portfolio of existing characters and stories are active in producing theme parks as a way for fans to interact with these stories and make the audience feel part of the action. Disneyland was the model for the modern theme park. The park is designed to sustain a feeling of involvement in a setting completely removed from the everyday world, although today, with technology and digital devices, this is changing, and we are looking towards ways of enhancing the experience through these technologies and moving into experiences that require less space to deliver on fantasy and thrills.

Types of Rides

Parks often have the choice of classic (clone) rides or customized or specially built rides. Clone rides are themed for a new location. Custom rides are designed for individual parks based on their specific needs and preferences,

Thrill rides: Roller coasters, water rides, extreme thrill rides (drop towers, pendulum rides)
Circular rides: Spinning (scrambler, tilt-a-whirl)
Family rides: For everyone (Ferris wheels, bumper cars, carousels, swings)
Transportation rides: Trains, monorails and sky rides

Dark rides: Enclosed, heavily themed rides, storytelling. A good ride has strong choreography, timing and effects.
Kiddie rides: Pint-sized bumper cars, miniature Ferris wheels and bouncy towers are favorite kiddie rides, along with bumper boats and circular rides themed as motorcycles, race cars and spaceships.
Pay-per-ride attractions: Extreme rides such as bungee jumps, go karts and skydiving simulators.

Visual Development and Art Direction for Theme Parks

Interview with Andy Sklar, entertainment designer and creative director

Andy Sklar has worked as a visual development artist and art director for the theme park and entertainment industry for 30 years at Walt Disney Imagineering, and currently works for Universal Studios.

How does the fantasy component play into your approach to environmental design while thinking about the practicality?

Framing is an effective starting point. How we frame something, be it a literal, structural or knowledge of the pre-existing story, is a helpful approach to visual story-telling. Much like a book cover invites you to pick it up and read it. The framing of a place is the emotional invitation to step into this world or narrative.

Walt Disney had the famous philosophy called *Wiener on a stick* about moving people through space. Drawing people into an experiential scene by having a visually appealing iconic element that pulls you, the audience, forward. The Castle at the end of Main Street USA is the ideal example, or Space Mountain sitting on the horizon of Tomorrowland.

There is always the operational reality that one must be plugged into as an environmental designer. A paved pathway may have to look the part of a storybook road, but it also has to be the right width for a fire truck and or a parade route to get through. As a creative thinker and designer in environmental design, you naturally must be tethered in reality.

In a theme park, people connect to their favorite characters and stories. What are some of the most important things you think about in creating this experience?

A winning character and licensing brand is a great starting point for designers, as your audience already has the emotional connections. Making the experience engaging and inspiring takes concerted thinking outside the box sometimes.

◄ Figure 8.19

Illustration by Andy Sklar, Magic Station. Source: Andy Sklar (used with permission).

There can be a learning curve for both the licensor and the designer in working together and being open to what the characters and the places they inhabit are and what are the possibilities. It is not enough to slap on characters to a façade and make sure color call outs are correct from a said style guide. You have to build experiences that adhere to the character's story, but also engage the guest in a unique experience that is unforgettable.

What are the most important things to think about in character design that will appeal to different markets?
What are the story and context of the character? Is it educational or entertainment driven? What world do they come from? When you know the story, the graphic elements all come into play. Shape, color, texture, patterns and materials inform the design.

Your work is an excellent example of keeping your own style and translating it into different projects, including children's books, greeting cards, figurines and environments. What are some key things you think about to keep your work consistent but adjust to the project?
Having a style can work for or against you. It is how I draw things – people like it, but it isn't always the right fit for the client when working as a commercial artist.

Sometimes you have to dial back style for a simple, direct and informative approach. Ultimately, it is working with both style and structure that makes the work most effective, I find for myself. You hope to find that opening within the project that speaks to you. With that window, you can bring your approach to the table and hope it translates.

Your work appeals to kids and adults alike. What in the style, characters and the stories that you tell in your work makes it cross over so well?
Artists and illustrators that have a timeless appeal inspire me personally.

I enjoy creating, expressing story and mood through color and shapes that I hope people might find fun and engaging. I personally find a lot of joy in making art. I hope that may speak to people too and comes through.[51]

Notes

1   Siegel, Tara J.
2   White, Deb 2017
3   White, Deb 2017
4   White, Deb 2017
5   White, Deb 2017
6   White, Deb 2017
7   White, Deb 2017
8   White, Deb 2017
9   Lee, Kevan 2017
10 The Family Dinner Project
11 Lanza, Michael 2017
12 Ginsburg, Kenneth 2007
13 Placemaking Chicago 2008
14 Food Explorers 2015
15 Atkinson, Mark 2017
16 Lanza, Michael 2017
17 Beckwith, Jay 2017
18 Sanders, Martin 2017
19 Voice of Play
20 Evergreen 2013
21 Urbanski, Matthew 2017
22 Kutska, Ken 2017
23 Bellali, Safir 2017
24 Urbanski, Matthew 2017

25 LandscapeStructures

26 White Hutchinson Leisure & Learning Group 2016

27 Westervelt, Eric 2014

28 Westervelt, Eric 2014

29 Kyle, Angela 2017

30 Siegal, James; Levner, Amy S. 2017

31 KaBOOM!

32 Fung, Alice 2017

33 Marinelli, Don 2010

34 Center For Green Schools

35 Zippin, Sam 2014

36 Steelcase Report 360° The Education Edition

37 Smith System

38 Fung, Alice 2017

39 Association of Science-Technology Centers

40 Louv, Richard 2017

41 Tang, Carol M. 2017

42 Shanklin, Michael; Kaye, Lauren; Kalmore, Shellie 2017

43 Munley, Mary Ellen 2012

44 Munley, Mary Ellen 2012

45 Piscitelli, Barbara; Anderson, David 2002

46 Munley, Mary Ellen 2012

47 Munley, Mary Ellen 2012

48 Russo, Cristina 2013

49 Hance, Jeremy 2011

50 White Hutchinson Leisure & Learning Group 2008

51 Sklar, Andy 2017

Chapter 9

Children's Media and Technology

Media and technology have had growing importance for the way kids play, learn, interact and are entertained. Educators, parents, health providers, content creators, designers and child development experts agree that media use can have a profound impact – both positive and negative – on learning, social development and behavior. As designers creating for children, how do we use all that technology has to offer to enhance development? What is the best way to prepare children for a technology-driven future? Our goal, as the creators of media content and devices, is to dial down the negative effects and boost up the positive ones.

In the last decade, the media and technology environments for children and youth in many cases have become completely embedded in children's lives. Children (and adults too) have learned to live without boredom, one of the roots of creativity and innovation. There are more ways to use media and more types of media to consume than ever before. Media is instantly accessible at any place and any time. The vast diversity of ways children and youth interact with media shows a huge variety in their preferences and patterns of use. Content created for children has diversified as the methods of distribution have expanded.[1]

We tend to lump the content (programs, communications, applications and interactions) with the technology (the software and devices). The truth is that the technology delivers a wow factor, but kids would not be spending so much time with it if it weren't for the engaging content.

Designers play a valuable role in delivering on the "cool" factor of technology while developing media content and devices that promote healthy child development. As creators, we explore broad modalities in design, including auditory, tactile and motion. However, we also need to question: When is the new way better than the old? What is the gain? When researching technology, center it around the appropriate usage. Question what is best done from a developmental perspective in the physical world and what is best executed through the use of technology. Does this technology create enhanced learning and playing for a deeper, longer-lasting experience?

◄ Figure 9.1

Playing, communicating and bonding with friends through technology is a priority for many young people.
Source: author photo.

Media Types

Media is used for making, communication, entertainment and education. In each of these categories, the level of expertise of those developing the content for TV shows, music, digital games, social networking, apps, computer programs, videos and websites varies, as does the quality. For content providers striving to make high-quality media for young people, the best way to have a positive impact is to understand what children need, want and understand at different ages and stages, and the role it plays in their lives. Seeking out high-quality media content for young people through editorial reviews and setting limits on how much time is spent with media are two good places to start.[2]

Today, we live in a world of screen-based media – but that is quickly changing, and in the very near future the screen will become less dominant, replaced by seamless technology. It will be integrated into our everyday.

Today, children's use of media can be broken down into these areas:

Passive consumption: Sitting and watching
Interactive consumption: Active media responds to the user
Communication: Interacting with others
Content creation: Tool for making

Educational Media

For designers, educators and content creators who are eager to use media to deliver educational programming, this genre offers a means of creatively producing the optimal combination of media to support learners. Books, television, computers, mobile devices and even toys are all delivery systems for the content. It is captivating for individuals and large groups and promotes attentiveness. Educational media is common in the 2–4 age group. As kids get older, the time spent on educational media for pleasure goes down.[3] Amy B. Jordan, an educational TV expert at Annenberg School of Communication, says: "To be truly educational you have to have a very narrow audience. TV is for mass audiences."[4] Educational content, therefore, is often designed to encourage developmental milestones within appropriate age groups. An educational property can teach a child about a particular subject or can help a child develop a specific skill. A child's learning and development are associated with interacting with the content.

Edutainment/Infotainment Media

Edutainment includes content that is primarily educational but has incidental entertainment value. Some programming and games may be explicitly designed with education in mind, while others may have secondary entertainment value. Many toys are "edutainment" products, as they have both educational and entertainment value. They entertain and at the same time teach children literacy, numerical, conceptual or motor skills. *Sesame Street* is a good example of edutainment. Infotainment is a form of media where the goal is to both entertain and inform. For example, the television show *Animal Planet* teaches in an entertaining format.

Entertainment Media

Entertainment media is media consumed for fun or amusement. Although there may be a moral to the story, the primary goal is to get kids engaged in the story, laughing, scared or excited. Most of this content is passive, such as viewing TV, movies or YouTube. Theater, dance, movies and video games are all forms of entertainment. Today, more and more entertainment media requires interaction from the viewer. Most forms of entertainment, such as music and movies, are developed to suit children's interests and offer meaning connected to their daily lives.

Interactive Media

Social networks, digital games and apps are all interactive. Kids are growing up with a touchscreen at their fingertips, and expect technology to respond intuitively to their needs and enhance and augment their experience of the real world. Children today are the first who have only ever known this world of digital, wireless connectivity, social media and apps. The strength of an interactive context is control. The more kids can control an interactive experience, the more they respond to it. Designing for open-ended experiences is a challenging task for designers, as we need to envision and program the possibilities.

Shuli Gilutz, children's UX design strategy and research consultant, highlights that designers remain focused on the content and experience, not the technology used. She explains:

> A good media product for children should be both entertaining and educational (similarly to a good media product for adults). We remember good movies, games and experiences if we had a meaningful experience, learned something and enjoyed it. This is true for good design for children, whether it is digital or not, interactive or not. The term Educational had been hijacked to refer to school pedagogy and content, and lies way beyond reading, writing and arithmetic. Children learn from anything they do: any game they play, any book they read, and show they watch. The question is: What *do* they learn? Which values? Which content? Which behaviors?
>
> As designers we can create wonderful experiences for children that will entertain, allow for creativity, social connections, curiosity about the world, learn new skills and content, and create wonder and joy. The most popular digital platform for kids today is YouTube, which allows for all this. It's basically the biggest library in the world at the fingertips of children, and they find all this content there: games, music, movies, TV shows, DIY, jokes, tutorials, etc. They also create their own content and put it out there, being creative and innovative. The most promising technologies will support these passions and take them further, while allowing for a safe and age-appropriate environment, similar to great physical playgrounds.[5]

Design Prompts: Media

- In your project, ask: What *do* they learn? Which values? Which content? Which behaviors?
- Can you design a single experience that ranges from entertainment to education?
- Design an educational experience that uses humor to convey the core concepts.
- Design an entertainment experience that sneaks in learning.
- Develop a character that portrays educational concepts through personalities.

Ethics and Media Use for Kids and Teens

Advice from Shuli Gilutz, children's UX design strategy and research consultant

In today's world there are many new ethical issues that designers need to consider related to children and technology. In a conversation with Shuli Gilutz, she suggested some important things to consider in our designs:

When designing technology for kids, companies tend to leave ethics for the legal departments, rather than incorporating them in the design process. This can not only create products that will ultimately create risks for children, and fail, but also, we miss design opportunities to create creative and unique solutions for these challenges. Designers should be aware of all legal requirements (COPPA, GDPR, etc.), but also about current research and best practices regarding safety online, privacy concerns, online advertising vulnerabilities and addiction designs. Each one of

these can be addressed with good design, to protect and inform children about their environment, and to contribute to their media literacy and give them the tools to be smarter online users in environments that do not provide these safety rails.

An example is incorporating principles of privacy by design (PbD) within the design process (which is also required by the EU 2018 General Data Protection Regulation (GDPR)). These design principles add responsibility to the empathy in user-centered design processes by creating proactive and preventative designs that by default allow the user a safe environment, where their data is protected end-to-end throughout the use lifecycle, with visibility and transparency. This, of course, must be age appropriate, and must be tested with children to make sure that we're getting what we hope for; like testing physical playgrounds for safety.[6]

Developmental Timeline for Media Usage

How do we as human beings come to understand the world through the screen? What do kids understand at different ages? Technology exposes children to content they can't see in their everyday. With technology, children can access multimedia presentations, visit places around the world, and learn about the world beyond their home and community. Different media formats also offer diverse entry points for children with various preferences for learning.

Babies and Toddlers

We know that in the first 2 years, face-to-face involvement with other people, manipulation of the physical environment, sensory experiences and open-ended play are what this age group needs for positive development, although a substantial number of infants and toddlers are watching TV and DVDs and are interacting with mobile devices on a regular basis.[7] The growing brains of babies and toddlers who use digital technologies are now wired differently. What we don't know is what these differences will cause in the long run.

Even though by the age of 6 months children have depth perception and visual acuity as good as those of adults, when babies are observing the real world, we question what they are "seeing."[8] There is a lot of representation and symbolism in books and screen-based programming. Basic understanding of story montage requires personal memory, self-reflection and interpretation. None of this is typically developed in infants at 12 months, and they don't understand sequence. At

18 months, they could start to understand narrative, and at 24 months, they show preferences for a correct version that is not chopped up (edited).[9]

Media Design for Babies and Toddlers

- Will gain more in a shared media experience with adults
- Hear words strung together
- Can tell there is something different about reality and screen representation, although they try to grab things (as if they were real) on the TV and picture books
- Brains have not yet developed enough to put stories together
- Respond to engaging people on screen who talk to them, repeat things and answer questions
- Beware of claims that content can make kids smarter[10]

Preschoolers

Tablets, computers and television can help with school readiness, address educational inequality and directly target specific learning areas. Interaction with a device gives them a feeling of control and motivates them to strive to new levels of competency. Preschoolers have a hard time understanding what is happening on the screen. In a popcorn study by John Flavell of Stanford University, he explained: "Even though by age three they can tell the difference between real and stuffed animals, a three-year-old thinks a bowl of popcorn will spill if you turn over the TV upside down." A 4-year-old writing to Mr. Rogers asked how people get in and out of the TV. And another child thought Big Bird is just a costume and there is a real bird inside. Even though 6th graders are better at this, they sometimes have a hard time distinguishing real from fake, fantasy and reality.[11] Developmental psychologist Dan Anderson, who has consulted on many preschool children's television shows, says:

- Know and use understandable language, action and context
- Dialogue and action must be linked
- The characters' action should illustrate what the scene is about
- Edits/cuts in sequence do elicit looking, but make informed judgments about how to use them[12]

Media Design for Preschoolers

- Content should be all sweetness and light-hearted
- Feel comfortable when content relates back to their daily life
- Start to understand time passing, story, and narrative in a linear story
- Learn from a back-and-forth interaction with content on the screen
- Learn cause and effect by touching a button
- Fill in the gaps with their imagination
- Are working on pre-reading readiness, phonics, math skills, and blending story and comprehension

- Parents can help with understanding by prompting questions before, during and after the experience
- Consider multiple children interacting at the same time to promote social skills and peer-to-peer learning
- Design characters that are easy to distinguish. For example, dinosaurs and dragons may be too similar
- Interest and communication can be gained through stimulation of several senses
- Prioritize sound and visual icons in telling the story versus reading
- Animate characters, environments, letters and words

Scary Programming

Children at this age have difficulty deciphering between the real and imagined dangers. To many kids this age, seeing something is believing it is real. This is important for content creators to know, because kids might get scared of things that you would never think are scary. Michael Brody, a child psychiatrist in Bethesda, MD, and chairperson of the television and media committee of the American Academy of Child and Adolescent Psychiatry, says: "Scary programming exacerbates anxieties in real life."[13] Once a kid is scared, they will never watch it again, so content developers want to get it right.

Elementary School Children

School-aged children use all forms of media and technology for their homework, connecting with others, gaming and creating their own content. YouTube channels, watching TV, using social networking sites and listening to music are all being juggled, and many kids are using more than one medium at once, sending texts while watching TV, or listening to music and posting social media updates. There is a lot of debate about children getting enough quiet time and space that is necessary to think, dream and wonder. Technology presents opportunities for kids to be inspired to engaged.

What do kids understand at this age? There is a lot of confusion, and they are still trying to figure things out. Abstract thinking does not develop fully until the teenage years. Without logic and abstract reasoning to understand themes, children take away a very different idea of what they see. What do they think when they see goblins and wizards? It depends on the individual child, their exposure and experience.

Media Design for Elementary School Age Children

- Use diverse visual media (illustrations, videos, photos) to describe content instead of words
- Use simple text. Scale can express a message, or graduated type can show a hierarchy
- Leverage knowledge children have from school and in their life
- Use humor where appropriate
- Design characters and content to be a bit more mature

- Provide customization capabilities
- Design for playification, not just gamification
- Make media social

Youth

Tweens/teens use it for everything: listening to music, watching TV, texting, gaming, filling out their college applications, making their own movies, interacting with their friends and blogging. Teens' approach to technology results in little distinction between digital and physical experiences. Neither is seen as inferior, and they switch between the two with ease. They are more selective about choosing content that they watch that represents their interests and identity. A more "serious" focus on computers emerges in middle school. Kids carry around cell phones and are freer to range farther from home.

Media Design for Youth

- Make it fast
- Provide self-expression capability
- Support positive social networking
- Provide increasingly "grown-up" content
- Design for intrinsic motivation
- Use more cultural inspiration
- Design for like-minded communities
- Meet them in their space

American Academy of Pediatrics on Screen time

The American Academy of Pediatrics continually revise their recommendations on screen time for children at different ages – for the latest guidelines, see www.aap. org/en-us/Pages/Default.aspx

Pros and Cons of Media and Tech Use

An aspect of a design that could be a pro for one user can turn into a con for another depending on the individual child or age groups.

Young Children

Pros

- Benefits of video chat and communication tools
- Children who may be geographically or economically disadvantaged have greater exposure to content
- Speaks to different intelligences and senses

- Promotes school readiness and problem-solving skills
- Practice the capacity for visual attention, hand–eye coordination, spatial skills and fine motor skills
- Expands horizons outside of daily activities

Cons

- Time reduced in physical world and engaging in sensory, active and creative play
- Decreased family time and parent–child interactions for building connections and learning
- Diminished time with learning people skills, peer-to-peer play
- Diversion from the present moment. Television, computers or digital games are used as a substitute teacher or electronic nanny
- Tech tools are expensive. Can the same learning happen in more affordable ways?

Older Children

Pros

- Practice in multi-tasking skills, complex reasoning and decision-making
- Encourages sharing of positive ideas, communication and collaboration
- Families enjoy using media together; may lead families to spend more time together, not less
- They feel in control
- May motivate them to strive for new levels of competency and skill development
- Exposes them to diverse viewpoints, experiences, cultures and opportunities
- Opportunity to learn anytime and anywhere

Cons

- Easy access to information gives less practice in problem-solving and critical thinking
- Multi-tasking may be harming the ability to remain focused
- Media and technology use is a source of tension for many families
- Ethics that go into machine learning, security of kids' data, privacy, tracking kids, legal and financial risks
- Posting material that could harm their reputation and self-esteem
- May have exposure to inappropriate material, online predators and physical molestation
- In face-to-face interaction, the experience or message is qualitative. You judge it by what is said or facial expressions. With social media, messages and images are now quantitative: number of likes, favorites or retweets

◀ Figure 9.2

Media and technology can take kids away from being in the moment and social learning.
Source: author photo.

◀ Figure 9.3

Technology can be used for making and sharing.
Source: author photo.

Parental Assistants

When designing to make kids' lives better, we often design for the parents too by offering parents support, advice, and updates on child development and childhood today. Designing for the parent is essential for all ages, but especially for those with younger kids who are not yet wired into social media or are unable to navigate information themselves. Many modern families have fewer support networks in the family and community and are raising children while juggling the responsibilities of work and managing their household. According to the Stylus report *Decoding digital parent* Feb 2016,[14] parenthood is becoming a digital affair as consumers turn to apps, devices and networks. Not trusting major marketers, a growing number of parents access the internet via mobile. It is the first port of call for information.

Supporting and Educating Parents about Play

Interview with Anna Yudina, director of marketing and communications, The Genius of Play (www.thegeniusofplay.org)
The Genius of Play offers parents and caregivers information to understand the developmental benefits of play, and helps families incorporate more play into their lives.

How did the research strategy drive your design process?
Before launching, we conducted an extensive study to understand parental attitudes about play and the challenges parents face when trying to balance play with other priorities. We also reviewed secondary research and interviewed experts in different fields, including child psychology, pediatrics and education. We discovered that time was a significant issue for today's families. We also learned that parents often felt insecure about their parenting skills and overwhelmed by information. Our solution was to present complex play research in a fun and easy-to-consume format while offering hands-on play ideas, expert advice and other resources.

How did you go about developing the online branding and interface design across media platforms?
It all started with a single landing page that housed our first video and infographic. We ran a short promotion to share the content on social media, and the reaction was immediate and very positive. Our next step was to introduce The Genius of Play™ animated characters through a series of fun and short videos that became the basis for our brand identity.

How did you go about building your audience?
As a non-profit initiative, we have a small budget, so we need to be creative. We have relied heavily on the viral power of social media to help spread our message and grow our following. We also use public service announcements, which provide reach that we couldn't

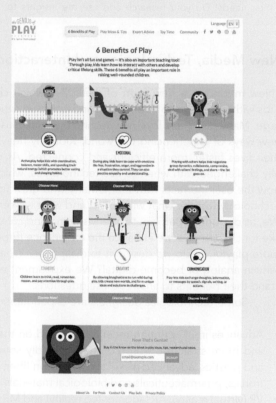

▲ Figure 9.4

"The characters were embraced by the audience and quickly became a key part of our brand's identity. We also introduced colors and icons to serve as visual cues for our 'Six benefits of play' framework and to complement the characters." Anna Yudina.
Source: Genius of Play (used with permission).

possibly afford using traditional media. Our relationships with parenting influencers and partnerships with other non-profit organizations help us extend and amplify our own communications. We attend events and conferences throughout the year and have recently convened an expert panel to draw the media's attention to the state of play in our society. At the end of the day, there is no single platform or tool – it's all about an integrated approach.

Do you have any suggestions for a designer creating a communications strategy?
First and foremost, know your audience and understand their needs! Do your research and use the insights to inform your initial strategy. You may even want to test your strategy on a small scale before investing time and resources into a bigger effort. Once you launch the program, be open to feedback and continue to evaluate results and make ongoing adjustments. Finally, understand that you can't do it alone! Identify and find like-minded individuals and organizations that will support you and advocate for your cause. Ultimately, it takes a movement to make a real social impact, but it can certainly be done with enough passion and vision.[15]

New Media, Technologies and Interactions

"New media" is a new form of literacy in which children need to be fluent. Kids without access to information and communication tools are at a disadvantage. Many designers love new technologies because they hold the promise of new ways of experiencing the world. Kids rely heavily on technology to make their lives faster, more efficient and connected. As designers, we are continually challenged with when to use technologies, which technologies to use and how the user interacts with them. Many who grew up in a world without digital technology wonder what will be lost and what will remain the same for children. We, as designers, participate in envisioning and creating what the future will be like for children. Do we enhance the experience through making the interaction more playful, or do we design it to disappear? What in the experience should remain analog or become digital? We experiment with technologies to understand their benefits to answer these questions.

Some technologies designers are experimenting with today are:

- Advances in 3D printing recently centered on the creation of new materials and more straightforward consumer-friendly software. In addition to starch and plastics, advanced nickel alloys, carbon fiber, glass, conductive ink, electronics, pharmaceuticals and biological materials are in the near future.
- VR (virtual reality), AR (augmented reality) and MR (mixed reality) are currently in everything: messaging, entertainment and learning. Worlds that are currently being produced through animated renderings will start to move into live action.
- Leveraging data: Information from different sources in a beneficial way will provide more customized experiences from retail to education to entertainment. Contextual media will respond and adapt to location, time of day, or whether to create targeted and immersive experiences. Ethical considerations with regard to collecting data about children and privacy are currently being challenged.

- AI: Advanced machine learning makes smart machines appear "intelligent" by enabling them to understand concepts in the environment and to learn. Instead of a fixed set of attributes, a smart machine can change its future behavior.
- Robots, autonomous vehicles and virtual personal assistants will take on roles that kids can't do on their own today and provide information when they need it.
- The Internet of Things, with the greater cooperative interaction between devices, entertainment and learning, can be integrated into everyday experiences.
- Wearables or invisible integration: The experience of the screen will open up into seamless experiences between the physical and digital worlds. The TV, internet, video games, tablets and cell phones will evolve into seamless technology. Our realities will be mixed.
- Drones are no longer just playthings but are being used to transport many things too difficult to reach through traditional transportation methods.
- Tactile tech: Advanced tactile input and emotional feedback from technology – from virtual reality (VR) to touch-sensitive haptics – enables users to experience worlds beyond their immediate environment in lifelike detail, nurturing new levels of empathy. Designers are experimenting with "soft technologies," natural interaction, haptic feedback, and design that guides, rather than dictates.[16]

▶ Figure 9.5

3D printed car: 3D printing provides designers with new methods for production and opportunities for co-creation with children, and children with new opportunities for making their ideas.
Source: author photo.

UX Strategy and Interaction Design

Interview with Maggie Hendrie, chair of the interaction design program at ArtCenter College of Design and a consultant in UX strategy and interaction design

How do you envision seamlessness from physical to digital play evolving?

We are moving away from screen-only technologies into more natural interfaces. From point and touch to voice, sensing, and tangible that can engage the whole body and create opportunities for unstructured play and other toy/game opportunities. Physical (tangible) computing will be necessary as the enabling technology and materials develop and become cheaper and more available. Seamlessness will come from new playful materials that respond to use, cross platforms and know who the player is. Not all interaction is "connected" to the web, but toys are increasingly being integrated into the Internet of Things.

Which technologies do you see as having the most promise for play in the coming years?

There is a lot of innovation in VR, but the ergonomics are not currently adapted for kids. AR has already shown new play experiences. Voice recognition software such as Siri and wearables has a lot of potential.

Are there specific ethical, social impact and sustainability issues in interaction design?

Ethical Issues

"Many students think they have nothing to lose by posting things online and then they Google themselves and have second thoughts."

- Digital capitalism: The predominant business model for digital is advertising through micro-targeting and profiling. This raises critical questions about children's rights to privacy.
- Ethics that go into machine learning: AI is not neutral; algorithms strategically encode social, economic and political points-of-view. They model the user experience just as much as they capture data on players' behavior.

- The security of kids' data and personal history: Many tech companies go through growth and acquisition, and their terms of use change frequently.
- Tech in the built environment can also be surveillance and very difficult to opt out of.
- Digital etiquette: What are the rules of engagement? How do we play together? There will also be questions of how play dates, classrooms and public spaces manage different user and parental desires for privacy.

Social Impact

The digital divide: We need to move towards net neutrality. Access to products, services and content that are the same across the board. As we move into subscription-based media, there will be a further division; those that can't afford it are the target, just like poor nutrition and digital advertising.

Sustainability

Digital data is not a zero footprint. The cloud is a place. A hot place with servers and storage architecture. Electronics are designed for obsolescence.

What should interaction designers think about when designing for children?

- Sensory experience as it relates to physical engagement, haptic feedback and design.
- New kinds of interactions, physical and digital, such as twisting, touching, moving, singing, communicating not just with the digitally enabled device but with other players.
- Transform a game into a toy and make it more interactive. A game has more rules; a toy is more tactile, open-ended and interactive.
- Recording and sharing others' experiences and stories.
- Design for open-ended play.
- Immersive narratives extend and augment stories and characters.
- Personalization.[17]

Sustainability and Ethics in Digital Media

There are many sustainability issues involving materials and power use and disposal of electronic devices. Many other issues include ethics, media usage and appropriateness for children.

- Privacy
- Everything digital requires power. Consider storage and reduce battery and power use.
- Design for reduced processing speed.
- Encourage e-recycling.
- Discourage planned obsolescence; design for upgrading.
- Create quality screen time – educational and edutainment content.

- Avoid negative messages such as gun violence and over-sexualization.
- Design using age appropriateness.
- Provide positive role models within various character traits, backgrounds and body types.
- Avoid gender and racial bias in product concepts and messaging. Consider multicultural perspectives. Create characters as positive role models. Avoid negative character traits.
- Reduce the "digital divide" and "app gap."
- Advertise appropriately.

Technology for Social Impact: Digital Divide

In recent years, the speed, flexibility and affordability of rapidly evolving digital technology have enabled millions of young people in developing countries to join the digital world. Increasingly, technology is being seen as a powerful development tool. It is used in youth-focused targets in global education, livelihoods and health. Deb Bauer, director of Dell Giving says:

> We believe that access to technology brings young people into contact with the broader world, opening up access to education and vocational training in a cost-effective way. What we've learned is that it isn't enough to merely provide the hardware, it's the quality of the wrap-around services – the teacher training, maintenance of technology, reliability of power, which provides the long-term benefits and this is one of the learnings we've been taking forward.[18]

Mariana Prieto, a design for social impact consultant, suggests:

> Technology has been a great tool for development. We now have mHealth tools that give access to healthcare to people in remote areas and drones that deliver vaccines in Rwanda. However, there is nothing more complex than simplicity. Then finding the most elegantly simple idea that can have a significant impact. Sometimes technology can help us achieve this. As designers, we need to be careful to not fall for new technology just for the excitement of it, when the most straightforward solution may be much more robust with old technology that is long lasting. In areas with limited power and electricity, technology may be difficult.

New technology is an exciting tool, but we need to make sure we don't abuse it and produce solutions that are resilient and long-term, even if they don't seem new and tech sexy. The upside to this is that many countries are leapfrogging old technology and building amazing systems using the latest technology that work so much better. An old example of this is the cellphone. Many countries never had significant landline coverage, but they leapfrogged to the mobile phone and have now been using mobile money for years, which other more developed nations are just catching up on.[19]

Characters, Storytelling and Narrative

Technology can be mysterious and magical to kids, although many kids love it because of the content. Children have an innate love of stories and characters that teach them about life, their emotions and others. Art director Scott Allen explained:

> Animals, creatures, fellow humans, aliens ... create an instant connection for the audience. Kids (and people of all ages) develop emotional connections with individual characters over time. Keep in mind *your* perception of a character isn't necessarily the same as someone else's perception of that character.

People have personal attachments.

Designers and creators of content use narrative as the optimal form of communication, making our messages engaging, memorable and understandable. The brain is wired to think and process information best in story form.[20] Children's early abilities to evaluative narratives are a strong predictor of further learning.[21] When designing for digital interaction, it is important that storytelling and emotion should not be separated from the interaction. They both must be developed at the same time.

When creating a project that includes storytelling, it is helpful to have a foundation in the basic principles of storytelling and characters. In its most basic form, a story is an event or sequence of events, set in a time and place. Meaningful events emerge through the chronological account and happen to someone over time. Most stories include a problem or tension, and the resolution of that problem. Literary devices such as rhyme or poetic prose with language and images help kids think in and retain stories in a memorable way. The three-act structure developed by Aristotle is a basic form of storytelling. This includes:

Act 1: The setup (the setting, the characters and relationships, the conflict)
Act 2: The middle is the confrontation (obstacles, conflict)
Act 3: The end: The resolution.

Genres of Stories

The same stories are told over and over in different ways. Classic stories cross the barriers of time and are updated for the modern world, including contemporary themes. The goal of the storyteller is to make the audience care about these characters and their predicament:

The great journey
Adventure – the quest
Rags to riches
Overcoming the monster
Rebirth
Coming of age
The fantastic and impossible
The sacrifice
The epic battle
The fall from grace
Love
Fate, revenge
The trick
Mystery

Character Archetypes

Characters in media play an important role in children's social and emotional development. Creating characters that serve as positive role models can instill prosocial behavior. Negative characters can show children potential outcomes of negative behavior. Behavioral scientist Susan Weinschenk recommends: "Consider archetypes in character development and design. It is instinctual to react to babies, puppies, kittens, because of their large eyes, they are soft, have funny movements, and make high-pitched sounds."[22] Art director and illustrator Douglas Day explains his approach:

> Archetypes are helpful to get the ball rolling. I look at opposites – the hero and anti-hero. Or several characters in a group. The bully, the rebel, the guardian and the victim. It is challenging to come up with unique ideas. I think about who is looking at this? Is there a fresh way of looking at this that hasn't been done?[23]

Classic Character Archetypes

The Hero: The hero (protagonist) has an objective, and encounters and overcomes obstacles along the way to achieving this goal. Heroes help children overcome challenges in the real world.

The Mentor: The mentor is usually old, and has magical abilities or greater knowledge than others. Mentors help heroes along their journeys, generally by showing them how to help themselves.

The Everyman: An average person who faces extraordinary circumstances and is just trying to get through a difficult situation they have little control over.

The Innocent: They are pure, although often surrounded by dark circumstances. They have not become jaded by the corruption and evil of others.

The Villain: The villain is often evil and wants to stop the hero from achieving their goal. They are power hungry. The villain's main vice will parallel the hero's chief virtue.[24]

Lessons to Learn from Superheroes

Through superhero stories, kids identify with characters who faced a threat and can learn about some basic life lessons. Gerard Jones explains in his book *Killing Monsters: Why Children Need Fantasy, Superheroes, and Make-Believe Violence* that a lot can be learned from superheroes:

- Achievement feels good

- Goals are achievable through commitment
- Clear choices must be made
- Sometimes conflict is useful
- Sometimes shattering old ways is necessary
- Loss and defeat are survivable
- Risk has its rewards[25]

Character Bonding

Howard Sklar is a post-doctoral researcher at the University of Helsinki. "As anyone who has watched an engaging film or read an engaging novel knows, we invest ourselves deeply in the experience of living with those characters," Sklar says. "We tend to respond to them as though they were real individuals."[26] Jackie March of the University of Sheffield suggests that media characters can serve as transitional objects for children – soothing them like a teddy bear and always there for them.[27]

Others are concerned about the influence of characters in children's lives today. Children often act out what they have seen in media in pretend play and real life. Susan Linn, author of the book *Consuming Kids*, argues that "children's reliance on characters is holding kids back from imagining their own world and make-believe figures."[28] In the documentary *Consuming Kids*, Michael Rich, MD MPH at Children's Hospital Boston, explains: "Growing up can be scary, and characters give stability. They offer continuity and attachment to touchstones in their lives as well as constants that make them feel loved and comfortable." He is concerned about how companies leverage this powerful attachment to make money.[29]

Building Narratives

- Character: Create and develop characters by writing bios. Where do they live? What do they eat? What are their favorite things to do? Least favorite? Why? Connect the characters to each other. Who are his/her friends? Enemies? Why? Explicit details will help with your character design and story development.
- Setting: It's essential to set the stage for the story correctly. Brainstorm where the story will take place and include a list of details (natural and man-made environments and objects) about the setting.

▲ Figure 9.6

Children's media today plays a role in influencing childhood from storytelling (narrative) play, fandom, social interaction among children and reflections of identity.
Source: author photo.

- Action: Think through the scenario of the actions of the character through a timeline or sequences of movements, motivations and interactions. Storyboarding is a great tool.
- Problem: Brainstorm a list of possible problem topics and solutions on a chart with the problem on the left side and the solution on the right side. Which ones are most intriguing and make sense for the style of your story?
- Resolution: This is often the most challenging part of a story. Develop a clear and distinct ending. Create a few ways to end a story and then choose the best one.

Designer's Tips on Character Design

- Visually strong and exciting characters get people's attention.
- Exaggerated features help the viewers to identify character traits. Which qualities will you exaggerate and what will you play down?
- Colors can help communicate a character's personality. Typically, dark colors such as black, purples and grays depict negative traits. Light colors such as white, blues, pinks and yellows express innocence, good and purity.

- Human or other. Animals, aliens or abstract characters distance the viewer from the character because they are somewhat disconnected from human experience
- Does gender play a role in your character, or are they gender neutral?

Creating Depth and Backstory

- Humanity is necessary to make people care.
- Develop personality. How do they react to situations? What are their quirks? How do they move?
- Emphasize ups and downs of emotions in facial features. Are they understated or exaggerated?
- Know their history. How did events and experiences affect who they are? What are their flaws? What are their goals? Dreams?
- Develop their world. Make it consistent with who they are and the story being told.
- Translation of media. What opportunities do you have to deepen the same story across platforms? What would they do in a book? TV? Video game? Movie? Toys?[30]

In an interview with creative director Alena St. James, she emphasized that characters and stories that connect to kids' lives and their own experiences resonate with them. She offered some suggestions to think about in characters and story that will appeal to different ages.

Little Kids – 6 and under

- Use simple graphic shapes
- Bolder and broader use of elements – fewer details
- Funny, silly designs

Middle Childhood – 7–10

- Grouping kids to play with others versus solo, having adventures with friends
- Take characters out of an average day and have them do amazing things
- Have the audience imagine and fill in details
- Characters are independent and are involved with peers
- Learning more about the world, values, how to handle themselves
- Better verbal skills and attention spans and have much more complex forms of play
- Silliness comes into play

Youth – 11+

- They take more risk, to prove themselves
- Audience wants to relate to adventurous and dynamic characters

- In one sense they want to grow up, and in another, it's hard to let go
- Even more complex forms of play
- More influence from their peers
- They can feel fear – but act anyway
- Monsters can be destroyed
- Self-assertion is powerful
- Merely being, they are heroic[31]

Character Design and Visual Storytelling

Interview with Douglas Day
Doug Day has experience in entertainment design and consumer products development, having worked for successful children's brands including Disney, Sesame Workshop and Warner Brothers.

How do characters and stories create connections for children?
The continuity of the characters and brand makes kids feel comfortable in what they can expect. For example, Disney has stringent guidelines of dos and don'ts on what they want to see; Sesame Workshop always has an intelligent approach.

Think subconsciously about the character. What kids see could support trust, self-esteem, competence. It can also give them a sense of pause to question the world.

What techniques do you use in your characters to convey emotion (happy, sad, angry, afraid, disgusted and surprised)?
I loved acting and stand-up comedy as a teen and in college. It strongly influences my work today. The connection works for character development, designing body poses, body language and facial expressions.

Do you consider how kids relate to humor to portray funny in your designs?
Humor is essential to kids. To best convey funny with no words, I look at slapstick comedy. Charlie Chaplin expresses humor with simple poses and body movement. Is there a particular gesture or way of looking? The Three Stooges and Marx Brothers use exaggeration. Villains are easily recognizable.

Do you approach characters and visual storytelling differently across media? Markets? If so, how?
Education is the most rigid. Curricular guidelines define what is acceptable.

Apps are interesting because of the small screen, and they are alive and moving. Simple artwork, bold shapes that promote clear communication is better than detailed line art.

In books, I can add subtleties. Readers put in their imaginations so don't need to detail it or be as expressive, just hints.

Entertainment: Particularly with animation, more is better. Bigger, flashier and tons of little details.

Edutainment: You always need to bring something new to keep it interesting. On *Sesame Street*, simple is better. In *Maya and Miguel* and *Dragon Tales*, diversity themes keep it interesting. Children enjoy learning about the customs.

Style Guides

Style guides are the visual representation of a brand, entertainment product or aesthetic of a company. They are used to keep the identity, characters and stories consistent across platforms. In a style guide, include:

- Character turnarounds
- Expressions
- Poses
- Synopsis
- Written descriptions
- Who do they resemble that is familiar?
- Design elements – borders – frames – patterns
- Colors of character
- Personalities and color

- Most appropriate consumer products (standard – paper – bedding – clothing – toys)
- Packaging color breakdown
- What colors will sell well – try to get away from stereotypes such as pink and blue

- Phrases they might say
- Type treatment[32]

Formats/Platforms

This section takes a look at different formats for media and their role in children's lives.

Books and e-Books

When reading books either physically or digitally, kids learn about the world outside of their immediate connections. Through stories, they connect to popular characters. They fantasize, consume creativity and use their imaginations in worlds that don't but could maybe someday exist. As children read, they develop emotions and reading ability, and question morals.

Parents are told to start reading to babies as soon as they are born. Even though they do not understand stories, the sounds of the words and the images on the page help them to pick up language and reading later on. When children are toddlers, preschoolers and early grade schoolers, the interaction between parent and child over a book or e-reader can be a strong interactive bonding experience as well as strengthening learning. Parents are in charge when reading to non-readers. An advantage of e-books is that kids can control the experience. As children get older, rates of reading among all children, especially teens, have fallen precipitously in recent years. Reading is the least common activity on multi-purpose tablets or small devices among all the options children have. They are more likely to have used a smartphone, iPod Touch, or tablet to watch a video, play games and use apps than they are to have read books on them. Many children multi-task while reading, such as having music or television on in the background, or flipping web pages as they are reading.[33]

It is important for designers and storytellers to think about how technology has changed and will continue to change the way children read. Children still love sitting down with "books" and seeing the story unfold. However, today, they also may interact with words and images on a screen, and the story may be read aloud to them. Electronic devices, including "learning" tools, offer multimedia experiences and blur the line between books and toys. Through digital technology and interaction with stories, the nature and depth of the reading experience for children and youth for both entertainment and education can be enhanced. Personalized experiences and non-linear narratives can help with learning to read, expand reading comprehension and change how we view stories.

Children's Book Formats and Genres

The book business is divided into two major markets: the trade and the educational market. A trade book is created for the general public audience, while the education market might be for libraries, used in the classroom or textbooks. There are also limited-edition books, mass-market paperbacks and the gift market. Traditionally, authors write books and a proposal is submitted to a publisher by the author or through an agent, depending on the publisher's requirements. If you are working on a book project, it is helpful to know which market you are creating the book for, what the traditional formats are, and how to best position it within a publisher's brand. There are also many opportunities today for self-publishing. Here is a list of different formats for children's books. Each has a diversity of genres, specific market, page and word count. Some of the categories share the same target user.

Children's Book Formats

- Baby books
- Toddler books (1–3 years)
- Picture books (4–8 years)
- Early picture books (4–6 years)
- Easy readers (easy-to-read) 6–8 years
- Transition books (early chapter books) (6–9 years)
- Chapter books (7–10 years)
- Middle grade (8–12 years)
- Young adult (12 years and up)
- A new age category (10–14) is emerging, especially with young adult nonfiction.[34]

Children's Book Genres

- Contemporary realistic fiction
- Historical fiction
- Nonfiction or information books
- Biography
- Traditional literature
- Poetry
- Modern fantasy
- Science fiction

Book Awards and Lists

I love to include children's books in my research to learn how authors have communicated the topic I am designing for to children. If you are looking for acclaimed examples of children's literature, read the books on these award lists:

- The Newbery Medal
- The Caldecott Medal
- The *New York Times* Children's Bestseller list
- Publishers' Weekly

Creating Engaging Stories and Books for Children

Advice from Marla Frazee, children's book author and illustrator

Marla Frazee was awarded a Caldecott Honor for *All the World* and *A Couple of Boys Have the Best Week Ever*, and the *Boston Globe* Horn Book Award for Picture Book for *The Farmer and the Clown*. She is the author-illustrator of *Roller Coaster*, *Walk On!*, *Santa Claus the World's Number One Toy Expert*, *The Boss Baby, and Boot & Shoe*, and the illustrator of *It Takes a Village* by Hillary Clinton, as well as many other books.

How do you get yourself in the head of the viewers/ readers/kids when you are creating your books? What are some of the things you think about?

I have strong memories of my feelings from when I was about 3 until 10 years of age. I remember trying to overcome the fear of doing things for the first time, all sorts of struggles with friends, and not being brave enough to ask the adults around me for what I needed. I know that kids (and adults for that matter) will connect to those feelings because we all had them.

Picture books are an emotional medium, and they should conjure up some intense emotion, whether it be funny, serious or downright silly. The hope is that a picture book will take the reader on an emotional journey.

What are some techniques that you use to develop the personality in a character visually?

Time in is the big thing. I always start out with a character in a somewhat stereotypical generic fashion. As I think about and draw that character, I get to know who they are. They may remind me of someone I know. I build a backstory in my head about them; their quirks, background and unique personality. Then, hopefully, they become who they are meant to be.

What are some things you think about when adding humor that kids will enjoy to your stories?

Humor is one of the hardest things. With *Boss Baby*, from the moment I got the idea for the book, I knew it was funny. But when I begin to write and illustrate it, I kept killing the humor. It took many drafts and revisions to make it work. If you are going for funny, you have to keep it frothy. If you load it down too much, you kill all of the bubbles.

What do you think about when developing a visual style for a specific story?

My style is like my personality; sometimes I wish I could change it! But it doesn't work that way. It always emerges on its own accord. When I approach a new project, I think about the story that needs to be told. Everything is in

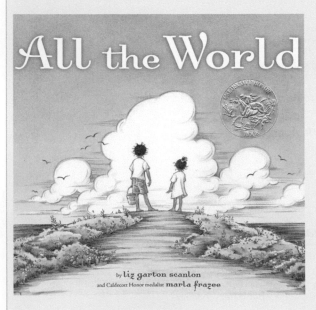

◄ Figure 9.7

All the World.
Source: Marla Frazee (used with permission).

Comic Books and Graphic Novels

Comic books are short stories with varying issues that tell a story over many issues. Graphic novels are much longer and tend to be much more complex. They have their storylines wrapped up in one or two books. I asked author and animation historian Michael Mallory how comics have changed over the last 70 years from the past to today. He explained:

> The single biggest change is that the medium became respectable. Now it is a genuine art form. It is a driving force of so much other entertainment and merchandising. In 1954 a Congressional committee tried to shut down comics for contributing to juvenile delinquency. Comics had to be approved and made safe again. The 90s ended the rating system, and now it is monitored by the companies for language, violence, sex and smoking. Most parents of kids who like comics don't go to comic book stores, so they don't know what kids are buying.

We discussed why there seems to be more graphic violence in comics today and the same comic characters have got darker and meaner. He explained: "The industry needs to keep topping itself. They are looking for something new, so they push it further and further."[36]

We also discussed the role the art style/writing style plays in the comic book development. Michael Mallory explained:

> The character is whom the reader follows so the writing is very important. The reader identifies with them on some level. You have to have a story and style of writing that fits. It is a visual medium. The artwork and character have to move. In legendary comic book artist Jack Kirby's work, nobody stood still. The action and motion need to propel the reader from panel to panel.[37]

There are many submarkets in the comic book industry, even with the same character property. Michael Mallory explained:

There are tighter niches in demographics and age divisions. I went to a comic book store and found 18 variations of Batman. The publishers are making non-violent cartoony versions of the same comics for younger kids to bring in younger readers. They are rounded, old-style Disney.

We then discussed how comic books help kids with their social and emotional development. Michael Mallory specified:

People connect to and identify with the characters. Kids understand the character dynamics of the group. Seeing them interact with each other models ways for kids to interact. For example, the Fantastic 4 argue and get mad. Kids could connect to that. They always come back and are welcomed and accepted. Characters and stories give kids a starting point for their imagination. What kid didn't enjoy running around with a towel wrapped around them like a cape and pretending to be Superman?[38]

Why People of All Ages Like Superman

Another example is how Superman with deep character can appeal to different developmental stages for different reasons, from a child to an adult.

- Preschoolers like his superpower and that he is above pain

- Early grade schoolers like him because he is strong
- Preteens like his secret identity
- Teens like that he is a protector of the community and the fictional history
- Adults like his innocence, that he goes through corruption and comes out clean[39]

◀ Figure 9.8

The comic book format engages visual readers.
Source: author photo.

Music

Exposing children to music during early development helps them learn the sounds and meanings of words, patterns and rhythms. They learn to play with language and humor and silliness. Dancing to music helps children build motor skills while allowing them to practice self-expression and strengthens memory skills. It can be used to enhance learning other subjects outside of music. It also provides kids with joy.

Although children still learn classic songs such as "Wheels on the Bus" and "Old McDonald," children's music has become sophisticated for the younger set. You will find Hollywood bands in every genre from bluegrass to metal playing daytime gigs at malls or concert venues with original creative music that speaks to issues in their lives. They are also introduced to the musical tastes of their parents from a very young age, and you may hear kids at three telling you how much they love the Beatles or country music. Soundtracks to blockbuster films with top talent are always on the bestseller music lists.

Kids may choose music by an artist they identify with or one who seems to embody their generation's trials and tribulations. In some cases, teens use music as a safe way to rebel against their parents and develop their own separate identity.

Some adolescents use music to deal with loneliness and to take control of their emotional status or mood. Music also can provide a background for romance and serve as the basis for establishing relationships in diverse settings. Adolescents use music in their process of identity formation, and their music preference provides them with a means to achieve group identity and integration into the youth culture.[40]

Sometimes, violent music can be an outlet that allows teens to release their anger or frustration without getting into trouble. The key to whether the music is good or bad often depends on how the teen lets this type of music affect them. Female adolescents are more likely than male adolescents to use music to reflect their emotional state, in particular when feeling lonely or

▶ Figure 9.9

Music offers ways for kids to express and explore emotions.
Source: author photo.

"down." Male adolescents, on the other hand, are more likely to use music as a stimulant, as a way to "boost" their energy level or to create a more positive image of themselves.[41]

Children's and Family Films

Children's films are made primarily for kids – although their parents may take them to the movies or view them with them at home. Most are non-offensive, wholesome and entertaining. Family films are made for a broader audience to include both children and adults. Genres of kids' and family films are live action, animation, musicals, fantasy, adventure, comedy, drama, and based on a book. When creating a movie for kids, think about content that is developmentally appropriate for a broad range of children.

Many parents look for movies of high quality. Does the storyline make sense? Are the characters well rounded? Is the dialogue believable? Parents, especially of younger children, look for educational content that teaches through essential messages and role models. Scenes with anxiety-stirring action such as chases or a looming disaster are particularly difficult for some children. Merchandising and children's movies are strongly linked.

Movie ratings provide parents with advance information about the content of movies to help them determine what's appropriate for their children. Some parents bring kids to NC-17 movies without thinking about it. Others take them very seriously. In recent years, adult content has moved into the PG-13 category. Creators of content for children and families need to consider the movie ratings.[42]

Live Action or Animation

With the technology available today, it is possible to shoot a live action movie with imaginary elements that are produced in special effects. It is also possible to write an animated movie that is close to realism. So how do you decide? Animation tends to favor the needs of stories about fantastical creatures (talking animals, robots, space aliens) and spectacle, and live action tends to favor the needs of stories that depend on seeing the faces of real people. If you are unsure whether you would like your story to be animated or live action, watch some movies that are made in both genres: *101 Dalmatians*, *Jungle Book*, *Robin Hood*. What does each movie have to offer? How does that compare with what you would like to achieve?[43]

Creating Engaging Family and Children's Movies

Claudia Puig, president, Los Angeles Film Critics Association
Claudia is a film critic, writer and journalist, and film consultant. She is a commentator for National Public Radio (NPR). As a bilingual Latina, Claudia is passionate about inclusion and a frequent speaker on diversity issues.

What can designers for kids and youth learn from watching films for their research?
They can learn from what engages children, what draws their attention, makes them laugh, gives them joy and a sense of comfort. For instance, kids loved the Minions in *Despicable Me*. Those creatures are brightly

colored, simply designed, have laughably odd voices and say funny things. They run around, pop in and out, and interject with comments, and kids are drawn to them. Why? Largely because they're like children in their energy and funny, non-sequitur comments, but are funnier and brighter to look at and never intimidating or scary.

There is also a lot to be learned about environments and design from watching children watch movies and their reaction to humor, character design and story development.

What do you think kids take away from movies at different ages?

- Toddlers and preschoolers: Toddlers take away some pretty basic information – good, bad, scary, safe, etc. Preschoolers might take away information that's a bit more defined – pretty, ugly, animal versus human versus plant life. They're a bit more aware of and interested in the world around them, but a lot of them are fixated on something favorite – for example, princesses, dinosaurs, kitties, trains, etc.
- Younger kids: Younger kids can take away messages and themes as well as action and storylines. They can discern the difference between a truly good guy versus a person pretending to be a good guy or a person who's villainized but perhaps is only misunderstood. The nuance begins to grow.
- Tweens and teens: Tweens and teens take more away from movies than their younger counterparts. Tweens are focused on the ages just above theirs, to learn how to act and what life will be like for them a few years down the road. They don't want to seem "babyish." They'd rather seem more mature. For teens, things begin to change dramatically. They're preparing to separate and be independent, and they're more rebellious and focused on others their own age above all. Movies made for them need to tap into their sense of being misunderstood by parents and authority figures, their growing sense of adulthood. Their hormones are kicking in, and they're thinking not just about romance and sex, but about where they fit into their social structure – i.e., school, clubs, athletics and music.

What roles do movies play in kids' lives?

Movies can play a significant role in kids' lives. Early on, they enjoy imaginative play and fantasy. As they get older, they look to movies to help them establish and reinforce their sense of identity. This can vary among kids. Some are much more connected to movies and TV. Others might connect more with characters they experience through reading. But movies can overlap, such as in the case of the Harry Potter book adaptations, or the Twilight trilogy.

What role do movies that have crossed the generations play in family bonding? For example: *Star Wars.*

They play an important role and become part of family traditions and connect family members in their affection for characters, excitement over the story and specific points in its telling. They help with family cohesion and create memorable shared experiences that can sometimes get repeated in future generations.

Do you have something to add about appropriate/inappropriate content and the importance of ratings?

I think it's essential that parents think seriously about appropriate and worthwhile content for their kids, especially taking into account the maturity level and interest of their specific children. Some 6-year-olds can handle a bit more tension than others. But violence and sexualization is something to be avoided in kids' viewing, as much as possible. With the prevalence of violent video games, TV and movies, some kids can get saturated by violent fare, at a very impressionable age. The MPAA [Motion Picture Association of America] ratings are helpful guidelines, but I think it's also worth checking out other sources to determine if something is appropriate for your child. Common Sense Media (commonsensemedia.org) is a useful resource that looks more deeply at family and assesses each movie on several factors, whereas the MPAA board, which is made up mostly of parents, rates movies on a broad basis (violence, nudity, bad language, drug use, etc.) The more research a parent does, beyond the rating system, the better.

If you had your wish list of what you would like to see content creators creating movies about for young children, what would they be?

- More movies showing women in non-traditional roles and unusual jobs, and important careers.
- More diversity in family fare in all content so that children grow up with less of a sense that white people, particularly white men, are in charge of things, and women tend to be aiding and facilitating, but not creating or solving.[44]

Children's Television

Kids' programming has been around since television was invented, and design has always been a part of it. Designers can learn about creating content for television and creating products and experiences for the real world by watching and critiquing television. Behavioral scientist Susan Weinschenk suggested: "Designers can learn a lot about characters by studying *Sesame Street*. Each character has specific colors, textures, voice, size, or shape. The eyes visually grab attention." Sesame Workshop[45] is also a research organization that produces research and content along with the Joan Ganz Cooney Center[46] on child development and media usage and a helpful resource for designers to refer to.

Television today has many channels available specifically for kids of different age groups. With so much content being made, there is also a huge range in the quality of the content.

Today's television experience and storytelling are in flux as content moves off the television set onto other screen-based digital devices and, in the future, immersive environments. Some trends in children's television include cross-platform storytelling, content mobility, event-based viewing, content delivery optimization and binge watching. Andra L. Calvert, professor of psychology at Georgetown's children's digital media center, explained: "Television tells a good story. When children are clicking around on a screen whatever narrative exists is interrupted."[47]

The link between television and advertising is constantly changing due to the way it is consumed. Author and animation historian Michael Mallory explained specifically how this relates to cartoons getting a green light to begin with:

> Advertising has always been a part of it. Today shows are not known for their sponsors as they were in the past. Although today you need to partner with a fast food chain for promotion for your show or feature film. Because of this much of American animation is for kids.

- Where and how kids watch TV has shifted from the TV set to mobile devices, with on-demand programming continually increasing.
- Broadcast television is the most accessible and widely used platform for educational content among lower-income children.
- Educational TV is the most popular genre among young children aged 2–4, and entertainment television starts to increase from age 5 to 8.
- Parental attitudes towards TV heavily influence the child's use. Parents who watch a lot of television have kids who watch a lot of TV.
- Children who watched a fast-paced television cartoon performed significantly worse on the executive function tasks than children in the other two groups when controlling for child attention, age, and television exposure.[48]
- Background television disrupts children's playtime. Parents interact less with the TV on. When parents play with children, the maturity level rises.

- TV shortens play and attention span.[49]
- Negative effects of exposure to violence, inappropriate sexuality and offensive language.
- Educational content from the Discovery channel and programs like *Animal Planet* expand children's exposure to the world.

Children's Television Act and Children's Online Privacy Protection Act

The FCC's Children's Television Act was drafted to enhance television for young viewers. The Children's Online Privacy Protection Act (COPPA) is the leading authority online and tries to ensure that parents remain in control of kids' data.

Designing Children's Media

Interview with Cameron Tiede, artist, illustrator and designer, who has created art and characters for toys, entertainment companies and products.

How does the medium you are designing for affect the way you approach the work?
Every medium has its strength – use it to your advantage. For example, in entertainment design, games and interactive are limited in graphics file sizes and need to be simplified where they can load in a short timeframe. The designs reflect the process.

What are some of the most significant issues that you think designers should be using their creativity and skills for when designing for kids?
Kids have too much in their world that is dumbed down to them. They want to be challenged, they want to think, and it just needs to be done in a creative way that makes it enjoyable. Designers could be using their creativity to help teach others, especially kids, to think and be creative.

What is the show *Thinkheads* about?
It's a social commentary on the not too distant future gone wrong with consumerism and over-consumption.

It features a family that challenges the norm and plays to the idea of creating by reducing, reusing and recycling. They have fun solving problems by making things and showing others a better way.

Can you describe your brainstorming process for the character design and look and feel of the show?
I draw and draw and draw. My brain is working as I do this. I pick out interesting visuals to begin to form a narrative. The aesthetics of the show are just an extension of my art/design style. Once the show got picked up by a production company to develop it, there were a few refinements, but most of the designs stayed very close to my originals.

Anything else of importance in this project that we can learn from and bring to other projects?
Timing and luck play a role. The show didn't make it to the screen. While great design is important, many other things are critical to making it happen. The valuable lesson here is not to give up.

▲ Figure 9.10

The *Thinkheads* have fun solving problems by making things and showing others a better way.
Source: Cameron Tiede (used with permission).

Visual Storytelling for Family and Children's Media

Interview with Adriana Galvez and Russell Chong

Adriana Galvez is an animation artist whose credits include Fox's *King of the Hill* and *The Simpsons*. Russell G. Chong is an artist/conceptual designer whose work includes Lucasfilm Animation on *The Clone Wars* and at Warner Bros. on *Batman the Animated Series*.

How is visual storytelling different across media?

Adriana: Today the characters and project are developed as brands with sequels already in mind. Everything now works across multiple platforms, mobile devices, TV, merchandise, toys. Platforms like YouTube and virtual reality have opened up the possibilities and even created a new Wild West for short films. VR and 360, in particular, are requiring a new mindset in storyboarding and directing these mediums that envelop the player in a moving environment.

Russell: Today animation is taken more seriously across the board. The special effects in movies; visual effects artists make dreams and nightmares for adults, and for younger people, animation has become super realistic in a cute, cartoony, sophisticated way. Right now the best movies are on TV, because you have more time to develop characters and situations as opposed to movies that only have a short format.

Russell, the environments you create take people to another world and allow them to live through their fantasies. Why is this important to kids? How does the fantasy component play into your approach to design?

Russell: I want to bring the viewer a sense of wonder by creating places they haven't seen or felt before and not spell things out. I think, what I have not seen? What needs to be seen? What needs to be felt? Although you could create something so new that it could alienate the audience. There is a fine line between alienating the audience or creating something they are comfortable and familiar with.

Adriana, you have worked on some of the most popular animated series, including *King of the Hill* and *The Simpsons*. What is it about these stories and humor in these shows that make it so appealing to kids and adults?

The creators and writers are very clever at layering and interweaving the storylines for kids and adults with different levels of humor, simultaneously. *King of the Hill* is very heartwarming, although still with some outrageously funny situations. It makes people feel good; it has heart. *Simpsons* is more whimsical and crazy. It takes more advantage of the medium with its surreal couch gags, political satire and themes. But also, in the end, it also has heart and is also family oriented. The work on these shows is a collaboration, from the actors and artists all working off the scripts. They are amazing feats of teamwork.

Russell, what was your favorite part of being a *Star Wars* artist?

Working within the *Star Wars* universe was always a dream of mine, and I finally got to be a part of it. The most creative aspect was coming up with new alien cultures and civilizations. All cultures have different ways of thinking, and they apply it to their architecture and technology. I got to design things for fictional beings and then extrapolated what this would look like. And of course, designing new *Star Wars* spaceships is always great fun!

Adriana, what role does color play in animation? Are there other considerations?

The color palette conveys the show's personality and sense of time and place. The art director or color director needs to think it all through with the creator and director following the series and the individual episode's needs. The color will influence and guide the emotions of the viewer, setting the mise-en-scène. The characters' personality can be influenced by the colors it wears or colors of the environment it lives in. There are also technical aspects of making sure the characters don't blend into the background. There are also special effects and dramatic sequences where color plays an active role. Like if the characters go to a black light party or a spooky noir sequence, or maybe a flashback sequence to a place in the past.

What else would you like creatives to think about when designing for kids?

Russell: Design with a way to inspire wonder. Wonder brings inspiration, curiosity and evolution. The sense of wonder and curiosity isn't there anymore for kids, and we need it. Don't stay within the confines and limits of what is already seen or known. Show people something new and different. Use the imagination!

Adriana: Inspire kids to get excited and help them want to explore and care for our world and be playful. Get them to participate, beyond being passive consumers. When you interact with something, you learn from it. Inspire them to make things and tackle creativity without fear.[50]

Digital Devices

Computers, tablets and online connectivity are becoming increasingly necessary to ensure that educational opportunity is open to all children, regardless of their economic status. Using digital tools for school assignments, grades, tutorials, researching papers and writing essays is now a key component of children's education.

The internet exposes children to valuable new ideas and information. Computers/mobile devices offer children new/interesting means of self-expression. Thanks to the wealth of information available on the internet, teachers note that students can find information for class assignments more easily now than in the past. They sometimes stumble across information in online videos or on websites that sparks their curiosity or turns out to be useful in classroom discussions. The web offers students a chance to experience social and cultural issues from a variety of perspectives. Children are encouraged to use tech tools to create projects and collaborate in groups regionally, nationally and internationally. Games and educational content allow children to develop spatial skills.[51]

Positive activities kids use digital tools for:

- Doing homework
- Playing educational games
- Looking up things they are interested in
- Making art or music or doing something creative
- Working with other students
- Connecting with teachers and friends
- Writing stories or blogs

Communications – Social Media

We are just now starting to see the first group of teenagers who have gone through their entire adolescence with many different platforms to interact with, including social media. Social media is not a way of communicating, but communication itself, and by using these tools, they are rewriting the rules of engagement. They seek involvement in a brand's development (and vice versa) and expect invisible digital networks that provide them with access to everything they need, wherever they are.[52]

The benefits of social media include:

- Allows people to keep/create connections you would never otherwise be able to because of distance/timing
- It is easy to find communities of people with similar interests (and which would otherwise be very difficult)
- With apps such as Snapchat, people can set up posts to disappear, and they can feel free to be silly and themselves without fear of it living forever somewhere

Communications media – whether pen and paper, phone calls, Facebook or Twitter – can exacerbate or alleviate the perils of teenage life. Social media facilitates new ways of expressing typical adolescent developmental needs, such as the need for connection and validation from peer groups. Today, it allows opening private exchanges for an entire school to see, adding photos and videos to words, allowing an entire community the chance to comment on what is seen or heard or said online, and maintaining a permanent record of all those interactions.[53] Time spent with media could subtract from face-to-face time, so heavy media users would forfeit opportunities to deepen empathy by conversing and learning from human facial and vocal cues. Does it make them feel more connected or more isolated? Better about themselves, or more depressed?

Social Psychological Theories Relating to the Socializing Influence of Media

Social Comparison Theory
People have an innate motivation to evaluate themselves through comparison with others. Upward social comparison can motivate people to improve themselves.[54]

Social Cognitive (Learning) Theory
People learn behavior and values by modeling others, noting what is deemed socially acceptable. This can happen by observing people in the real world or through mass media and social media.[55]

Cultivation Theory
Repeated exposure to consistent themes (in television) can lead viewers to internalize those perspectives and accept media portrayals as representations of reality.[56]

Super-Peer Theory
The media can act as a powerful peer, making risky behaviors and aesthetics seem normative.[57]

Digital Games

Kids play games for educational purposes, for entertainment and for creativity (drawing, making music or making videos). There are many apps based on TV or movie characters or branded content. Many children have access to a handheld game player or smartphone. Older children spend more time with digital games than younger children do (under 8). Boys are more likely than girls to play console video games. Gaming and learning applications equip children to pay attention to details in order to complete activities. The preoccupation many children have can mean they are focusing on simultaneous activities to achieve the goals of the game. Many apps and computer activities are created to encourage children to complete a level to unlock the next level. Using goal-oriented gamified learning is becoming more popular in and out of school because of its ability to motivate and engage children and youth. Informality is at the center of it. With play spaced learning, kids don't need to go to any place at any time to have a learning experience – there are more between places to learn.

Genres of Games

- Action games
- Adventure
- Action adventure games
- Role-playing
- Strategy games
- Multiplayer games
- Casual games
- Educational games

Advice from a Play Designer and Researcher from Toca Boca

Interview with Chris Lindgren, kids' play and research manager, and Petter Karlsson, play designer

Toca Boca, located in Stockholm, Sweden, creates digital toys and everyday products that are filled with fun and silliness that kids from any corner of the world can instantly relate to. Through research and testing, apps and products are designed from their perspective. There are no rules, no boundaries and no "right" or "wrong" way to do anything.

How does the Toca Life apps series fit into a child's life?

The Toca Life apps are role-playing games where children play with the everyday life. Occasionally out of the ordinary things will happen such as going on vacation, to the doctor, or a new sibling will arrive. The apps are like a play set or dollhouse where they have characters, props and environments that kids get to experiment with. It is multi-touch, so they can play on their own or with others.

The apps put the child in the director's chair, where they are in control of how the story plays out and act out different roles. They are open, and a child can expand and test limits if they want to that may not be appropriate at home or in school. They play with reality.

Can you give a few examples of what you consider in child development in the design of this app series?

Earlier we worked with the age span 3–9-year-olds. Our new apps are focusing on 6–9-year-olds. For the younger kids who are just learning to process experiences, the apps are more straightforward and repetitive. The apps for the older kids have more complicated environments and interactions. They use more references to youth popular culture. They all allow the children to express themselves and be creative.

What type of testing do you do with kids, and how does it affect the development of the product?

We have two main types of testing that align with the milestones on the project. With every milestone, we test and evaluate. We film them during the test and write a summary afterward. If we have questions for the kids, we ask those outside of the testing session.

Exploratory testing allows us to better understand and figure out what is interesting about the theme we are working with. We observe kids playing with other toys or have a co-creation session around the theme. For example, if we are doing an app about pets, we would want to find out what is attractive to kids about pets. Sometimes the testing confirms what we already know; other times we are surprised. When we were testing the topic of vacation, we found out that family is significant on vacation and learned a lot about different kinds of families.

Later on, in our play observations, we test new mechanics, features, themes and user experience. The kids may be playing with a prototype that showcases a new game mechanic or an almost complete prototype or a complete prototype. We may have two kids next to each other, and by sharing the tablet they create a conversation, and we find out what they are thinking. We want to know where they go first, what is intuitive, and what are the unexpected things they do. Most importantly, we want to know if it is fun.

What do you see as the benefits for children who play with your apps?

It is hard for kids to be in control of their world. In a digital world, they are open, free and feel powerful.

We design in things that are not as easy to play with or impossible to play in the real world and use humor. So, for example, they can throw everything in the scene in the garbage can, they can flush pets down the toilet, or they can put 20 hats on their head.

How do you go about story and character development in your apps?
We have characters and environments that we set up the stage and use prompts, and the kids create. We have a very open environment, and everyone on the team contributes and thinks about how to make it fun. We think about how different kids would play with it, from the ones who just want to create everyday storytelling to playing in more extreme exploring the boundaries of the system. Everyone adds to the suggestions for the features, jokes, artwork, and it organically evolves. Big ideas will be discussed for final decisions.

The characters are all designed to be diverse, and we have a structure that we use to make sure we have included all skin, body types, genders, cultures and special needs. Here are our seven aspects of diversity:

Physical characteristics
Cultural characteristics
Functional diversity
Body shape
Gender
Family structure
Age[58]

▶ Figure 9.11

The Toca Boca apps spark kids' imagination with open-ended play, and they can play seriously or be silly.
Source: Toca Boca (used with permission).

Gamification of Everyday Life

Today, designers in industries outside of gaming are now taking the science, fun and addicting elements behind what makes gaming so compelling and applying them to real-world activities. Gaming design theories are being used to engage people to play more, learn more and be healthier. They are also being used in marketing to engage people with brands in non-traditional ways to ultimately buy more.

Embedding games into everyday activities encourages kids to spend more time on things that they have to do but don't necessarily want to do, such as homework and chores. Through studying game mechanics, designers can use techniques for gamification to help kids develop compared with what many parents are concerned about, which is wasting time on a video game.

By looking at the goals and objectives and a clear pathway to achieve them, kids better understand why they are doing the activity and feel the accomplishment that leveling up gives them. Why do kids/youth like games? They are challenging, they use creativity, they get to socialize with friends. They have a unique narrative, they are rewarding, and the amount of risk involved is manageable.

Points, badges and leaderboards are the most basic examples of gamification. Many people in design find these too simplistic, and their inclusion is not what truly satisfies what makes us like games. According to Yu-kai Chou, core drivers include:

- Social influence: Kids feel like you are part of something bigger
- Sense of accomplishment by improving and leveling up
- Empowerment of creativity and feedback
- Ownership and possession – improve it, protect it, get more
- Scarcity and impatience – you want it because you can't have it
- Unpredictability and curiosity[59]

Interaction designer Brian Boyl explains:

> The misconception of many designers about gamification is that a product interface or experience needs to be turned into a video game. Most of the time the results usually stand out as being contrived. Making a gamified experience that looks, acts and feels like a video game is not the correct way to think about it. What they need to do is understand the role of their product and adapt it the context of their project.
>
> For example, Strava, a website and mobile app used to track athletic activity such as cycling and running via GPS, uses gamification to engage its user base. They could have approached it as a video game, but they didn't. Instead, the feedback, graphics, use of gamification is executed in a way that makes sense in the type of product that they are offering – performance and competitive athletic system. The game is built around segments that you bike or run. The interface looks like performance statistics – not a video game – next to your Personal Record and times on the right just where your place is. It is very clearly a competitive game but not ....[60]

Digital in the Playground

Interview with Dr. Nis Bojin about Biba

Nis is a designer, strategist and writer in games and new media. Biba is a series of mobile games for parents to use with their children on the playground.

How did you put together your research strategy?

First, we wanted to get a sense of what Biba could offer if there was a product–market fit, and if so, if the product would fulfill its objectives. Our initial research strategy moving forward consisted of the following questions and associated methods:

Phase 0: Is there indeed a problem to be addressed and what is it? (Literature review/interviews)

Phase 1: Could Biba work in addressing this problem? Was there precedent/research/theory to back up the efficacy of such a product? (Literature review)

Phase 2: Did parents want it/would they use it? (Focus groups and ethnographic research)

Phase 3: Would parents and kids enjoy using it? (User testing–participant observation–survey)

Phase 4: Was the product easy to use? (User testing–participant observation–survey)

Phase 5: Was the product working in achieving its aim of getting kids more active? (User testing via quantitative measures w/control group)

Phase 6: Did the product work in the wild concerning achieving the intended outcomes? (Embedded longitudinal trials with families w/control group)

During this time we collaborated (and continue to collaborate) with Child Psychology faculty at Simon Fraser University to ensure we were approaching each of these research questions as rigorously and impartially as possible.

▶ Figure 9.12

Mom and daughter playing with Biba in a playground. Source: Biba (used with permission).

Did you use co-design/participatory design research methods in creating the project?

We needed to iterate quickly, so we committed to rapid prototyping assisted with frequent family user testing, but did not go the participatory design route. From this process, we concluded to focus on "refereed" play; the caregiver is leading a play session with their child/children, holding the screen-based device at all times. Also in the Biba product was the notion of "embodied play," a child is using their body during gameplay, and we invite this through the concept of "incidental activity." Put plainly, this means that a game's premise naturally asks a child for something that would require physical movement, and thus, the physical engagement is entirely natural and intuitive for a child.

Your product is for 3–9-year-olds with a broad age group of different developmental needs. How did you design the content to cater to each age group?

Preschool 3–5

Children of preschool age are already eager to explore their physicality in a playground space without requiring much prompting. So we decided to offer an experience for the parent that already aligned with what they were doing on the playground with their child and their digital device. Our preschool product was thus Biba GoSeek: a hide-and-go-seek variant for the playground in which the parent chooses a themed frame and then takes a picture of their child and customizes photos when they 'find' them. Given that parents are typically equipped with screen-based devices on the playground and already use these devices to take pictures, we thought to try and simply align ourselves with these behaviors for parents of 3–4-year-olds.

Early elementary 6–8

Our aim here was to maintain an intrinsic-first approach to our designs, such that the role-play of the premise presented is the appealing aspect of gameplay. We were highly cognizant of the fact that associating extrinsic rewards with a gameplay experience such as this could ultimately result in over-justification effects, and we wanted to avoid this at all costs. Biba, and the physical activity it incites, is designed to be fun for its own sake.

Later elementary 8–10

This is where kids tend to become a lot more concerned with physical mastery and testing themselves against (and with!) their friends – which means sociality comes to the fore here as well. So we wanted to cultivate physically testing forms of play that permitted peer benchmarking either collaboratively or competitively in fostering the development of a child's physical competencies. In achieving this, we designed games that privileged measurable outcomes, e.g., a driving game where kids test for best times; a relay game where pairs of kids collaborate to last the greatest number of rounds possible; and a baseball game where kids make choices about what pitches or hits to pick in outsmarting their opponent, followed by a challenge to use equipment and tag back in to "home base" before the other. Scores and benchmarks can then be compared and bested on repeat visits, and if parents are using our Play Tracker app, they can create activity profiles that let them track the specific physical accomplishments of their children.

What do you see as the future of analog/digital play in the built environment in the next 15 years?

I think technology will continue to embed itself in everything we interact with; the things we wear, ride and use. If we thought today's children were digitally literate, I don't think we've seen anything yet. So I do foresee commonplace wearables with non-invasive sensors in things like clothing and shoes that can be used to interface with basic near-field type communications and so on. The big breakthrough aspect of this, I think, will be a stronger relationship between the outcomes of outdoor play and activity and indoor, digital play.[61]

Notes

1   Common Sense Media 2015
2   Common Sense Media *Reviews*
3   Common Sense Media 2013
4   Guernsey, Lisa 2007
5   Gilutz, Shuli 2018
6   Gilutz, Shuli 2018
7   Archer, Karin 2017
8   American Optometric Association
9   Guernsey, Lisa 2007
10 Guernsey, Lisa 2007
11 Guernsey, Lisa 2007
12 Guernsey, Lisa 2007
13 Guernsey, Lisa 2007
14 Stylus.com 2016
15 Yudina, Anna 2017
16 Cearley, David 2016
17 Hendrie, Maggie 2017
18 Kelly, Annie 2013
19 Prieto, Mariana 2016
20 Weinschenk, Susan M. 2011
21 Haden, C.A.; Haine, R.A.; Fivush, R. 1997
22 Weinschenk, Susan (interview October 18 2017)
23 Day, Douglas 2017
24 Scribendi
25 Jones, Gerard 2003

26   Sklar, Howard 2009
27   Guernsey, Lisa 2007
28   Linn, Susan 2004
29   Rich, Michael 2008
30   Pixar
31   St. James, Alena 2017
32   Day, Douglas 2017
33   Common Sense Media May 12 2014
34   Backes, Laura
35   Frazee, Marla 2017
36   Mallory, Michael 2017
37   Mallory, Michael 2017
38   Mallory, Michael 2017
39   Jones,, Gerard 2003
40   American Academy of Pediatrics 2009
41   American Academy of Pediatrics Council on Communications and Media
42   Motion Picture Association of America
43   Alcott, Todd 2008
44   Puig, Claudia 2018
45   Sesame Workshop
46   Joan Ganz Cooney Center
47   Guernsey, Lisa 2007
48   Lillard, Angeline S.; Peterson, Jennifer 2011
49   Elkind, David 1981
50   Galvez, Adriana; Chong, Russell 2017
51   Digiparenthood
52   Stylus.com
53   boyd, danah 2007
54   Festinger, Leon 1954
55   Bandura, A. (1986; 2001)
56   Gerbner, George (1994)
57   Common Sense Media 2015
58   Lindgren, Chris; Karlsson, Petter 2018
59   Chou, Yu-kai 2015
60   Boyl, Brian 2017
61   Bojin, Nis 2017

Bibliography

Adawaiah Dzulkifi, Mariam; Faiz Mustafar, Muhammad (2013) *The Influence of Colour on Memory Performance*. Retrieved from www.ncbi.nlm.nih.gov/pmc/articles/PMC3743993/.

Age of Montessori (2018) ageofmontessori.org. *Your Child's Developmental "Windows of Opportunity"*. Retrieved from http://ageofmontessori.org/your-childs-developmental-windows-of-opportunity/.

Alcott, Todd (2008) *Screenwriting 101: Animation vs. Live Action*. Retrieved from www.toddalcott.com/screenwriting-101-animation-vs-live-action.html.

Allen, Nathan (October 17, 2017) Interview.

Allen, Scott (March 30, 2017) Interview.

Amatullo, Mariana (November 29, 2016) Interview.

American Academy of Pediatrics *Benefits of Breastfeeding* (2017) Retrieved from www.aap.org/en-us/advocacy-and-policy/aap-health-initiatives/Breastfeeding/Pages/Benefits-of-Breastfeeding.aspx.

American Academy of Pediatrics (November 2009) *Impact of Music, Music Lyrics, and Music Videos on Children and Youth*. Retrieved from http://pediatrics.aappublications.org/content/124/5/1488.full.pdf.

American Association for the Advancement of Science (2018) *Growth Stages 1: Infancy and Early Childhood*. Retrieved from http://sciencenetlinks.com/lessons/growth-stages-1-infancy-and-early-childhood/.

American Montessori Society (2017) *Introduction to the Montessori Method*. Retrieved from https://amshq.org/Montessori-Education/Introduction-to-Montessori.

American Optometric Association (2017) *Infant Vision: Birth to 24 Months of Age*. Retrieved from www.aoa.org/patients-and-public/good-vision-throughout-life/childrens-vision/infant-vision-birth-to-24-months-of-age.

Amsterdam, B. (1972). Mirror Image Reactions before Age Two. *Developmental Psychobiology*, 5, 297–305. See *When the Self Emerges: Is That Me in the Mirror?* Retrieved from www.spring.org.uk/2008/05/when-self-emerges-is-that-me-in-mirror.php.

Archer, Karin (2017) *Infants, Toddlers and Mobile Technology*. Retrieved from http://scholars.wlu.ca/cgi/viewcontent.cgi?article=3031&context=etd.

Armstrong, Patricia. *Bloom's Taxonomy*. Retrieved 2001 from https://cft.vanderbilt.edu/guides-sub-pages/blooms-taxonomy/.

ArtCenter (2017) *About Design-Based Learning*. Retrieved from www.artcenter.edu/teachers/about_dbl.php.

ASCD (2017) *The Whole Child Approach*. Retrieved from http://wholechildeducation.org.

Association of Science-Technology Centers Urban Library Council (2014) *Learning Labs in Libraries and Museums*. Retrieved from www.imls.gov/sites/default/files/publications/documents/learninglabsreport_0.pdf.

Association of Waldorf Schools of North America (2017) *Waldorf Education: An Introduction*. Retrieved from https://waldorfeducation.org/waldorf_education.

Atkinson, Mark (February 3, 2017) Interview.

Backes, Laura (2017) *Understanding Children's Book Genres*. Retrieved from www.right-writing.com/genres.html.

Bandura, A. (1986) *Social Foundations of Thought and Action: A Social Cognitive Theory*. Prentice Hall.

Bandura, A. (2001) Social Cognitive Theory of Mass Communication. *Media Psychology*, 3, 265–298.

Becerra, Liliana (January 24 2017) Interview.

Beck, Melinda (November 27 2017) Interview.

Beckwith, Jay (April 22 2017) Interview.

Beisert, Fridolin (October 24 2017) Interview.

Bellali, Safir (March 17 2017) Interview.

Belton, Teresa (September 23 2016) *How Kids Can Benefit from Boredom*. Retrieved from https://theconversation.com/how-kids-can-benefit-from-boredom-65596.

Bennett, Katherine (February 28 2017) Interview.

Benoit, Diane (2004) *Infant–Parent Attachment: Definition, Types, Antecedents, Measurement and Outcome*. Retrieved from www.ncbi.nlm.nih.gov/pmc/articles/PMC2724160/,

Benyamin, Beben (May 22 2015) cited in *Nature vs. Nurture Debate: 50-Year Twin Study Proves It Takes Two to Determine Human Traits*. Retrieved from www.medicaldaily.com/nature-vs-nurture-debate-50-year-twin-study-proves-it-takes-two-determine-human-334686.

Berger, Warren (2016) *A More Beautiful Question: The Power of Inquiry to Spark Breakthrough Ideas*. Bloomsbury.

Berk, Jed (November 16 2017) Interview.

Beststart.org (2017) *Supporting Development*. Retrieved from www.beststart.org/OnTrack_English/2-support.html.

Bethune, Kevin (2017) *The 4 Superpowers of Design*. Retrieved from www.ted.com/talks/kevin_bethune_the_4_superpowers_of_design.

Bethune, Kevin (February 5 2018) Interview.

Bingham Newman, Ann (October 21 2017) Interview.

Bojin, Nis (April 19 2017) Interview.

Borysiewicz, Ania (April 19 2017) Interview.

Bostain, Penny (April 27 2017) Interview.

boyd, danah (2007) *Why Youth (Heart) Social Network Sites: The Role of Networked Publics in Teenage Social Life*. MIT Press. Retrieved from www.danah.org/papers/WhyYouthHeart.pdf.

Boyl, Brian (November 15 2017) Interview.

Bradley, Heather (February 16 2015) *Design Funny*. How Books. Retrieved from www.howdesign.com/design-books/3-serious-reasons-design-funny/.

Brown, Stuart (2017) US Coalition of Play Conference on the Value of Play: Where Design Meets Play, April 2–5 2017.

Brucken, Carolyn; Cooper, Alban; Wilson, Sarah (October 2 2017) *Play Exhibit at the Autry Museum of the American West*. Interview.

Bruno, Oliveiero; Novelli, Luana (2017) Retrieved from www.ncbi.nlm.nih.gov/pmc/articles/PMC3217667.

Burns, Will (August 7 2017) *Should Education Focus Less on the Creative Arts, More on the Art of Creativity?* Retrieved from www.forbes.com/sites/willburns/2017/08/07/should-education-focus-less-on-the-creative-arts-more-on-the-art-of-creativity/#2dd5a07c5b12.

Bushnell, Brent (January 2017) Interview.

Carey, Susan (1986) *Cognitive Science and Science Education*. Retrieved from http://edci670.pbworks.com/w/file/fetch/59138742/Carey_1986.pdf.

Carr-Gregg, Michael (2017) *Five Greatest Challenges Facing Parents of Teens Today*. Retrieved from www.theparentingplace.com/behaviour-and-discpline/five-greatest-challenges-facing-parents-of-teens-today/.

Carter, Christine (September 16 2008) *7 Ways to Foster Creativity in Your Kids.* Retrieved from https://greatergood.berkeley.edu/article/item/7_ways_to_foster_creativity_in_your_kids.

Cearley, David (January 15 2016) *Top 10 Technology Trends for 2016.* Retrieved from www.forbes.com/sites/gartnergroup/2016/01/15/top-10-technology-trends-for-2016/#10ca28c05ae9.

Center for Green Schools (2017) Retrieved from http://centerforgreenschools.org/.

Center for Parenting Education (2017) *Benefits of Chores.* Retrieved from https://centerforparentingeducation.org/library-of-articles/responsibility-and-chores/part-i-benefits-of-chores/.

Center on the Developing Child (2017) Harvard University *Brain Architecture.* Retrieved from https://developingchild.harvard.edu/science/key-concepts/brain-architecture/.

Center on the Developing Child (2017) Harvard University *Executive Function & Self- Regulation.* Retrieved from https://developingchild.harvard.edu/science/key-concepts/executive-function/.

Cherry, Kendra (August 15 2017) *What Is Nature vs. Nurture?* Retrieved from www.verywellmind.com/what-is-nature-versus-nurture-2795392.

Chick, Nancy (2018) *Metacognition: Thinking about One's Thinking.* Retrieved from https://cft.vanderbilt.edu/guides-sub-pages/metacognition/.

Chick, Nancy (2018) *What Are Learning Styles?* Retrieved from https://cft.vanderbilt.edu/guides-sub-pages/learning-styles-preferences/.

Child Art. (2018) Retrieved from https://en.wikipedia.org/wiki/Child_art.

Chou, Yu-kai (March 1 2015) *Gamification and Behavioral Design.* Retrieved from http://yukaichou.com/gamification-examples/octalysis-complete-gamification-framework/.

Christensen, Tanner (May 18 2015) *The Differences between Imagination, Creativity, and Innovation.* Retrieved from http://creativesomething.net/post/119280813066/the-differences-between-imagination-creativity.

CliffsNotes (2018) *Cognitive Development Ages 7–11.* Retrieved from www.cliffsnotes.com/study-guides/psychology/development-psychology/physical-cognitive-development-age-711/cognitive-development-age-711.

Color Matters (2017) *Why Are School Buses Yellow?* Retrieved from www.colormatters.com/color-matters-for-kids/why-are-school-buses-yellow.

Common Sense Media *Reviews* (2018) Retrieved from www.commonsensemedia.org/reviews.

Common Sense Media (November 1 2012) *Children, Teens, and Entertainment Media: The View from the Classroom.* Retrieved from www.commonsensemedia.org/research/children-teens-and-entertainment-media-the-view-from-the-classroom.

Common Sense Media (2013) *Zero to Eight, Children's Media Use in America 2013.* Retrieved from www.commonsensemedia.org/research/zero-to-eight-childrens-media-use-in-america-2013.

Common Sense Media (May 12 2014) *Children, Teens, and Reading.* Retrieved from www.commonsensemedia.org/research/children-teens-and-reading.

Common Sense Media (January 21 2015) *Children, Teens, Media, and Body Image.* Retrieved from www.commonsensemedia.org/research/children-teens-media-and-body-image.

Common Sense Media (October 3 2015) *The Common Sense Census: Media Use by Tweens and Teens.* Retrieved from www.commonsensemedia.org/research/the-common-sense-census-media-use-by-tweens-and-teens.

Complexity Labs (October 17 2016) *Divergent and Convergent Thinking.* Retrieved from http://complexitylabs.io/divergent-convergent-thinking/.

Consumer Products Safety Commission (CPSC) (2017) *Art and Craft Safety Guide*. Retrieved from www.cpsc.gov/PageFiles/112284/5015.pdf.

Consumer Products Safety Commission (CPSC) (2017) *Small Parts for Toys and Children's Products Business Guidance*. Retrieved from www.cpsc.gov/Business--Manufacturing/Business-Education/Business-Guidance/Small-Parts-for-Toys-and-Childrens-Products.

Corsaro, William A. (November 29 2017) *Peer Culture*. Retrieved from www.oxfordbiblio graphies.com/view/document/obo-9780199791231/obo-9780199791231-0010.xml.

Cron, Robert (March 8 2017) Interview.

Csikszentmihalyi, Mihaly (September 1 2017) Interview.

Dam, Rikke (2017) *Raymond Loewy's The MAYA Principle*. www.interaction-design.org/literature/article/design-for-the-future-but-balance-it-with-your-users-present.

Daniels, Natasha (2016) *Childhood Fears by Age*. Retrieved from www.anxioustoddlers.com/worries-by-age/.

Davidson, Jill Camber (2009) *Wisconsin's Model Academic Standards for Nutrition*. Retrieved from https://dpi.wi.gov/sites/default/files/imce/team-nutrition/pdf/nestandards.pdf.

Davol, Leslie (February 6 2017) Interview.

Day, Douglas (April 24 2017) Interview.

Day, Kelly (April 15 2015) *11 Ways Finland's Education System Shows Us That "Less Is More"*. Retrieved from https://fillingmymap.com/2015/04/15/11-ways-finlands-education-system-shows-us-that-less-is-more/.

Deitz, Rena (April 12 2017) Interview.

Dell'Antonia, K.J. (April 3 2010) *Preschoolers Know All about Brands*. Retrieved from www.slate.com/articles/double_x/doublex/2010/04/preschoolers_know_all_about_brands.html.

del Moral Pérez, M. Esther; Guzmán Duque, Alba P.; Fernández García, L. Carlota (January 2018) *Game-Based Learning: Increasing the Logical-Mathematical, Naturalistic, and Linguistic Learning Levels of Primary School Students*. Retrieved from http://multipleintelligencesoasis.org/wp-content/uploads/2018/02/Game-Based-Learning-Increasing-the-Logical-Mathematical-Naturalistic-and-Linguistic-Learning-Levels-of-Primary-School-Students.pdf.

Demetrios, Eames (May 1 2017) Interview.

Digiparenthood (2017) *10 Benefits of Exposing Young Children to Modern Technology*. Retrieved from https://digiparenthood.wordpress.com/2013/08/23/10-benefits-of-exposing-young-children-to-modern-technology/.

Doheny, Kathleen (2017) *Autism Cases on the Rise; Reason for Increase a Mystery*. Retrieved from www.webmd.com/brain/autism/searching-for-answers/autism-rise.

Doherty, William (1999) cited in Senior, Jennifer *All Joy and No Fun*. Ecco; Reprint edition 2015.

Donaldson, O. Fred (December 5 2012) *The Positive Power of Play*. Retrieved from https://childrenscreativity.wordpress.com/2012/12/05/the-positive-power-of-play/.

Doyle, William (March 18 2016) *Why Finland Has the Best Schools*. Retrieved from www.latimes.com/opinion/op-ed/la-oe-0318-doyle-finnish-schools-20160318-story.html.

Dube, Marie-Catherine (September 1 2017) Interview.

Dziengel, Ana (April 21 2017) Interview.

Eccles, Jacquelynne S. (1999) *The Future of Children – Fall 1999. Middle Childhood*. Retrieved from https://pdfs.semanticscholar.org/6d23/31eed233d80c076305010522f9357a2cc114.pdf.

Educational Development Center (2013) *Preventing Bullying*. Retrieved from http://preventingbullying.promoteprevent.org/what-bullying.

Edwards, Carolyn P.; Knoche, Lisa; Kumru, Asiye (2001) *Play Patterns and Gender.* Retrieved from http://digitalcommons.unl.edu/cgi/viewcontent.cgi?article=1610&context=psychfacpub.

Eliot, Lise (2010) *Pink Brain, Blue Brain: How Small Differences Grow into Troublesome Gaps.* Mariner Books.

Elkind, David (1981) *The Hurried Child.* Perseus Publishing; 3rd edition 2001.

Empowered by Color *The Color Black.* Retrieved from www.empower-yourself-with-color-psychology.com/color-black.html.

Eton, Sarah Elaine (December 31 2010) Retrieved from https://drsaraheaton.wordpress.com/2010/12/31/formal-non-formal-and-informal-learning-what-are-the-differences/.

Evergreen (2013) *Landscape and Child Development*; Second edition. Retrieved from www.evergreen.ca/downloads/pdfs/Landscape-Child-Development.pdf.

Feder, Karen (September 15 2017) Interview.

Festinger, Leon (May 1 1954) *A Theory of Social Comparison Processes.* Retrieved from http://journals.sagepub.com/doi/abs/10.1177/001872675400700202.

Fields, R. Douglas (April 1 2011) *Why We Prefer Certain Colors.* Retrieved from www.psychologytoday.com/blog/the-new-brain/201104/why-we-prefer-certain-colors.

Food Explorers (2015) Retrieved from www.thefoodexplorers.co.uk/150th-market-celebration.

Frans, Peter (June 11 2014) *Effective Problem Solving & Decision Making.* Retrieved from www.linkedin.com/pulse/20140611165242-65881436-effective-problem-solving-decision-making/.

Frazee, Marla (August 23 2017) Interview.

Frost, Joe (February 29 2016) *Balancing Safety and Challenge in Playground Design.* Retrieved from https://issuu.com/penchura/docs/balancing_safety___challenge_in_pla.

Frost, J.L.; Wortham, S.C.; Reifel, S. (2010) *Play and Child Development.* Pearson Allyn Bacon Prentice Hall. Updated on July 20 2010. Retrieved from www.education.com/reference/article/gender-differences-play/.

Fry Kasuba, Terri (January 20 2017) Interview.

Fuentes, Patricio (September 1 2017) Interview.

Fung, Alice (April 17 2017) Interview.

Gallardo, Ramil (2014) *Physical and Motor Development of Children and Adolescents.* Retrieved from www.slideshare.net/rhamylle13/physical-and-motor-development-of-children-and-adolescents.

Galvez, Adriana; Chong, Russell (April 25 2017) Interview.

Gardner, Howard (2017) *Theory of Multiple Intelligences.* Retrieved from http://multipleintelligencesoasis.org/.

Geller, David (2017) *When Will My Child Be Ready to Shower Instead of Bathe?* Retrieved from www.babycenter.com/404_when-will-my-child-be-ready-to-shower-instead-of-bathe_70665.bc.

genderspectrum (2017) *Dimensions of Gender.* Retrieved from www.genderspectrum.org/quick-links/understanding-gender/.

Gerbner (1994) *Cultivation Theory.* Retrieved from https://en.wikipedia.org/wiki/Cultivation_theory.

Gill, Tim (2010) *Nothing Ventured … Balancing Risk and the Benefits of the Outdoors.* Retrieved from www.englishoutdoorcouncil.org/wp-content/uploads/Nothing-Ventured.pdf.

Gilutz, Shuli (March 15 2018) Interview.

Ginsburg, Kenneth R. (January 2007) *The Importance of Play in Promoting Healthy Child Development and Maintaining Strong Parent–Child Bonds.* Retrieved from http://pediatrics.aappublications.org/content/119/1/182.

Gladwell, Malcolm (2002) *The Tipping Point: How Little Things Can Make a Big Difference*. Bay Back Books.

Goodheart-Willcox Publisher (2017) Chapter 3: *Observing Children: A Tool for Assessment*. Retrieved from www.hasdk12.org/cms/lib/PA01001366/Centricity/Domain/915/CC%20ch%203%20study%20guide.pdf.

Goodwin, Bill (April 7 2017) Interview.

Goodwin, Paula (April 18 2017) Interview.

Gopnik, Alison (2009) *The Philosophical Baby: What Children's Minds Tell Us about Truth, Love, and the Meaning of Life*. Bodley Head; First edition.

Gopnik, Alison (2011) *TED Talk: What Do Babies Think?* Retrieved from www.ted.com/talks/alison_gopnik_what_do_babies_think.

Gopnik, Alison; Meltzoff, Andrew N.; Kuhl, Patricia K. (2000). *The Scientist in the Crib: What Early Learning Tells Us about the Mind*. William Morrow Paperbacks.

Graber, Evan G. (2017) *Physical Growth of Infants and Children*. Retrieved from www.msdmanuals.com/home/children-s-health-issues/growth-and-development/physical-growth-of-infants-and-children.

Gray, Peter (December 4 2008) *The Value of Play II: How Play Promotes Reasoning* Retrieved from www.psychologytoday.com/blog/freedom-learn/200812/the-value-play-ii-how-play-promotes-reasoning.

Gray, Peter (January 26 2010) *The Decline of Play and Rise in Children's Mental Disorders*. Retrieved from www.psychologytoday.com/blog/freedom-learn/201001/the-decline-play-and-rise-in-childrens-mental-disorders.

Gray, Peter (April 7 2014) *Risky Play: Why Children Love It and Need It*. Retrieved from www.psychologytoday.com/blog/freedom-learn/201404/risky-play-why-children-love-it-and-need-it.

Gray, Peter (October 31 2016) *The Culture of Childhood: We've Almost Destroyed It*. Retrieved from www.psychologytoday.com/blog/freedom-learn/201610/the-culture-childhood-we-ve-almost-destroyed-it.

Gray, Peter (February 6 2018) *Freedom to Learn: The Roles of Play and Curiosity as Foundations for Learning*. Retrieved from www.psychologytoday.com/blog/freedom-learn/200811/the-value-play-i-the-definition-play-gives-insights.

Guernsey, Lisa (2007) *Screen Time*. Basic Books.

Haberman, Mike (March 6 2015) *Future Friday: How to Distinguish between Fads, Micro Trends, Macro Trends and Megatrends*. Retrieved from http://omegahrsolutions.com/2015/03/future-friday-how-to-distinguish-between-fads-micro-trends-macro-trends-and-megatrends.html.

Haden, C.A.; Haine, R.A.; Fivush, R. (1997) *Developing Narrative Structure in Parent–Child Reminiscing across the Preschool Years*. Retrieved from www.ncbi.nlm.nih.gov/pubmed/9147838.

Hampton, Natalie (January 8 2018) Interview.

Hance, Jeremy (June 7 2011) *Do Kids Learn Anything at Zoos?* Retrieved from https://kidsnews.mongabay.com/2011/06/do-kids-learn-anything-at-zoos/.

Hanson, Cynthia (2017) *Only Children: Why "Only" Isn't Lonely*. Retrieved from www.parents.com/pregnancy/considering-baby/another/only-children/.

Harms, William (2013) *Children's Complex Thinking Begins Forming before They Go to School*. Retrieved from https://news.uchicago.edu/article/2013/01/23/children-s-complex-thinking-skills-begin-forming-they-go-school.

Harris, J.R. (1998) *The Nurture Assumption: Why Children Turn out the Way They Do*. Free Press.

Hayes, Dayle (2015) *5 Ways to Promote a Positive Body Image for Kids*. Retrieved from www.eatright.org/resource/health/weight-loss/your-health-and-your-weight/promoting-positive-body-image-in-kids.

Health Fitness Revolution (June 3 2015) *Top 10 Health Benefits of Youth Sports.* Retrieved from www.healthfitnessrevolution.com/top-10-health-benefits-youth-sports/.

Hearron, P.F.; Hildebrand, V. (2009) Excerpt from *Guiding Young Children*, 2009 edition, pp. 62–67. Merrill, an imprint of Pearson Education Inc.

Heimann, Roleen (September 5 2017) Interview.

Hein, George E. (October 15–22 1991) *Constructivist Learning Theory: The Museum and the Needs of People.* Retrieved from www.exploratorium.edu/education/ifi/constructivist-learning.

Hendrie, Maggie (February 28 2017) Interview.

Hines, Melissa (2015) Cited in *What's Wrong with Pink and Blue?* Retrieved from http://lettoysbetoys.org.uk/whats-wrong-with-pink-and-blue.

Hodges, Charlie (March 21 2018) Interview.

Hoecker, Jay L. (March 23 2016) *Is Baby Sign Language Worthwhile?* Retrieved from www.mayoclinic.org/healthy-lifestyle/infant-and-toddler-health/expert-answers/baby-sign-language/faq-20057980.

Holloway, Ani (Fall 2016) Mindfulness Lecture. Glenoaks Elementary School, Glendale, CA.

Holman, Cas (August 28 2017) Interview.

Holohan, Meghan (September 1 2016) *Children as Young as 3 Have Poor Body Image, Talk of Dieting Says Recent Survey.* Retrieved from www.today.com/health/children-young-3-have-poor-body-image-talk-dieting-says-t102453.

Huggamind (2017) *Infant Brain Stimulation.* Retrieved from www.huggamind.com/highcontrast.php#highcontrast.

Ideonomy (2017) *638 Primary Personality Traits.* Retrieved from http://ideonomy.mit.edu/essays/traits.html.

IMDb (2008) *Consuming Kids, The Commercialization of Childhood.* Documentary. Retrieved from www.imdb.com/title/tt1337599/.

inclusive playgrounds.org (2018) Retrieved from www.inclusiveplaygrounds.org/2play/overview.

Indian Express (July 30 2015) *India to Surpass China by 2022 to Become World's Most Populous Nation.* Retrieved from http://indianexpress.com/article/india/india-others/india-to-surpass-china-by-2022-to-become-worlds-most-populous-nation/.

International Labour Organization (2018) *Child Labour.* Retrieved from www.ilo.org/global/topics/child-labour/lang--en/index.htm.

Jambhekar, Shri (October 13 2017) Interview.

Jenkin, Matthew (December 2 2015) *Tablets out, Imagination in: The Schools That Shun Technology.* Retrieved from www.theguardian.com/teacher-network/2015/dec/02/schools-that-ban-tablets-traditional-education-silicon-valley-london.

Joan Ganz (2017) Cooney Center Retrieved from http://joanganzcooneycenter.org/.

Johnson, Chandra (September 7 2015) *How Important is Gender in Children's Toys and Play?* Retrieved from www.deseretnews.com/article/865636087/How-important-is-gender-in-childrens-toys-and-play.html.

Johnson, Christie, and Wardle (2005) *The Importance of Outdoor Play for Children.* Retrieved from www.communityplaythings.com/resources/articles/2010/outdoor-play.

Jones, Gerard (2003) *Killing Monsters: Why Children Need Fantasy, Superheroes, and Make-Believe Violence.* Basic Books.

J. Paul Getty Museum (2017) *Grade-by-Grade Guide to Building Visual Arts Lessons.* Retrieved from www.getty.edu/education/teachers/building_lessons/guide.html.

Kable, Jenny (February 10 2010) *The Theory of Loose Parts.* Retrieved from www.letthechildrenplay.net/2010/01/how-children-use-outdoor-play-spaces.html.

KaBOOM! (2017) *Creating Play Everywhere*. Retrieved from https://kaboom.org/playability/play_everywhere/playbook.

Kardefelt-Winther, Daniel. The Talkoot event, January 19–21 2018, Helsinki, Finland.

Katz, Michelle (October 25 2017) Interview.

Kawata, Jessie (April 18 2017) Interview.

Kelly, Annie (June 17 2013) *Technology Can Empower Children in Developing Countries – If It's Done Right*. Retrieved from www.theguardian.com/sustainable-business/technology-empower-children-developing-countries.

Kennedy-Moore, Eileen (February 26 2012) *Children's Growing Friendships: How Children's Understanding of Friendship Changes and Develops with Age*. Retrieved from www.psychologytoday.com/blog/growing-friendships/201202/childrens-growing-friendships.

Kennedy-Moore, Eileen (October 1 2013) *Teaching Children to Read Emotions: Emotional Understanding through Books*. Retrieved from www.psychologytoday.com/blog/growing-friendships/201310/teaching-children-read-emotions.

Kennedy-Moore, Eileen (June 30 2015) *Do Boys Need Rough and Tumble Play? Why Do Boys Wrestle, Rough House, and Play Pretend Fighting?* Retrieved from www.psychologytoday.com/blog/growing-friendships/201506/do-boys-need-rough-and-tumble-play.

Keyes, Richard (October 9 2017) Interview.

KidsHealth *Encouraging your Child's Sense of Humor*. Retrieved from http://kidshealth.org/en/parents/child-humor.html#.

Kids Matter (2017) *Learning Positive Friendship Skills*. Retrieved from www.kidsmatter.edu.au/mental-health-matters/social-and-emotional-learning/social-development/learning-positive-friendship.

Kids Matter (2017) *Motivation and Praise*. Retrieved from www.kidsmatter.edu.au/mental-health-matters/social-and-emotional-learning/motivation-and-praise.

Kids Matter (2017) *Resilience*. Retrieved from www.kidsmatter.edu.au/mental-health-matters/social-and-emotional-learning/resilience.

Kids Matter (2017) *Why Culture Matters for Children's Development and Wellbeing*. Retrieved from www.kidsmatter.edu.au/sites/default/files/public/KM%20C1_Cultural%20Diversity_Culture%20Matters%20for%20Development.pdf.

Kid Sense *Strength and Endurance* (2017) Retrieved from https://childdevelopment.com.au/areas-of-concern/gross-motor-skills/strength-and-endurance/.

Kid Sense *Visual Perception* (2017) Retrieved from https://childdevelopment.com.au/areas-of-concern/fine-motor-skills/visual-perception/.

Kim, Jin Hyung (October 6 2017) Interview.

Kunter, Yesim (October 13 2017) Interview.

Kutner, Lawrence (2017) *Humor as a Key to Child Development*. Retrieved from https://psychcentral.com/lib/humor-as-a-key-to-child-development/.

Kutska, Ken (September 6 2017) Interview.

Kyle, Angela (April 1 2017) Interview.

Land, George (2011) *The Failure of Success*. TEDxTuscon. Retrieved from www.youtube.com/watch?v=ZfKMq-rYtnc&feature=youtu.be.

LandscapeStructures (2017) *Children Need to Play, Learn and Grow Together, Side by Side*. Retrieved from www.playlsi.com/en/playground-design-ideas/inclusive-play/inclusive-play-commitment/.

Lane, Carla (2017) *The Distance Learning Technology Resource Guide*. Retrieved from www.tecweb.org/styles/gardner.html.

Lanza, Michael (April 4 2012) *Playborhood*. Free Play Press.

Lanza, Michael (August 11 2017) Interview.

Larmer, John; Mergendoller, John R. (2010) *Seven Essentials for Project-Based Learning*. Retrieved from www.ascd.org/publications/educational_leadership/sept10/vol68/num01/Seven_Essentials_for_Project-Based_Learning.aspx.

Laughery, Kenneth R. (1993) Everybody Knows or Do They? *Ergonomics in Design*, 1, 8–13.

Learning Design (2017) *Drawing Development in Children*. Retrieved from www.learningdesign.com/Portfolio/DrawDev/kiddrawing.html.

LearningRx (2017) *4 Cognitive Stages for Child Development*. Retrieved from www.learningrx.com/4-cognitive-stages-for-child-development-faq.htm.

Lee, Kevan (August 23 2017) *The Healthiest Way to Work: Standing vs. Sitting and Everything in Between*. Retrieved from https://open.buffer.com/healthiest-way-to-work-standing-sitting/.

LEGO Serious Play (2017) www.lego.com/en-us/seriousplay.

Leslie, Alan M. (1987) *Pretense and Representation: The Origins of "Theory of Mind"*. Retrieved from http://citeseerx.ist.psu.edu/viewdoc/download?doi=10.1.1.694.2866&rep=rep1&type=pdf.

Levine, Paul (August 22 2017) Interview.

Liberman, Isabelle Y.; Shankweiler, Donald; Liberman, Alvin M. (1989) *The Alphabetic Principle and Learning to Read*. Retrieved from www.haskins.yale.edu/sr/SR101/SR101_01.pdf.

Lillard, Angeline S.; Peterson, Jennifer (October 2011) *The Immediate Impact of Different Types of Television on Young Children's Executive Function*. Retrieved from http://pediatrics.aappublications.org/content/128/4/644.full.

Lindgren, Chris; Karlsson, Petter (January 21 2018) Interview Toca Boca.

Linn, Susan (2004) *Consuming Kids*. The New Press.

LoBue, Vanessa (March 29 2016) *When Do Children Develop Their Gender Identity?* Retrieved from http://theconversation.com/when-do-children-develop-their-gender-identity-56480.

Louv, Richard (March 10 2017) Interview.

Madden Ellsworth, Lindsay (2017) *Role of 4-H Dog Programs in Life Skills Development*. Retrieved at http://dx.doi.org/10.1080/08927936.2017.1270596.

Mallory, Michael (August 24 2017) Interview.

Mansfield, Tessa (April 17 2017) Interview.

Mardell, Ben (2018) *Key Learnings from the Pedagogy of Play Project*. Retrieved from https://vimeo.com/270655162.

Marinelli, Don (2010) *A Conversation with Don Marinelli*. Retrieved from www.youtube.com/watch?v=MDYcsjIyP54.

Martin, Francois (October 15 2017) Interview.

Mayden, Jason (February 24 2018) Interview.

McGhee, Paul (2017) *How Humor Facilitates Children's Intellectual, Social and Emotional Development*. Retrieved from www.laughterremedy.com/articles/child_development.html.

McGhee, Paul (2002) *Understanding and Promoting the Development of Children's Humor*. Kendall/Hunt.

McLeod, S.A. (2015). *Observation Methods*. Retrieved from www.simplypsychology.org/observation.html.

McLeod, Saul (2008) *Information Processing*. Retrieved from www.simplypsychology.org/information-processing.html.

McLeod, Saul (2012) *The Zone of Proximal Development and Scaffolding*. Retrieved from www.simplypsychology.org/Zone-of-Proximal-Development.html.

McLeod, Saul (2014) *Lev Vygotsky*. Retrieved from www.simplypsychology.org/vygotsky.html#MKO.

McLeod, Saul (2015) *Jean Piaget's Theory of Cognitive Development.* Retrieved from www.simplypsychology.org/piaget.html#schema.

MediaSmarts (2017) *How Marketers Target Kids.* Retrieved from http://mediasmarts.ca/digital-media-literacy/media-issues/marketing-consumerism/how-marketers-target-kids.

Mental Health Daily (2017) *At What Age Is the Brain Fully Developed?* Retrieved from http://mentalhealthdaily.com/2015/02/18/at-what-age-is-the-brain-fully-developed/.

Mersch, John (2017) *Tween Child Development 9–11 Years Old.* Retrieved from www.medicinenet.com/tween_child_development/article.htm.

Mgbemere, Bianca; Telles, Rachel (2013) *Types of Parenting Styles and How to Identify Yours.* Retrieved from https://my.vanderbilt.edu/developmentalpsychologyblog/2013/12/types-of-parenting-styles-and-how-to-identify-yours/.

Michel Carter, Christine (December 21 2016) *The Complete Guide to Generation Alpha, the Children of Millennials.* Retrieved from www.forbes.com/sites/christinecarter/2016/12/21/the-complete-guide-to-generation-alpha-the-children-of-millennials/#19328f303623.

Michelon, Pascale (2006) *What Are Cognitive Abilities and Skills, and How to Boost Them?* Retrieved from http://sharpbrains.com/blog/2006/12/18/what-are-cognitive-abilities/.

Micklo, Stephen J. (September 22 1995) *Developing Young Children's Classification and Logical Thinking Skills.* Retrieved from www.tandfonline.com/doi/abs/10.1080/00094056.1995.10522639.

Milestones on the Road to Reading Success (2018) Retrieved from www.treloar.org.uk/media/filer_public/25/b5/25b5626c-de1b-43f4-a17a-9820ce803670/reading_milestones.pdf.

Mindful Schools *Why Mindfulness Is Needed in Education* (2018) Retrieved from www.mindfulschools.org/about-mindfulness/mindfulness-in-education/.

MindsetWorks (2018) *Decades of Scientific Research That Started a Growth Mindset Revolution.* Retrieved from www.mindsetworks.com/science/.

Moats, Louisa; Tolman, Carol (2009) *Speaking Is Natural; Reading and Writing Are Not.* Retrieved from www.readingrockets.org/article/speaking-natural-reading-and-writing-are-not.

Moges, Bethel; Weber, Kristi (May 7 2014) *Parental Influence on the Emotional Development of Children.* Retrieved from https://my.vanderbilt.edu/developmentalpsychologyblog/2014/05/parental-influence-on-the-emotional-development-of-children/.

MoMA (2017) *Century of the Child.* Retrieved from www.moma.org/interactives/exhibitions/2012/centuryofthechild/.

Montes, Devin (December 12 2017) Interview.

Montimore, Terry (April 7 2017) Interview.

Morin, Amanda (2017) *5 Ways Kids Use Working Memory to Learn.* Retrieved from www.understood.org/en/learning-attention-issues/child-learning-disabilities/executive-functioning-issues/5-ways.

Morin, Amanda (2017) *Math Milestones: What to Expect at Different Ages.* Retrieved from www.understood.org/en/learning-attention-issues/signs-symptoms/age-by-age-learning-skills/math-skills-what-to-expect-at-different-ages.

Morrison, Mary Kay (2007) *Using Humor to Maximize Learning.* R&L Education.

Motion Picture Association of America *Film Ratings* (2017) Retrieved from www.mpaa.org/film-ratings/.

Mott, Tom (January 18 2017) Interview.

Mumper-Drumm, Heidrun (December 2 2016) Interview.

Munley, Mary Ellen (April 2012) *Early Learning in Museums.* Retrieved from www.si.edu/Content/SEEC/docs/mem%20literature%20review%20early%20learning%20in%20museums%20final%204%2012%202012.pdf.

Murphy Paul, Annie (November 1 2015) *Peer Pressure Has a Positive Side.* Retrieved from www.scientificamerican.com/article/peer-pressure-has-a-positive-side/.

Nakano, Mari (January 20 2017) Interview.

National Disability Authority Centre of Excellence in Universal Design (2017) *The 7 Principles.* Retrieved from http://universaldesign.ie/What-is-Universal-Design/The-7-Principles/.

National Education Association (2017) *Preparing 21st Century Students for a Global Society.* Retrieved from www.nea.org/assets/docs/A-Guide-to-Four-Cs.pdf.

National Institute of Mental Health (2017) *Autism Spectrum Disorder.* Retrieved from www.nimh.nih.gov/health/topics/autism-spectrum-disorders-asd/index.shtml.

National Scientific Council on the Developing Child (2008) *The Timing and Quality of Early Experiences Combine to Shape Brain Architecture.* Retrieved from http://developingchild.harvard.edu/wp-content/uploads/2007/05/Timing_Quality_Early_Experiences-1.pdf.

Navarro, Joe (April 9 2010) *Body Language Essentials for Your Children – For Parents.* Retrieved from www.psychologytoday.com/blog/spycatcher/201004/body-language-essentials-your-children-parents.

New York Times (March 13 1997) *Balloons Made of Latex Pose Choking Hazard.* Retrieved from www.nytimes.com/1997/03/13/garden/balloons-made-of-latex-pose-choking-hazard.html.

New York Times (December 23 2012) *Guys and Dolls No More.* Retrieved from www.nytimes.com/2012/12/23/opinion/sunday/gender-based-toy-marketing-returns.html.

New York Times (October 18 2015) *Move Over, Millennials, Here Comes Generation Z.* Retrieved from www.nytimes.com/2015/09/20/fashion/move-over-millennials-here-comes-generation-z.html.

Nicolopoulou, Ageliki (2009) *Using a Narrative- and Play-Based Activity to Promote Low-Income Preschoolers' Oral Language, Emergent Literacy, and Social Competence.* Retrieved from www.ncbi.nlm.nih.gov/pmc/articles/PMC4391821/.

Noxon, Christopher (February 14 2017) Interview.

Observing, Recording, and Reporting Children's Development, CRI Preschool Assessment Instrument (2017) Retrieved from http://laffranchinid.faculty.mjc.edu/Ch5.pdf

Oddleifson Robertson, Katrin (2017) *The Arts and Creative Problem Solving.* Retrieved from www.pbs.org/parents/education/music-arts/the-arts-and-creative-problem-solving/.

Ormrod, Jeanne (2011) *Educational Psychology: Developing Learners.* Retrieved from http://teachingasleadership.org/sites/default/files/Related-Readings/LT_Ch2_2011.pd Jeanne Ormrod's Educational Psychology: Developing Learners.

Oswalt, Angela (2017) *Early Childhood Emotional and Social Development: Identity and Self Esteem.* Retrieved from www.gulfbend.org/poc/view_doc.php?type=doc&id=12766.

Oswalt, Angela (2008) Retrieved from www.mentalhelp.net/articles/early-childhood-emotional-and-social-development-identity-and-self-esteem/.

Otto, Ali (September 28 2017) Interview.

Pan, Hsinping (November 20 2017) Interview.

Pangrazi, Robert P.; Beighle, Aaron (2016) *Dynamic Physical Education for Elementary School Children.* Pearson.

Pashler, Harold; McDaniel, M.; Rohrer, D.; Bjork, R. (2008) Learning Styles: Concepts and Evidence. *Psychological Science in the Public Interest* 9, 105–119.

Passaro, Jamie (June 31 2015) *The Chemicals in a Plastic Doll*. Retrieved from www.theatlantic.com/health/archive/2015/07/whats-in-my-daughters-doll/399914/.

Passman, R.H. (1987) Attachments to Inanimate Objects: Are Children Who Have Security Blankets Insecure? *Journal of Consulting and Clinical Psychology*, 55, 825–830.

Pessoa, Luiz (2009) *Cognition and Emotion*. Retrieved from www.scholarpedia.org/article/Cognition_and_emotion.

Petridis, Alexis (March 20 2014) *Youth Subcultures: What Are They Now?* Retrieved from www.theguardian.com/culture/2014/mar/20/youth-subcultures-where-have-they-gone.

Piaget's Theory of Cognitive Development (2018) Retrieved from https://en.wikipedia.org/wiki/Piaget%27s_theory_of_cognitive_development.

Piscitelli, Barbara; Anderson, David (2002) *Young Children's Perspectives of Museum Settings and Experiences*. Retrieved from www.researchgate.net/publication/253933420_Young_Children%27s_Perspectives_of_Museum_Settings_and_Experiences.

Pixar (2018) *Character Design*. Retrieved from http://pixar-animation.weebly.com/character-design.html.

Placemaking Chicago (2008) *What Is Placemaking?* Retrieved from www.placemakingchicago.com/about/.

Play Therapy (2017) Retrieved from https://en.wikipedia.org/wiki/Play_therapy.

Poesch, Eric (January 26 2017) Interview.

Poliakoff Struski, Valerie (November 20 2017) Interview.

Pontrella, Caitlin (August 12 2017) Interview.

Poole, Carla; Miller, Susan A.; Booth Church, Ellen (2017) *Ages & Stages: How Children Build Friendships*. Retrieved from www.scholastic.com/browse/article.jsp?id=3747174.

Prieto, Mariana (December 19 2016) Interview.

Principles of Child Development and Learning That Inform Developmentally Appropriate Practice (2017) Retrieved from www.ce71a.com/New_University_School/principles_of_child_development.htm.

Provine, Robert (2001) *Laughter: A Scientific Investigation*. Penguin Books; Reprint edition (December 1 2001).

Puig, Claudia (January 15 2018) Interview.

Rechner, Erin (April 19 2017) WGSN Interview.

Rich, Michael (2008) Retrieved from documentary, *Consuming Kids*, 2008.

Robinson, Ken (2009) *The Element: How Finding Your Passion Changes Everything*. Penguin Books.

Rockbrook Camp for Girls (2017) *Why Camp Is Great for Children*. Retrieved from www.rockbrookcamp.com/parents/children-camp-great/.

Rodriguez, Shirley (December 8 2017) Interview.

Romm, Cari (October 10 2016) *Little Kids Use Dreams to Figure Out Their Real Lives*. Retrieved from www.thecut.com/2016/10/what-do-babies-and-little-kids-dream-about-animals-mostly.html.

Russo, Cristina (March 11 2013) *Can You Worry about an Animal You've Never Seen? The Role of the Zoo in Education and Conservation*. Retrieved from http://blogs.plos.org/scied/2013/03/11/zoo-education/.

Ryan, David B. (2017) *How Does Peer Pressure Influence Teen Purchasing Choices?* Retrieved from http://living.thebump.com/peer-pressure-influence-teen-purchasing-choices-8191.html.

Sample, Ian (2014) *Talking to Babies Boosts Their Brain Power*. Retrieved from www.theguardian.com/science/2014/feb/14/talking-to-babies-brain-power-language.

Sanders, Martin (November 10 2017) Interview.

Savage, Jennifer S.; Fisher, Jennifer O.; Birch, Leann L. *Parental Influence on Eating Behavior* (2017) Retrieved from www.ncbi.nlm.nih.gov/pmc/articles/PMC2531152/.

Scholastic *20 Ways to Boost Your Baby's Brain Power*. Retrieved from www.scholastic.com/parents/resources/article/thinking-skills-learning-styles/20-ways-to-boost-your-babys-brain-power.

Schor, Edward L. (2004) *Caring for Your School-Age Child: Ages 5 to 12*. American Academy of Pediatrics.

Science Buddies (2017) *Steps of the Scientific Method*. Retrieved from www.sciencebuddies.org/science-fair-projects/project_scientific_method.shtml.

Scribendi (2017) *5 Common Character Archetypes in Literature*. Retrieved from www.scribendi.com/advice/character_archetypes_in_literature.en.html.

Seidi, Amanda; Cristia, Alejandrina (2012) *Infants' Learning of Phonological Status*. Retrieved from www.ncbi.nlm.nih.gov/pmc/articles/PMC3487416/.

Self-Concept (2017) Retrieved from https://en.wikipedia.org/wiki/Self-concept.

Separation Anxiety (April 18 2017) Retrieved from www.psychologytoday.com/conditions/separation-anxiety.

Sesame Workshop (2017) Retrieved from www.sesameworkshop.org.

Shaffery, Joseph (2016) *10 Ways to Design for the Human Brain*. Retrieved from www.invisionapp.com/blog/design-for-the-human-brain/?utm_campaign=Weekly+Digest&utm_source=hs_email&utm_medium=email&utm_content=25249127&_hsenc=p2ANqtz-_FFtvR5jRvvbOu9vAbxg7CtTi7R5DTlLo5awIgHQAShHGH-FjbfXptOJs6-s2Wjkut2tl6iWcwAXWe2dPCNc7SZ1ng9g&_hsmi=25249127.

Shane, Brian (December 28 2012) *Schools Use Smart Devices to Help Make Kids Smarter*. Retrieved from www.usatoday.com/story/news/nation/2012/12/27/students-use-cellphones-as-part-of-classroom-lessons/1794883/.

Shanklin, Michael; Kaye, Lauren; Kalmore, Shellie (November 13 2017) Interview.

Shiffrin, Richard M.; Nosofsky, Robert M. (1994) *Seven Plus or Minus Two: A Commmentary on Capacity Limitations*. American Psychological Association Inc. Retrieved from http://pubs.cogs.indiana.edu/pubspdf/21879/21879_1994_rms-rmn_pr_seven.pdf.

Shinskey, Jeanne (July 26 2016) *Why Can't We Remember Our Early Childhood?* Retrieved from https://greatergood.berkeley.edu/article/item/why_cant_we_remember_our_early_childhood.

Shouse, Benjamin (2011) *Why Don't We Remember Being Babies?* Retrieved from www.livescience.com/32963-why-dont-we-remember-being-babies.html

Siegal, James; Levner, Amy S. (October 25 2017) Interview.

Siegel, Tara J. (2010) *Quality Environments for Children: A Design Development Guide*. Retrieved from www.liifund.org/wp-content/uploads/2011/03/LIIF-Quality_Environments_for_Children-2010.pdf?

Sklar, Andy (January 25 2017) Interview.

Sklar, Howard (2009) *Believable Fictions*. Retrieved from https://blogs.helsinki.fi/hes-eng/volumes/volume-5/believable-fictions-on-the-nature-of-emotional-responses-to-fictional-characters-howard-sklar/.

Sleep Foundation (2017) *Children and Sleep*. Retrieved from https://sleepfoundation.org/sleep-topics/children-and-sleep.

Small, Meredith F. (1998) *Our Babies, Ourselves. How Biology and Culture Shape the Way We Parent*. Anchor Books.

Smith System (2017) *Enhancing Classroom Learning Environments by Design*. Retrieved from https://smithsystem.com/design-principles/.

Sparks and Honey (2017) Retrieved from web.sparksandhoney.com/.

Speech Therapy Centres of Canada (2013) *Children and Age-Appropriate Attention Spans*. Retrieved from www.speechtherapycentres.com/children-and-age-appropriate-attention-spans/.

Stages of Writing Development. Retrieved from www.mecfny.org/wp-content/uploads/2015/06/StagesofWritinghandout.pdf.

Stanford Children's Health *Cognitive Development*. Retrieved from www.stanfordchildrens.org/en/topic/default?id=cognitive-development-90-P01594.

Steelcase Report 360° The Education Edition (2015) Retrieved from www.steelcase.com/content/uploads/2015/07/SE-360-Education-Magazine.pdf.

Steinberg, Laurence (2008) Retrieved from http://headsup.scholastic.com/students/peer-pressure-its-influence-on-teens-and-decision-making.

St. James, Alena (March 15 2017) Interview.

Strittmatter, Neeti (March 30 2016) Interview.

Stromberg, Joseph (January 28 2015) *9 Surprising Facts about the Sense of Touch*. Retrieved from www.vox.com/2015/1/28/7925737/touch-facts.

Study.com (2017) *Factors Influencing Motor Development*. Retrieved from https://study.com/academy/lesson/factors-influencing-motor-development.html.

Stylus (February 2016) *Decoding Digital Parenting*. Retrieved from www.stylus.com/.

Surbhi, S. (January 3 2017) *Differences between Inductive and Deductive Reasoning*. Retrieved from https://keydifferences.com/difference-between-inductive-and-deductive-reasoning.html.

Sweet, Elizabeth (December 21 2012) *Guys and Dolls No More?* Retrieved from www.nytimes.com/2012/12/23/opinion/sunday/gender-based-toy-marketing-returns.html.

Sweetland Edwards, Haley (December 2 2015) *Why Mark Zuckerberg Wants to Spend on Personalized Learning*. Retrieved from http://time.com/4132619/mark-zuckerberg-personalized-learning/.

Talking Point (2017) *11–17 Years*. Retrieved from www.talkingpoint.org.uk/ages-and-stages/11-17-years.

Tang, Carol M. (2017) Interactivity Conference, May 3–5 2017, Pasadena, CA.

Tarkett (2017) *Children's Perceptions of Color and Space*. Retrieved from https://media.tarkett.com/docs/BR_TNA_HD_KIDS_STUDY.PDF.

Teach Me to Talk! (February 12 2008) *First 100 Words – Advancing Your Toddler's Vocabulary with Words and Signs*. Retrieved from http://teachmetotalk.com/2008/02/12/first-100-words-advancing-your-toddlers-vocabulary-with-words-and-signs/.

Teacher Tap (2018) *Project, Problem, and Inquiry-based Learning*. Retrieved from https://eduscapes.com/tap/topic43.htm.

The Center for Early Childhood Education, Eastern Connecticut State University (2017) *TIMPANI Toy Study*. Retrieved from www.easternct.edu/cece/timpani/.

The Education Alliance (2002) Individualist and Collectivist Perspectives on Education, *The Diversity Kit*. The Education Alliance at Brown University.

The Family Dinner Project (2017) Retrieved from https://thefamilydinnerproject.org/.

The GiggleBellies *What Colors Do Kids Prefer?* (November 1 2013) Retrieved from http://blog.thegigglebellies.com/2013/11/what-colors-do-kids-prefer-the-science-of-playful-colors/.

The Human Memory (2017) *Memory Recall/ Retrieval*. Retrieved from www.human-memory.net/processes_recall.html.

The LEGO Foundation (2017) www.legofoundation.com/ko-kr/research-and-learning/foundation-research.

The National Council of Teachers of Mathematics (2000) *Principles and Standards for School Mathematics*. Retrieved from www.nctm.org/uploadedFiles/Standards_and_Positions/PSSM_ExecutiveSummary.pdf.

The Peak Performance Center (2017) *Types of Attention*. Retrieved from http://
thepeakperformancecenter.com/educational-learning/learning/process/obtaining/
types-of-attention/.

The Reggio Approach (2017) *What Is The Reggio Emilia Approach?* Retrieved from www.
aneverydaystory.com/beginners-guide-to-reggio-emilia/main-principles/.

Thompson, Alison (2004) *Developing Fundamental Movement Skills*. Teachers' guide.
CCEA. Retrieved from http://ccea.org.uk/sites/default/files/docs/curriculum/area_of_
learning/physical_education/FMS_teacher_guide.pdf.

Tichnor-Wagner, Ariel (August 3 2016) *A Global Perspective: Bringing the World into
Classrooms*. Retrieved from www.edweek.org/tm/articles/2016/08/03/a-global-
perspective-bringing-the-world-into.html.

Tiede, Cameron (January 22 2018) Interview.

Turrell, Andrew (2011) *Selective Attention and User Experience*. Retrieved from http://
uxmag.com/articles/selective-attention-and-user-experience.

UNICEF *Early Childhood Development: The Key to a Full and Productive Life* (2017)
Retrieved from www.unicef.org/dprk/ecd.pdf.

UNICEF *Gender Action Plan 2018–2021*. Retrieved from www.unicef.org/gender/files/
2018-2021-Gender_Action_Plan-Rev.1.pdf.

UNICEF *Innovation in Education* (2017) Retrieved from www.unicef.org/education/bege_
73537.html.

United Nations *Sustainable Development Goals* (2017) Retrieved from www.un.org/
sustainabledevelopment/sustainable-development-goals/.

United Nations Human Rights Office of the High Commissioner (November 20
1989) *Convention on the Rights of the Child*. Retrieved from www.ohchr.org/EN/
ProfessionalInterest/Pages/CRC.aspx.

United States Department of Agriculture (2017) *Expenditures on Children by Families,
2015*. Retrieved from www.cnpp.usda.gov/sites/default/files/crc2015.pdf.

University of Oregon Center on Teaching and Learning *Phonemic Awareness* (2017)
Retrieved from http://reading.uoregon.edu/big_ideas/pa/pa_what.php.

Urbanski, Matthew (April 27 2017) Interview.

Voice of Play (2017) *Benefits of Play*. Retrieved from http://voiceofplay.org/benefits-of-
play/.

Voss, Angie (2017) *Social and Emotional Aspects of Sensory Processing Challenges*.
Retrieved from http://asensorylife.com/social--emotional-factors.html.

Waterlow, Lucy (April 12 2013) *Traditional Playtime Now Over by Age 7: British Children
Abandoning Dolls and Toys in Favour of Gadgets and Computers Earlier Than
Before*. Retrieved from www.dailymail.co.uk/femail/article-2308038/Traditional-
playtime-age-7-British-children-abandoning-dolls-toys-favour-gadgets-computers-
earlier-before.html.

Weber, Steven (2017) *What Is the Purpose of the American High School?* Retrieved from
http://edge.ascd.org/blogpost/what-is-the-purpose-of-the-american-high-school.

WebMD.com (2017) *Your Teen's Sexual Orientation and Gender Identity –
Topic Overview*. Retrieved from www.webmd.com/parenting/tc/
your-teens-sexual-orientation-and-gender-identity-topic-overview#1.

Weinschenk, Susan (October 18 2017) Interview.

Weinschenk, Susan M. (2011) *100 Things Every Designer Needs to Know about People
(Voices That Matter)*. New Riders.

Weinschenk, Susan M. (2011) *The Secret to Designing an Intuitive UX*. Retrieved from
https://uxmag.com/articles/the-secret-to-designing-an-intuitive-user-experience.

Weise, Marin L.; Adolph, K.E. (2000) Locomotor Development. In L. Balter (Ed.),
Parenthood in America: An Encyclopedia. ABC-CLIO.

Weissberg, Roger (2016) Retrieved from www.edutopia.org/blog/
    why-sel-essential-for-students-weissberg-durlak-domitrovich-gullotta.
Westervelt, Eric (August 4 2015) *Play Hard, Live Free: Where Wild Play Still
    Rules.* Retrieved from www.npr.org/sections/ed/2015/08/04/425912755/
    play-hard-live-free-where-wild-play-still-rules.
White, Deb (February 9 2017) Interview.
White, Rachael E. (2012) *The Power of Play.* Retrieved from www.childrensmuseums.org/
    images/MCMResearchSummary.pdf.
White Hutchinson Leisure & Learning Group (2008) *History of Children's Entertainment
    and Edutainment Centers.* Retrieved from www.whitehutchinson.com/leisure/
    historychildren.shtml.
White Hutchinson Leisure & Learning Group (June 25 2016) *Children's Outdoor Play
    Areas & Adventure Play Gardens.* Retrieved from www.whitehutchinson.com/leisure/
    gardenchildren.shtml.
Winger, Dan (February 15 2017) Interview.
Wiser, Dana (April 21 2017) Interview.
Wolchover, Natalie (June 29 2012) *Your Color Red Really Could Be My Blue.* Retrieved
    from www.livescience.com/21275-color-red-blue-scientists.html.
Yamaguchi, Dice (October 1 2017) Interview.
Young, Sarah (2017) *Batman Effect: Playing Make Believe Could Help Boost Your Child's
    Productivity, Finds Study.* Retrieved from www.independent.co.uk/life-style/batman-
    effect-playing-make-believe-child-productivity-tasks-concentration-study-dora-
    explorer-bob-a8112291.html.
Zippin, Sam (May 28 2014) *How Does Your School's Physical Environment Affect
    Students?* Retrieved from www.schooldude.com/community/discover/blogs/
    how-does-your-schools-physical-environment-affect-students.
Zopf, Jini (November 8 2016) Interview.

Index

Note: Please note that page numbers in *italic* represent text within figures.

child psychology 114, 405

childhood amnesia 210–211, *210*

childrearing practices 130

Children's Advertising Review Unit (CARU) 115

Children's Design Guide 17–19, *19*

Children's Online Privacy Protection Act (COPPA) 115, 399, 425

Children's Television Act 425

choking hazards 76, 89, 185

Chong, Russell 427

citizenship 17, 219, 330, 342, 379, 382

city play spaces 370, *371*

classroom design/classroom furniture design 377–379, *378*

clothing, children's 179–180, 241–242, 255; design process 51, 72–73, 74–76, 79, 87

co-creation 17, *20*, 22, 120, 133, 139, 167, *407*, 430

co-design 1, 15, 16, 22, 23, 39, 99, 101–103, *102*, 257, 433

cognitive development 192–231; Piaget's model of 43–44, 197–200, *198*, *199*; theories of 197–202, *198*, *199*, *201*

cognitive ergonomics 53, *54*

collaboration 329

collectivist perspective, on education 350

color 37, 52, 68, 69, 70, 75, 76–79, *77*; in animation 427; in character design 413, 424; in classroom furniture design 378; and co-design 163, 357; cultural associations with 77, 80; and developmental stages 81–82; emotional responses to 80–81; and functionality 79; and gender 242, 243; and markets 79–80; and parental opinions 81; and play therapy 249; prototypes 105, 106; Alena St. James on 92; science on 79; in style guides 415–416; in toy design 314, 318, 319

comic books 45, 47, 157, 419–420, *420*

communication 19, 30, 69, 102, 104, 110, 131, 137, 249; and 21st century skills movement 329; and cognitive development 202; and developmental stages 141, 142, 144; and humor 264, 266; language 225, 227, 229, 235, 325; non-verbal 219, 235, *236*, 236, 285, 288, 303; and play 277–278; and Reggio Emilia approach 340, 342; and social media 428–429; verbal 224, 235, 236; visual 70, 115, 174, 318; written 318

communications media 429

communications strategies 406

community play 308, *309*, 336–337

community spaces 319, 362–375, *363*, *364*, *366*, *369*, *371*, *372*, *373*, *376*

compensation, for children 39–40

competitive analysis 61, 131

concept testing 36–37, 105, 189, 283

conceptual models 200

consent forms 15

constructive play 99, 219; children's spaces 373, 374; play 285, 292, *293*, 296–299, *297*, *298*, *299*, 304, 312

constructivism 133, 197–201, *198*, *199*, *201*, 335–337

Consumer Product Safety Commission (CPSC) 56, 368, 369

consumer trends 46, 138; *see also* trend analysis

consumerism 45, 67, 255, 313, 425

controlled observations 40–43

Convention on the Rights of the Child 102, 107

convergent thinking 218, 219

cooking 186, 187, 248, 360

cool hunting 74

cooperative play 252, 281–284, *283*, *284*

COPPA (Children's Online Privacy Protection Act) 115, 399, 425

copyrights 62, 63, 64, 69, 109

CPSC (Consumer Product Safety Commission) 56, 368, 369

creative intelligence 218

creative play 170–171, 218–219, 220, 235, 304–306, *305*, *306*

creative process 101, 138, 192, *221*; *see also* idea generation/ideation

creativity, children's 17, 18, 97, *171*; and cognitive development 192, *212*, 218–225, *221*, *225*; role of play in 33, 135, 220–221

creativity, designer's 9, 58, 96

Creativity project 224–225

critical thinking 310, 403; children's spaces 373, 386–387, 389; cognitive development 219, 224; education 329, 341, 343, 346

Cron, Robert 308, *309*

cross-cultural perspectives 44, 46

Csikszentmihalyi, Mihaly 135, 136

cultivation theory 429

cultural identity 239, 240, *240*

cultural museums 386–389, *388*, *389*

cultural research 44–46

cultural shifts 257, 318–319, 320

culture, children's 21, 44, 45, 74, 75, 133, 135, 274–275

intelligence 195–197, *196*
interaction design/interactive design 54, 105, 209, 408
interactive media 11, 274, 397–403, *404*
interactive play 115
interests, of children 127, 129, 130; and alone time *264*; and children's media and technology 397, 402, 428; and children's spaces 354, 355, 359, 360, 370; and culture 243; and design process 7, 12, 24, 26, 29, 53, 66, 69; and games 308; and open-ended play 280; and self-identity 239; shared 45, 149, 152; and youth subcultures 255
interests, of kidults 157
interface design 64, 217, 405
intergenerational engagement 12, 134, 156, 158, 328
International Labour Organization (ILO) 87, 107
International Playground Safety Institute 83–84, 118, 368–369, *369*
International Rescue Committee 274, 351–352
Internet of Things (IoT) 139, 312, 407, 408
interview questions 14, 29
interviews 14, 15, 22–23, 27, 28–29, 41, 189, 200; case studies 60
introductions, session 41
invention, designing for 298–299, *298, 299*
inventory management 116
IoT (Internet of Things) 139, 312, 407, 408
IP (intellectual property) 61, 62–64
It's a bob 295, *295*

Jambhekar, Shri 53
Jones, Gerard 412
juvenile arthritis, exercise toys for children with 173–174, *174*

KaBOOM! 374–375, *376*
Kalmore, Shellie 385, *386*
Kardefelt-Winther, Daniel 17
Karlsson, Petter 430–431, *431*
Kasuba, Terry Fry 28, 100, 200
Katz, Michelle 62, 64
Kawata, Jessie 344–345, *345*
Kaye, Lauren 385, *386*
Keyes, Richard 76–77, 77–78, *77*, 79, 80, 81
Kidspace Children's Museum 385, *386*
kidults 155–158, *158*
Kim, Andrew 335
Kim, Jim Hyung 236, *237*
kitchen 53, 187, 200, 248, 337, 357, 360

Knorr, Caroline 238
Kunter, Yesim 23, 299, *300*, 314, 316
Kutska, Kenneth S. 83–84, 118, 368–369, *369*
Kyle, Angela 373, *373*

language 165, 174: cognitive development 194, 212, 215, 223, 225–229, *227, 230*; development stages 140, 141, 142, *142*, 146, 149, 151; *see also* body language
language play 225, 226, 342
Lanza, Michael 361–362, *362*, 365
Larmer, John 345–346
Leadership in Energy and Environmental Design (LEED) certification 52, 379
learning 7, 11, 194–195, *195*; academic 180, 333; accidental 12, 225, 326, 373; active 330, 346, 348, *373*, 377; and animals 261; and attention 330; and behavior 26, 207; and the brain 207; and color 79; and communities 363–364, 373, *373*; and the constructivist approach 198, *198*; contexts for 325–326, *326, 327*; to cook 187; and culture 46; definition of 324–325; Eames Demetrios on 49; design-based 346–347, *347*; developmental ways of 325; and digital play 309; and digital skills 224; to eat 185–186, *186*; and education 323, 335; and edutainment 265; emotional 232, 234–235, *234, 236*, 236–237, *237*; and empathy 32; and entertainment 398, 399, 400; formal 325, *326*; and the four pillars 17; game-based 348; through gamification 348–349, 429; and humor 264–265; individual 280–282; informal 12, 225, 326, 373; inquiry-based 346–347, *347*; and the intelligences 197; and loose parts 294; and manipulatives 295; and memory 209, 211; and metacognitive practices 217; and motivation 238; motor 168; and museums 388; and music and movement 222, 223–224; non-formal 325–326; off-line open source 132; open-ended 294; peer-to-peer 133–134, *134*, 401; personalized 347–348; and pets 261; in physical education 182; through play 275, 277–282, 315–316, 348–349, 352; and multiple play patterns 298, *298*; principles of 195; problem-based 346; and problem-solving 213; project-based 329, 345–346, 376; research as 43; self- 239; sensory 389; to sleep

Printed and bound by CPI Group (UK) Ltd, Croydon, CR0 4YY

21/10/2024

01777092-0002